24 0088232 3

IMPERIAL COLLEGE
THE[illegible]ARY

Geometric Methods in System Theory

NATO ADVANCED STUDY INSTITUTES SERIES

Proceedings of the Advanced Study Institute Programme, which aims at the dissemination of advanced knowledge and the formation of contacts among scientists from different countries

The series is published by an international board of publishers in conjunction with NATO Scientific Affairs Division

A	Life Sciences	Plenum Publishing Corporation
B	Physics	London and New York
C	Mathematical and Physical Sciences	D. Reidel Publishing Company Dordrecht and Boston
D	Behavioral and Social Sciences	Sijthoff International Publishing Company Leiden
E	Applied Sciences	Noordhoff International Publishing Leiden

Series C – Mathematical and Physical Sciences

Volume 3 – Geometric Methods in System Theory

519.71

Geometric Methods in System Theory

Proceedings of the NATO Advanced Study Institute held at London, England, August 27–September 7, 1973

edited by

D. Q. MAYNE and R. W. BROCKETT

D. Reidel Publishing Company

Dordrecht-Holland / Boston-U.S.A.

Published in cooperation with NATO Scientific Affairs Division

First printing: December 1973

Library of Congress Catalog Card Number 73–91206

ISBN 90 277 0415 5

Published by D. Reidel Publishing Company
P.O. Box 17, Dordrecht, Holland

Sold and distributed in the U.S.A., Canada, and Mexico
by D. Reidel Publishing Company, Inc.
306 Dartmouth Street, Boston, Mass. 02116, U.S.A.

All Rights Reserved
Copyright © 1973 by D. Reidel Publishing Company, Dordrecht
No part of this book may be reproduced in any form, by print, photoprint, microfilm, or any other means, without written permission from the publisher

Printed in The Netherlands by D. Reidel, Dordrecht

Contents

Introduction

Geometric Methods in System Theory

In automatic control there are a large number of applications of a fairly simple type for which the motion of the state variables is not free to evolve in a vector space but rather must satisfy some constraints. Examples are numerous; in a switched, lossless electrical network energy is conserved and the state evolves on an ellipsoid surface defined by $x'Qx$ equals a constant; in the control of finite state, continuous time, Markov processes the state evolves on the set $x'x = 1$, $x_i \geq 0$. The control of rigid body motions and trajectory control leads to problems of this type. There has been under way now for some time an effort to build up enough control theory to enable one to treat these problems in a more or less routine way. It is important to emphasise that the ordinary vector space-linear theory often gives the wrong insight and thus should not be relied upon.

On the other hand there are very good mathematical reasons for looking at control theory geometrically. In the study of optimal control the subtleties of the singular problem are dealt with most efficiently via a differential geometric technique. In stochastic control the study of problems where the diffusion is not fully elliptic is clarified by differential geometry. In stability theory geometric methods can lead to surprisingly delicate sufficient conditions. In all these areas differential geometry serves to overcome the limitations of the well known linear theory and enables one to analyse large classes of nonlinear problems without difficulty.

Work applying geometrical methods to control problems has been going on for some time now in widely scattered centres in Europe and the U.S.A. Our motivation in organising this Advanced Study Institute was to bring together some of the main contributors in the hope that a general evaluation and synthesis would take place. The Institute was lively, stimulating and broad in scope, fulfilling our expectations. These proceedings formed the basis for the talks that were given. The first five contributions are mostly expository and, taken together, provide the basic information and results assumed by most of the other contributions. The expository lectures are intended as a combination textbook and guide to the literature for the novice. The research papers explore in depth some of the more interesting questions of current interest.

It is a pleasure to thank NATO for their generous support for the Institute. We also acknowledge with gratitude the fellowships provided by the National Science Foundation of the U.S.A.

D.Q. Mayne R.W. Brockett

DYNAMICAL POLYSYSTEMS
and
CONTROL THEORY

C. LOBRY

U.E.R. Mathématiques et Informatique

Université Bordeaux I, 351, cours de la Libération
TALENCE (FRANCE).

In these lectures we shall see how it is possible to generalize many results on controllability of linear control systems to the nonlinear ones. A little use of basic definitions of differential geometry is made ; the necessary material used is exposed in appendix.

The basic paper on this subjet is R. Hermann paper "On the Accessibility Problem in Control Theory" exposed at the International Symposium on Nonlinear Differential Equations and Nonlinear Mechanics in 1961. In essence all my lectures are contained in this paper.

Not many papers where published in this spirit, with notable exceptions like Hermes (1) and Kucera (1) (2) before 1970. After 1970, a collection of papers appeared ; see for instance, Brockett, Elliott, Francis, Haynes-Hermes, Jurdjevic, Jurdjevic-Sussmann, Krener, Lobry, Stefan, Sussmann in the references. They have in common the fact that the pseudo-group of local diffeomorphisms generated by a family of vector fields on a manifold M is considered to be the relevant mathematical object to look at in the study of nonlinear control systems.

This point of view is exposed here. Not all the known results are exposed but only the basic ones ; for further development the reader is referred to the authors listed above. Chapter I is introductory ; Chapter II is mostly based on recent papers by Sussmann and Krener, and contains the basic results; Chapter III is devoted to applications to some problems of controllability.

I wish to apologize for the poor english of the notes.

I want to acknowledge both professors L. MARKUS and G. REEB for the encouragements they gave me and J. MARTINET for many mathematical discussions.

I thank professors R. BROCKETT and D. MAYNE who gave me the opportunity for these lectures

I thank also Mme Polzin who tried, with some success, to transform a badly hand written text into a well typed set of notes in a very short time.

I - DYNAMICAL POLYSYSTEMS AND CONTROL THEORY.

The idea of Dynamical Polysystems (say D.P.) goes back to Bushaw (1)(2). In the first § we give an abstract formulation, in the second one we discuss connections with control systems.

§1. Dynamical Polysystems

Let us define a control group. Let I be a non empty set and consider the set of all finite sequences $((t_1,i_1)(t_2,i_2)\dots(t_p,i_p))$ with values in $R\times I$; we now introduce a simplification rule :

a) every term of the form $(0,i_j)$ is suppressed

b) if $i_j = i_{j+1}$ the terms $(t_j,i_j)(t_{j+1},i_{j+1})$ are replaced by $(t_j + t_{j+1},i_j)$.

It is easy to see that by this procedure one obtains after a finite number of steps an irreductible sequence (which may be empty and in this case is denoted by 0) which depends only of the initial one.

I.1.1 - Definition : The set of all such irreductible sequences is denoted by G(I) and is called "control group" (modelled on I), elements of G(I) are called "controls".

The word "group" is justified by the following :

I.1.2 - Proposition : The mapping from $G(I)\times G(I)$ into G(I) which associates to the two controls s_1 and s_2 the irreductible sequence obtained after reduction of the sequence (s_1,s_2) defines a non commutative group structure on M . This law is denoted by :

$$(s_1,s_2) \longrightarrow s_1 s_2$$

The neutral element is the empty sequence and the inverse is denoted by s^{-1} .

Remark that the inverse of the control $((t_1,i_1),\dots,(t_j,i_j),\dots,(t_p,i_p))$ is the control $((-t_p,i_p),(-t_{p-1},i_{p-1}),\dots,(-t_j,i_j),\dots,(-t_1,i_1))$. One can also define the multiplication of a control by a real λ ; if the control is $s=((t_1,i_1),\dots,(t_j,i_j),\dots,(t_p,i_p))$ then by definition $\lambda.s$ is $((\lambda t_1,i_1),\dots,(\lambda t_j,i_j),\dots,(\lambda t_p,i_p))$.

I.1.3 - Proposition : The law defined above has the following properties :

$$1.s = s$$
$$(\alpha\beta).s = \alpha.(\beta.s)$$
$$\alpha(s_1 s_2) = (\alpha.s_1)(\alpha.s_2)$$

but in general :

$$(\alpha+\beta).s \neq (\alpha.s)(\beta.s) .$$

We say that a control is positive if every t_j of the sequence $((t_1,i_1),\ldots,(t_j,i_j),\ldots,(t_p,i_p))$ is positive ; we denote respectively by $G^+(I)$ and $G^-(I)$ the set of positive and negative controls. Exept if I is reduced to one element (in which case G(I) is R) the group G(I) is not the union of $G^+(I)$ and $G^-(I)$.

We topologize G(I) by the strongest topology for which the mappings :

$$(t_1,t_2,\ldots,t_p) \rightarrow t_1.s_1\ t_2.s_2,\ldots,t_p.s_p \ ;\ s_i \in G(I)\ ;\ p \in N\ ;$$

from R^p into G(I) are continuous.

We are also able to define a differentiable structure on G(I) by this way ; a mapping f from G(I) into a manifold M is smooth if and only if for every $s_1,s_2,\ldots,s_p$ in G(I) the mapping :

$$(t_1,t_2,\ldots,t_p) \longmapsto f(t_1.s_1,\ldots,t_i s_i,\ldots,t_p s_p)$$

is a smooth mapping.

I.1.4 - Definition : A Dynamical Polysystem controlled by G(I) on a manifold M is a mapping :

$$\Pi : G(I) \times M \longrightarrow M$$

which has the following properties :

i) $\Pi(0,x) = x \ ; \quad x \in M$

ii) $\Pi(s_1,s_2,x) = \Pi(s_1(\Pi(s_2,x)); \ x \in M \quad s_1 \in G(I) , \ s_2 \in G(I)$,

iii) the mapping Π is smooth.

The last statement means :

for every sequence $s_1 s_2,\ldots,s_p$ of elements of G(I) the mapping

$$(t_1,t_2,\ldots,t_p,x) \longrightarrow \Pi(t_1.s_1,\ldots,t_i.s_i,\ldots,t_p.s_p,x)$$

is a smooth mapping from $R^p \times M$ into M .

If the set I has only one element the definition reduces to the definition of a smooth dynamical system because there will not be possible confusion between elements of G(I) and elements in M we denote the action of $G(I)$ on M by :

$$\Pi(s,x) = sx$$

I.1.5 - Definition : The set :

$$G(I)x =' \{sx \ ; \ s \in G(I)\}$$

is the orbit of x , the set :

$$G^{+}(I)x =' \{ sx \ ; \ s \in G^{+}(I)\}$$

is the positive orbit of x and the set :

$$G^{-}(I)x =' \{ sx \ ; \ s \in G^{-}(I)\}$$

is the negative orbit of x .

For every x in M the formula :

$$X^{i}(x) = \frac{d}{dt} (\lambda(t,i)x)_{t=o}$$

defines a vector in TM_x and $x \to X^i(x)$ defines a vector field on M .

I.1.5 - Definition : The family $(X^i; \ i \in I)$ of vector fields defined above is the infinitesimal transformation of the Dynamical Polysystem.

Conversely let $(X^i; \ i \in I)$ be a family of complete vector fields on M ; the formula :

$$(t_1,i_1)(t_2,i_2),\ldots,(t_p,i_p)x = X_{t_1}^{i_1} \circ X_{t_2}^{i_2} \circ \ldots \circ X_{t_p}^{i_p}(x)$$

defines a Dynamical Polysystem controlled by $G(I)$ and its infinitesimal transformation is the family $(X^i; \ i \in I)$. Moreover if two Dynamical Polysystems have the same infinitesimal transformation they coincide. By this way, we see that (as in the case of smooth dynamical systems), a Dynamical Polysystem is equivalent to a family D of complete vector fields on M . We introduce now some notations.

Let D be a subset of $V(M)$, let $G(D)$ be the control group modelled on D and denote by the same symbol $G(D)$ the Dynamical Polysystem controlled by $G(D)$ whose infinitesimal transformation is the family D ; we call it the Dynamical Polysystem generated by D . If s is an element of $G(D)$ the mapping $x \to sx$ is a smooth mapping from M into M ; its derivative at point x is a linear isomorphism from TM_x into TM_{sx} and is denoted by $\xi \to Dsx.\xi$.

Remark : To a non complete vector field corresponds the notion of local Dynamical system. One can easily define the corresponding notion of local Dynamical Polysystem generated by family of vector fields which are not necessarily complete.

§2. Controlled systems considered as Dynamical Polysystem.

Consider on R^n (or on a manifold M) the control system :

$$\frac{dx}{dt} = f(x,t,u) \; ; \; x \in R^n \; , \; u \in U \subset R^p$$

and assume that f is smooth with respect to x,t,u and satisfies standard growth assumptions in order to ensure that for every piecewise smooth control $t \to \mathfrak{U}(t)$ defined on R the response :

$$x(t,t_o,x_o,\mathfrak{U})$$

is defined for every initial condition t_o, x_o and every t in R . Suppose we are concerned by a cost criterium of the type :

$$\int_{t_o}^{t} f^o(x(t,t_o,x,\mathfrak{U}),t,\mathfrak{U}(t))\,dt \; .$$

Then, we consider the set D of all vector fields in R^{n+2} defined by :

$$X \quad (x^o,x,x^{n+1}) = \begin{pmatrix} f^o(x,x^{n+1},\mathfrak{U}(x^{n+1})) \\ f(x,x^{n+1},\mathfrak{U}(x^{n+1})) \\ 1 \end{pmatrix}$$

where $\mathfrak{U}$ ranges over the set of all smooth mappings from R into U . The elements X of D are complete vector fields ; let G(D) be the D.P. generated by D .

Suppose $t \to \mathfrak{U}(t)$ is a piecewise smooth control, this means that $\mathfrak{U}$ has a finite number of switches on every compact interval, and on any maximal open interval without switch $\mathfrak{U}$ is equal to the restriction of some smooth mapping. So on the interval $[t_o,t_1]$ we have the following picture :

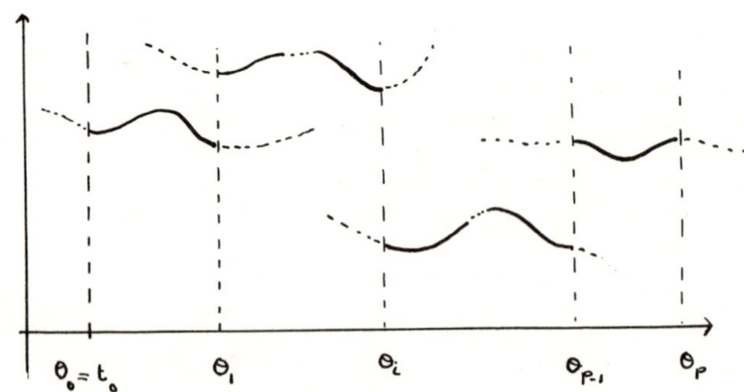

- A piecewise smooth control.

On each interval $]\theta_i,\theta_{i+1}[$, the control $\mathfrak{U}$ is the restriction of some smooth mapping $\mathfrak{U}_{i+1}$. Let us consider the control (in the sense of §1) of G(D) defined by :

$$s = ((\theta_p-\theta_{p-1},X^p),\ldots,(\theta_{i+1}-\theta_i,X^{i+1}),\ldots,(\theta_1-\theta_o,X^1)) .$$

I.2.1 - <u>Proposition</u> : The point :

$$\begin{pmatrix} y_o^o \\ y \\ y^{n+1} \end{pmatrix} = s \begin{pmatrix} 0 \\ X_o \\ t_o \end{pmatrix}$$

in R^{n+2} satisfies :

I.2.2

$$y^o = \int_{t_o}^{t_1} f^o(x(t,t_o,x_o,\mathfrak{U}),t,\mathfrak{U}(t))\,dt$$

$$y = x(t_1,t_o,x_o, \quad)$$

$$y^{n+1} = t_1$$

Proof of prop.I.2.1 : By definition of the action of s on the point $\begin{pmatrix} 0 \\ x_o \\ t_o \end{pmatrix}$ we obtain $s\begin{pmatrix} 0 \\ x_o \\ t_o \end{pmatrix}$ by integrating first the system :

$$\frac{dx^o}{dt} = f^o(x,x^{n+1},\mathfrak{U}_1(x^{n+1})) \quad x^o(0) = 0$$

$$\frac{dx}{dt} = f(x,x^{n+1},\mathfrak{U}_1(x^{n+1})) \quad x(0) = x_o$$

$$\frac{dx^{n+1}}{dt} = 1$$

up to time $\theta_1-\theta_o$. The last equation gives us :

$$x^{n+1}(t) = \theta_o+t = t_o+t \ , \ \theta_o \leq \theta_o+t \leq \theta_1$$

so in the interval considered we have :

$$\mathfrak{U}_1(x^{n+1}(t)) = \mathfrak{U}_1(\theta_o+t) = \mathfrak{U}(\theta_o+t)$$

and then :

$$x^o(\theta_1-\theta_o) = \int_{t_o}^{\theta_1} f^o(x(t,t_o,x_o,\mathfrak{U}),t,\mathfrak{U}(t))\,dt$$

$$x(\theta_1-\theta_o) = x(\theta_1,t_o,x_o,\mathfrak{U})$$
$$x^{n+1}(\theta_1-\theta_o) = \theta_1 \quad ;$$

starting from this point with vector field X^2 during time $\theta_2-\theta_1$ we find the cost at time θ_2, the response at time θ_2 and time θ_2 itself ; and so on ...

By this proposition we associate to a control in the usual meaning a "control" in the sense of an element of G(D) ; the converse is also true : to every element of G(D) one can associate in evident way a control in the usual sense for which relations I.2.2 of prop.I.2.1 are satisfied.

Now we have as immediate corollaries of prop. I.2.1.

I.2.3 - Proposition : The reachable set for control system (1) in the augmented space including time and cost criterium is the positive orbit :

$$G^+(D)\begin{pmatrix}0\\x_o\\t_o\end{pmatrix} \quad .$$

The reachable set at time t_1 is the section of $G^+(D)\begin{pmatrix}0\\x_o\\t_o\end{pmatrix}$ by the hyperplane defined by equation $x^{n+1}=t_1$.

From the classical theory one knows that a great number of interesting properties of control systems are related to the geometry of the reachable set (in the augmented space or not) (see Lee-Markus (1)) ; so the study of the structure of $G^+(D)$ gives us the opportunity to prove some results in control theory. The procedure given here to relate a control system to a dynamical polysystem is not unique. This was only an example (cf.III).

II - STRUCTURE OF TRAJECTORIES OF A D.P.

We consider on a manifold M the Dynamical Polysystem G(D) generated by some family D of complete vector fields. We prove (in §1) that G(D).x is a smooth manifold, the dimension of which is determined by the "infinite Taylor expansion of the system" at point x in the analytical case ; in §2 we prove that in the C^∞ case "almost all systems" are trivial in the sense that G(D).x is the whole space. The last paragraph is devoted to the study of $G^+(D).x$.

§1. The orbit G(D).x is a submanifold of M .

The natural topology of G(D).x is the topology induced by the topology of G(D) ; from the definition of the topology of G(D) it follows :

II.1.1 - A subset θ of G(D).x is open if and only if for every sequence $s_1,s_2,\dots,s_p$ of controls in G(D) the inverse image of θ under the mapping :

$$(t_1,t_2,\dots,t_p) \to t_1.s_1\ t_2.s_2\ t_3,\dots,t_p.s_p.x$$

is an open subset of R^p .

The orbit G(D).x is a hausdorff, arcwise connected space for this topology. The mapping $x \to sx$ is a diffeomorphism, thus if X is a vector field on M the formula :

II.1.2 $$sX(sx) = Dsx.X(x)$$

defines a new vector field sX .

II.1.3 - Definition : The closure of a family D of vector fields is the family :

$$\Delta(D) =' \{ sX ; s \in G(D) ; X \in D\} .$$

The closure of D is the smallest family which contains D and is G(D) invariant.

II.1.4 - Definition : The dimension of a D.P. G(D) at a point x is the dimension of the linear subspace of TM_x :

$$\mathcal{L}(\Delta(D)(x))$$

generated by the set :

$$\Delta(D)(x) =' \{ X(x) ; X \in \Delta(D)\} .$$

The first interesting thing is that the <u>dimension</u> is <u>constant along any orbit of</u> D . To prove this we first remark that, because G(D) is a group, it is enough to prove that for any x_o in M and any s_o in G(D) the dimension of G(D) at point $s_o x_o$ is greater than dimension at point x_o . In order to prove this last point we have to show that :

$$X \in \Delta(D) \Rightarrow Ds_o x_o . X(x_o) \in \mathcal{L}(\Delta(D)(s_o x_o))$$

by definition of $\Delta(D)$ the vector field X is of the form :

$$X = sY$$

for some s in G(D) and Y in D ; from this we get :

$$Ds_o x_o . X(x_o) = Ds_o x_o o Dsy_o . Y(y_o) \quad ; \quad y_o = s^{-1} x_o$$

$$Ds_o x_o . X(x_o) = Ds_o sy_o . Y(y_o)$$

which proves that $Ds_o x_o . X(x_o)$ is an element of $\Delta(D)(s_o x_o)$.

The linear subspace $(\Delta(D)(y))$ is a good candidate to be the tangent space $T(G(D)x)_y$ of G(D)x at point y . To prove this we shall construct a local parametrization at point y ; because group property there is no loss of generality in assuming y=x , thus :

II.1.5 - <u>Lemma</u> : Let p be the dimension of the system at point x, then there exist :

i) points $y_1, y_2, \ldots, y_p$ in G(D).x ,

ii) controls $s_1, s_2, \ldots, s_p$ in G(D) ,

iii) vector fields $X^1, X^2, \ldots, X^p$ in D ,

such that the mapping :

$$(t_1, t_2, \ldots, t_p) \to \varphi_x(t_1, t_2, \ldots, t_p) = s_p(t_p, X^p) s_p^{-1} \; s_{p-1} \ldots$$

$$s_i(t_i, X^i) s_i^{-1} \ldots s_2^{-1} s_1(t_1, X^1) y_1$$

restricted to a suitable neighbourhood of 0 in R^p is a local parametrization of rank p (see A.2.1) of G(D).x at point x .

Proof of lemma II.1.5 : To prove the lemma we have to prove that $_x(t_1, t_2, \ldots, t_p)$ belongs to G(D).x , which is evident by construction, that $\varphi_x(0,0,\ldots,0) = x$, which is also evident by construction, and that the rank of the mapping φ_x in the underlying euclidien space of M has maximum rank p at 0 . To prove the last point it is enough to prove that the mapping φ_x , as a mapping from R^p into M , has rank p at 0 . Choose a basis $\xi_1,\xi_2,\dots,\xi_p$ of $\mathfrak{L}(\Delta(D)(x))$ made of elements of $\Delta(D)(x)$; thus by definition :

$$\xi_i = s_i X^i(x) = Ds_i y_i . X^i(y_i)$$

for some controls s_i in G(D), some vector fields X^i in D and some points y_i in G(D).x ; i=1,2,...,p . For the mapping defined in the statement we have :

$$\varphi_x(0,0,\dots,0,t_i,0,\dots,0) = s_i(t_i,X^i)s_i^{-1}\, s_1 y_1 = s_i(t_i,X^i)y_i \;;$$

(see the picture) :

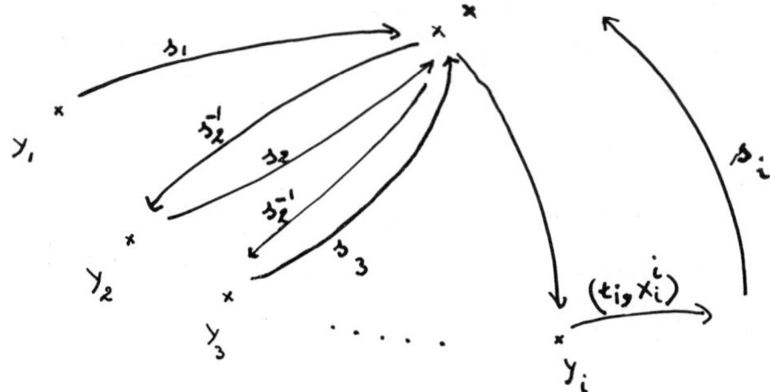

Then, we have :

$$\left.\frac{\partial_x}{\partial t_i}\right|_{t=0} = Ds_i y_i . X^i(y_i) \;,$$

which proves the lemma.

We now prove a lemma which will enable us to prove that the local coordinate systems defined above are compatible.

II.1.6 - Lemma : Let $(\varphi_x,\mathcal{V}_x)$ be a local parametrization of G(D).x at point x such that :

(h) $t\in\mathcal{V}_x \Rightarrow I_m(D\varphi_x(t)) = \mathfrak{L}(\Delta(D)(\varphi_x(t)))$.

Then, for any sequence $s_1,s_2,\dots,s_q$ of controls in G(D) and any point y in G(D).x the inverse image of $\varphi_x(\mathcal{V}_x)$ under the mapping :

$$(t_1,t_2,\dots,t_q) \to t_1.s_1\, t_2.s_2,\dots,t_q.s_q\, y$$

is an open subset of R^q .

Proof of Lemma II.1.6 : By a straightforward induction argument the problem reduces to mappings of the form :

$$t \to tsy \quad ;$$

but we have :

$$s=(t_1,X^1)(t_2,X^2)(\dots)(t_r,X^r)$$

thus the mapping considered is a composed mapping obtained from the two mappings :

$$\lambda_1,\lambda_2,\dots,\lambda_r \to (\lambda_1,X^1)(\lambda_2,X^2)\dots(\lambda_r,X^r)$$

$$t \to t\, t_1 \ , t\, t_2,\dots,t\, t_r$$

and the problem reduces to a mapping :

$$t \to \gamma(t) = s(t,X)y \quad , \quad X \in D \ .$$

By definition of the vector field sX , the mapping γ is a solution of the differential equation :

$$\frac{dz}{dt} = sX(z) \ .$$

From the property (h) we assumed on φ_x we know that the vector field sX is tangent to the manifold defined by $\varphi_x(\mathcal{V}_x)$; thus if $\gamma(t_o)$ belong to $\varphi_x(\mathcal{V}_x)$, the same holds for t close enough to t_o (see A.5.5). Lemma II.1.6 is proved.

We now prove the :

II.1.7 - Theorem (Sussmann (1)). Consider an orbit $G(D).x_o$ of some point x_o in M under the action of a dynamical polysystem G(D). For any x in $G(D).x_o$ choose a local parametrization $(\varphi_x,\mathcal{V}_x)$ as defined by lemma II.1.5.

1) The family $(\varphi_x,\mathcal{V}_x)_{x \in M}$ is an atlas of $G(D).x_o$
2) The topology induced by the atlas is the topology induced by G(D).

Proof of theorem II.1.7 : In order to prove 1), we have to prove that any two local parametrizations $(\varphi_x,\mathcal{V}_x)$ and $(\varphi_y,\mathcal{V}_y)$ of the atlas (see A.2.2, A.2.3) are compatibles. Let t be a point in $\mathcal{V}_x$; the image of the linear map $D\varphi_x(t)$ is generated by the vectors :

$$D\, s_p(t_p,X^p)s_{p-1}\dots s_i(y_i(t)).X^i(y_i(t)) \ ;$$

where :

$$y_i(t) = (t_i, X^i) s_i^{-1} s_{i-1} \dots s_2^{-1} s_1 (t_1, X^1) y_1$$

$$t_1, t_2, \dots, t_p) = t \quad .$$

Thus the image of $D\,\varphi_x(t)$ is exactly the subspace $\mathcal{L}\,(\Delta(D)(\,\varphi_x(t)))$ (because dimension is constant and equal to p along the orbit $G(D).x_o$), and then lemma II.1.6 proves that the two local parametrizations are compatible. We now prove 2). The topology defined by this atlas is strongest than the topology defined by G(D), because every local parametrization is the restriction of some mapping :

$$(t_1, t_2, \dots, t_q) \to t_1.s_1\; t_2.s_2 \;\dots\; t_q.s_q\; x_o \;;$$

conversely we know by lemma II.1.6 that the inverse image of $\varphi_x(\mathcal{V}_x)$ under such a mapping is an open subset of R^q , this proves (by I.1.1) that $\varphi_x(\mathcal{V}_x)$ is an open subset for the topology induced by G(D) ; this will be true for any open neighbourhood θ_x contained in $\mathcal{V}_x$ (because φ_x, θ_x is a local parametrization of the same type 1) and this achieves the proof because the subsets $\varphi_x(\theta_x)$ define a basis for the manifold topology of $G(D).x_o$.

II.1.8 - The family of subspaces $\mathcal{L}(\Delta(D)(x))$ is called a distribution (i.e. to each point x in M one defines a subspace $_x$ of tangent space TM_x) ; a manifold N is called an "integral submanifold" of a distribution $\mathcal{L}$ iff for every x in N the tangent space TN_x to N is precisely $\mathcal{L}_x$. From lemma II.1.6 one can derive that the submanifold $G(D).x_o$ defined by Th.II.1.7 is the unique maximal integral submanifold of the distribution defined by $\mathcal{L}(\Delta(D)(x))$. For further development on this subjet see Sussmann (1) Sussmann and Jurdjevic (1) and Stefan (1).

II.1.9 - The submanifold $G(D).x_o$ is not necessarilly regularly imbeded (A.4.1). For instance take the manifold T×R , where T is the torus. The torus is the space $[0,1] \times [0,1]$ where $\{0\}\times[0,1]$ is identified to $\{1\}\times[0,1]$ and $[0,1]\times\{0\}$ to $[0,1]\times\{1\}$; take on the torus the vector field $X=\begin{pmatrix}1\\ \alpha\end{pmatrix}$, α irrationnal, in this case the image of $t \to X_t(x_o)$ is a dense subset of T ; now, on T×R one takes the family D generated by the two vector fields $X^1 = \begin{pmatrix}1\\ \alpha\\ 0\end{pmatrix}$

and $X^2 = \begin{pmatrix} 0 \\ 0 \\ 1 \end{pmatrix}$; for the D.P. generated by D any orbit is of the form $\{X_t(x_o) ; t \in R\} \times R$, which is dense.

§2. The analytical and generic cases.

We denote by $[D]^\infty$ the smallest family, closed under lie-bracket operation, which contains D . We call it the Lie-closure of D .

II.2.1 - Definition : The rank of a Dynamical Polysystem G(D) at a point x , is the dimension of the linear space $\mathcal{L}([D]^\infty(x))$ generated by values at point x of the elements of the Lie-closure of its infinitesimal transformation.

II.2.2 - Proposition : The rank of a Dynamical Polysystem is smaller than its dimension.

Proof of Prop.II.2.2 : The Lie-closure of $\Delta(D)$ contains the Lie-closure of D ; from theorem II.1.7 and A.5.8, we know that if X and Y are in $\Delta(D)$, then their bracket $[X,Y]$ is also a vector field tangent to G(D).x , for any x in M . So the dimension of the linear space $\mathcal{L}([\Delta(D)]^\infty(x))$ is exactly the dimension of G(D) at point x . Thus the rank, which is the dimension of $\mathcal{L}([D]^\infty(x))$, is smaller than the dimension of G(D).x . It may happen that the inequality is strict for every point in M . For instance, take $M=R^3$ and the two vector fields :

$$X = \frac{\partial}{\partial x} + \Phi(x)\frac{\partial}{\partial z} \ ; \ Y = \frac{\partial}{\partial x} + \psi(x)\frac{\partial}{\partial y}$$

where Φ and ψ are smooth mappings such that :

$$x \leq +1 \Rightarrow \Phi(x) = 0 \ ; \ x > +1 \quad \Phi(x) > 0$$
$$x \leq -1 \Rightarrow \psi(x) > 0 \ ; \ x > -1 \quad \psi(x) = 0$$

in this case the rank is 2 at every point an the dimension is 3.

Thus in general the Lie-closure of D cannot characterise the dimension of the system ; the pathology in example above comes from the fact that the mappings Φ and ψ vanish identically on a half line ; this cannot occur in analytical and in generic case. The aim of this § is to prove it.

II.2.3 - Definition : A familly D of vector fields is "locally finitely generated" if for any point $x \in M$ there exists a familly X^i ; $i=1,2,\dots,p$ of vector fields of D such that, for any X in D, there is a neighbourhood $\mathcal{V}_x$ (which may depend on X) of x and smooth functions f_i such that :

$$X(y) = \sum_{i=1}^{p} f_i(y)\, X^i(y) \qquad y \in \mathcal{V}_x \;.$$

II.2.4 - Lemma : If D is a "locally finitely generated" familly of vector field closed under Lie-bracket operation, then the dimension at x of the Dynamical Polysystem generated by D is equal to the dimension of the space generated by values at x of elements of D (which in this case is the rank of the D.P.).

Proof of Lemma II.2.4 : The matter is to prove that for any s in G(D), any x in M and any X in D the vector $sX(sx) = Dsx.X(x)$ belongs to the linear space generated by vectors $X(x)$; $X \in D$. By a straigthforward induction argument
(★) this reduces to the case where $s = (\mathbf{s},Y)$, for some Y in D ; now, by usual argument based on connexity of $[0,\mathbf{s}]$ this reduces in proving :

Given X in D and x in M, there is $\varepsilon > 0$ such that for any $s, |s| < \varepsilon$, the vector $D(s,Y)x.X(x)$ is an element of $(D(x))$.

First we can choose a local coordinate system ; thus now the familly D is defined on R^n and x is the origin. Let $X^1, X^2, \dots, X^p$ be the vectors of definition II.2.3 ; there exists a neighbourhood $\mathcal{V}$ of 0 such that for every y in this neighbourhood one has :

$$[X^i, Y](y) = \sum_{j=1}^{p} f_{ij}(y)\, X^j(y) \;;$$

because D is assumed to be closed under brackets.
Choose ε so small that :

$$Y_s(0) \in \mathcal{V} \quad |s| < \varepsilon \;.$$

Denote by $V^i(s)$ the vector of R^n :

$$V_i(s) = (DY_s(o))^{-1}\, X^i(Y_s(o)) \;,\; i=1,2,\dots,p$$

by definition of the Lie-bracket (A.5.6) one has :

$$V_i'(s) = (DY_s(o))^{-1}\, [X^i, Y](Y_s(o)) \; i=1,2,\dots,p$$

(★) We parametrize the family D by D itself; thus by definition one has : $(s,Y)x = Y_s(x)$.

thus :

$$V_i'(s) = (DY_s(o))^{-1} \sum_{j=1}^{p} f_{ij}(Y_s(o))\, X^j(Y_s(o))$$

and by the linearity of $(DY_s(o))^{-1}$:

$$V_i'(s) = \sum_{j=1}^{p} f_{ij}(Y_s(o))\, V_i(s) \quad i=1,2,\dots,p \ .$$

The vectors $V_i(s)$ can be considered as the column vectors of an $n\times p$ matrice $V(t)$ which is solution of the differential equation :

$$V'(s) = V(s)^t.(f_{ij}(Y_s(o))) \ .$$

Let $(\phi(s))$ be the fondamental matrix of this system ; then :

$$V(s) = V(o) \ \ \phi(s)$$

$$DY_s(o)\ V(s)\ \phi^{-1}(s) = DY_s(o)\ V(o)$$

and thus, by definition of $V_i(s)$,

$$(X^1(Y_s(o)),X^2(Y_s(o)),\dots,X^p(Y_s(o)))\phi^{-1}(s) =$$

$$(DY_s(o).X^1(o),DY_s(o).X^2(o),\dots,DY_p(o).X^p(o)) \ ;$$

this proves that $DY_s(o).X^i(o)$ is a linear combination of elements $X^j(Y_s(o))$ of the set $D(x)$. We have prove that $DY_s(o).X^i(o)$ belongs to $(D(Y_s(o)))$ and thus by linearity, for any X in D the vector $DY_s(o).X(o)$ belongs to $(D(Y_s(o)))$; this proves the lemma.

II.2.5 - <u>Theorem</u> : If D is a family of analytic vector fields on an analytic manifold M, then the dimension of the orbit $G(D).x_o$ through any point x_o is the rank of the system at this point .

Proof of theorem II.2.5 : Take the Lie-closure $[D]^\infty$ of D . This family is closed under Lie brackets (!) and it follows from algebra that any family of analytic vector fields is "locally finitely generated". (This comes from the fact that the set of "germs" of vector fields at 0 is a Noeterian module on the ring of analytic functions ; see Lobry (2)).

So in the analytical case the dimension of $G(D).x$ is caracterised by the coefficients of power expansion at point x . In the algebraic case, if the family D has a finite number of elements, this can be done by a finite number of computations.

Now we want to prove that "almost all systems" have rank n (where n is the dimension of the manifold M) at any point x in M . In order to avoid little technical difficulties, we shall restrict to the case where M is an open subset of R^n ; the general case can be transcribed easily if one introduce Jet bundles.

II.2.6 - Theorem : The subset G of the set of pair of vector fields on an open subset U of R^n such that :

$$\left.\begin{matrix} X \in G \\ Y \in G \end{matrix}\right\} \Rightarrow \operatorname{rank}\left\{\begin{matrix} [X\ Y]^1(x), & [X\ Y]^2(x), \dots, [X\ Y]^{2n}(x) \\ [Y\ X]^1(x), & [Y\ X]^2(x), \dots, [Y\ X]^{2n}(x) \end{matrix}\right\} = n$$

where $[X\ Y]^i$ is defined by induction :

$$[X,Y]^0 = X$$
$$[X,Y]^{i+1} = \quad X,Y^{\,i},Y \quad ,$$

is an open subset for the norm $\|\ \|_{2n,K}$ (see B.2) and is dense for any norm $\|\ \|_{k,K}$, $k \in N$.

Proof of theorem II.2.6 : Let us consider the space of k-jets ($k \geq 2n$) of mappings from R^n into $R^n \times R^n$ (considered as pair of vector fields). Denote by $\overline{X}$, $\overline{Y}$ the k-jet of such a pair. The bracket operation induces natural mappings from $J^k_{n,2n}$ into R^n ; we denote them by :

$$(\overline{X}, \overline{Y}) \to [\overline{X}, \overline{Y}]^{(p)}$$
$$(\overline{X}, \overline{Y}) \to [\overline{Y}, \overline{X}]^{(p)} .$$

By definition $[\overline{X}, \overline{Y}]^{(p)}$ is the value at 0 of $[X , Y]^p$ where X and Y are two vector fields whose k-jet at 0 are respectively $\overline{X}$ an $\overline{Y}$; this definition does not depend on the choice of X and Y . This mappings are well defined for any p less than k ; thus the two mappings :

$$(x,\overline{X},\overline{Y}) \to G(x,\overline{X},\overline{Y}) = ([\overline{X}\ \overline{Y}]^{(1)}, [\overline{X},\overline{Y}]^{(2)}, \dots, [\overline{X},\overline{Y}]^{2n})$$

$$(x,\overline{X},\overline{Y}) \to H(x,\overline{X},\overline{Y}) = ([\overline{X},\overline{Y}]^{(1)}, [\overline{X},\overline{Y}]^{(2)}, \dots, [\overline{X},\overline{Y}]^{2n})$$

are well defined ; (actually they don't depend on x). Let $M_{n-1}(n,2n)$ be the subset of all matrices whose rank is strictly smaller than n-1 ; from the construction it is now clear that two vector fields X and Y are in G if at least one of the two following holds (see B.1 for notations) :

$$G(x,J_k(X,Y)(x)) \notin M_{\leq n-1}(n,2n)$$
$$H(x,J_k(X,Y)(x)) \notin M_{\leq n-1}(n,2n) \quad ;$$

thus the density part of theorem (openness is trivial) is a consequence of B.2.1, if we prove that :

$$S = G^{-1}(M_{\leq n-1}(n,2n)) \cap H^{-1}(M_{\leq n-1}(n,2n))$$

is contained in a finite number of submanifolds of codimension strictly greater than n. Let X_o and Y_o be the first components of a k-jet of pair of vector fields ; consider the covering of $J^k_{n,2n}$ in 3 subsets :

$$\Delta =' \{\overline{X}, \overline{Y} ; \overline{X}_o = \overline{Y}_o = 0\}$$
$$E_1 =' \{\overline{X}, \overline{Y} ; \overline{X}_o = 0\}$$
$$E_2 =' \{\overline{X}, \overline{Y} ; \overline{Y}_o = 0\}$$

The set Δ has codimension $2n$ in $J^k_{n,2n}$, then $U\times\Delta$ has codimension $2n$ in $U\times J^k_{n,2n}$. Thus it remains to prove that the restriction of $G^{-1}(M_{\leq n-1}(n,2n)) \cap H^{-1}(M_{\leq n-1}(n,2n))$ in the open sets $U\times E_1$ and $U\times E_2$ are contained in finite union of manifolds of codimension strictly greater than n ; but because G and H does not depend on x, **and because of the symmetry it** is enough to prove that the set :

$$W =' \{\overline{X},\overline{Y} ; \operatorname{rank}([\overline{X}\,\overline{Y}]^{(1)}, [\overline{X}\,\overline{Y}]^{(2)},\ldots,[\overline{X}\,\overline{Y}]^{2n}) = n\}$$

is contained in a finite number of submanifolds of E_2 of codimension strictly greater than n. Because we are in E_2 the vector $\overline{Y}_o$ is not 0 ; so it can be considered as the k-jet of some vector field Y which has no critical point at 0 ; by a smooth change of coordinate, which induces a linear diffeomorphism of $J^k_{n,2n}$, we can replace Y by the vector field $\frac{\partial}{\partial x_1}$; thus we can write with evident notations :

$$[\overline{X},\overline{Y}]^{(1)} = \frac{\partial \overline{X}}{\partial x_1},\ [\overline{X},\overline{Y}]^{(2)} = \frac{\partial^2 \overline{X}}{\partial x_1^2},\ldots,[\overline{X},\overline{Y}]^{(p)} = \frac{\partial^p \overline{X}}{\partial x_1^p} \ ;$$

the mapping from E_2 into $M(n,2n)$:

$$\overline{X} \to \left(\frac{\partial \overline{X}}{\partial x_1}, \frac{\partial^2 \overline{X}}{\partial x_2^2},\ldots,\frac{\partial^{2n} \overline{X}}{\partial x_1^{2n}}\right) = \Gamma(\overline{X})$$

has maximum rank ; by A.6.1 the set $M_{\leq n-1}(n,2n)$ is a collection of finite number of submanifolds of codimension $(n-k)(2n-k)$ n+1 ; the same is true for its inverse image by Γ in E_2. This achieves the proof of the theorem.

This result extend to a general manifold in the following :

II.2.7 - Theorem : The set of Dynamical Polysystems G(D) whose rank is n at every point of the manifold M contains an open dense subset for the C^k topology of whitney $(k \geq n)$.

So, almost all dynamical polysystems on a connected manifold are trivial, because if the rank is n there exists only one orbit for if it was not the case trajectories would define a covering of M by open subsets. To make reference to the notion of general position of linear control systems we introduce the :

II.2.8 - Definition : A Dynamical Polysystem G(D) on a manifold M of dimension n is in general position if the rank of G(D) is n at every point.

II.2.9 - Conclusion : In the smooth case, general position holds for almost all systems ; in analytic case general position holds for the restriction of G(D) to each of its orbits.

§3. Positive orbits.

We are now concerned by the positive orbits. One cannot expect the positive orbit to be a manifold, (take $D=\{\frac{\partial}{\partial x},\frac{\partial}{\partial y}\}$ in R^2!), so we will give only one topological result on positive orbits.

II.3.1 - Theorem : Let G(D) be a Dynamical Polysystem in general position on a manifold M ; then the interior of G(D).x is dense in G(D).x .

Proof of theorem II.3.1 : (Krener (1)) Take an open neighbourhood $\mathcal{U}$ of x ; because the rank of the system at x is n there exists at least X^1 in D such that $X^1(x)$ is not 0 ; thus the set $S_1 = \{X^1_{s_1}(x) \; ; \; 0 < s_1 < \varepsilon_1\}$ is a smooth manifold of dimension 1 if we choose ε_1 small enough ; moreover we choose it such that S_1 is contained in $\mathcal{U}$. There exists at least one point t_1 in $[0,\varepsilon_1]$ and an element X^2 of D such that $X^1(X_{t_1}(x))$ and $X^2(X_{t_2}(x))$ are independant for if not it would say that all the vectors of D are tangent to the submanifold of dimension S^1 and thus the rank on S^1 would be 1 which is impossible (if $n>1$!). Let S^2 be the two dimensional manifold :

$$S_2 = \{ X^2_{s_2} \circ X^1_{s_1}(o);\ 0 < s_2 < \varepsilon_2 \ ;\ t_1-\varepsilon_2 < s_1 < t_1+\varepsilon_2 \}$$

where ε_2 is chosen in such a way that :

i) $0 < t_1-\varepsilon_2 < t_1+\varepsilon_2 < \varepsilon_1$

ii) the maps $(s_1,s_2) \to X^2_{s_2} \circ X^1_s(o)$ is of rank 2 every where

iii) S_2 is contained in $\mathcal{U}$.

If n is strictly greater than 2 one can define in the same way S_3 , and by a trivial induction we prove existence of :

i) n vector fields $X^1\ X^2 \ldots X^n$ of D (not necessarilly distinct)

ii) n strictly positive numbers $t_1,t_2,\ldots,t_n$,

such that :

$$X^n_{t_n} \circ X^{n-1}_{t_{n-1}} \circ \ldots \circ X^1_{t_1}(x) \quad \in \mathcal{U}$$

and the mapping :

$$(s_1,s_2,\ldots,s_n) \to X^n_{s_n} \circ X^{n-1}_{s_{n-1}} \circ \ldots \circ X^1_{s_1}(x) = K_{\mathcal{U}}(s_1,s_2,\ldots,s_n)$$

has rank n at point $t_1,t_2,\ldots,t_n$; thus by inverse function theorem there exists a neighbourhood $\mathcal{V}_t$ of $(t_1,t_2,\ldots,t_n)$ such that :

$$s=(s_1,s_2,\ldots,s_n) \in \mathcal{V}_t \Rightarrow \quad s_1 > 0, s_2 > 0, \ldots, s_n > 0$$

and the set $K_{\mathcal{U}}(\mathcal{V}_t)$ is an open subset of $\mathcal{U}$. Thus we have proved the existence of an interior point of $G^+(D).x$ in the neighbourhood $\mathcal{U}$ of x . Let y be a point in $G^+(D).x$, choose a neighbourhood $\mathcal{W}$ of y . Let s be a control in $G^+(D).x$ such that $s.x = y$. Take $s^{-1}(\mathcal{W}) = \mathcal{U}$ in the preceding ; the point $s\,K_{\mathcal{U}}(t_1,t_2,\ldots,t_n)$ is obviously an interior point of

$G^+(D).x$. This completes the proof of the theorem.

If the hypothesis of general position is not assumed th.II.3.1 is not true ; take for instance in R^2 ; $D=\{\frac{\partial}{\partial x}; \frac{\partial}{\partial x}+\varphi(x)\frac{\partial}{\partial y}\}$ where the mapping φ is such that :

$$x \le 1 \Rightarrow (x) = 0$$

$$x < 1 \Rightarrow (x) \ne 0$$

and take the origin as starting point. Obviously one only needs that the rank is maximum only on a dense subset of M .

As a rule $G^+(D).x$ is not closed ; for instance take in R^3 the familly $D = \{(1-y^2)\frac{\partial}{\partial x}+\frac{\partial}{\partial y}+\frac{\partial}{\partial z} ; (1-y^2)\frac{\partial}{\partial x}-\frac{\partial}{\partial y}+\frac{\partial}{\partial z}\}$; (see Lobry (1)) . For further development on analytic case, see Sussmann and Jurdjevic (1), for connections with theory of integration of distribution see Sussmann (1), and for sufficient condition for closedness of $G^+(x)$ see Krener (1).

III - APPLICATIONS TO CONTROL THEORY.

We give some examples of applications to control theory of the techniques of I and II. This examples are given as an illustration, there are other ones yet known and no doubt that there are others to be found.

§1. Controllability of linear systems.

Let us consider the system :

III.1.1 $$\frac{dx}{dt} = A(t)x + b(t)u \quad x \in R^n \; ; \; |u| \le 1 \quad t \ge t_0$$

assume that $t \to A(t)$ and $t \to b(t)$ are smooth, and define the mapping :

$$D_A : C^\infty(R, R^n) \to C^\infty(R, R^n)$$
$$V \to D_A V$$

$$D_A V(t) \underset{\text{def}}{=} -A(t)\, V(t) + V'(t)$$

III.1.2 - Theorem : (Weiss (1)) The reachable set from $(0,t_o)$ at time $t_o + \varepsilon$ by peicewise smooth controls is a symmetric convex neighbourhood of 0 if (and only if in analytical case) the rank of the infinite system of vectors :

$$b(t_o)\ ;\ D_A b(t_o), \ldots,\ D_A^k b(t_o), \ldots$$

is equal to n .

Proof of th.III.1.2 : Consider on R^{n+1} the family D of two vector fields :

$$X = \begin{pmatrix} A(t)x + b(t) \\ \hline 1 \end{pmatrix} \qquad Y = \begin{pmatrix} A(t)x - b(t) \\ \hline 1 \end{pmatrix}$$

and compute the brackets according to formula A.5.7. Before doing this one remarks that from well known Jacobi identity (i.e. $[[X,Y],Z] + [[Y,Z],X] + [[Z,X],Y] = 0$) and a straightforward induction argument, in order to compute the rank of a dynamical polysystem generated by a family D of vector fields, one only needs to compute the brackets of the form :

$$[\ldots[[X^1,X^2],X^3],\ldots],X^q]\quad X^i \in D\ ;\ q \in N\ .$$

In our case :

$$[X,Y](x,t) = \left(\begin{array}{c|c} A(t) & A'(t)x+b'(t) \\ \hline 0 & 0 \end{array}\right) \begin{pmatrix} +A(t)\quad x - b(t) \\ \hline 1 \end{pmatrix} -$$

$$\left(\begin{array}{c|c} A(t) & A'(t)x-b'(t) \\ \hline 0 & 0 \end{array}\right) \begin{pmatrix} A(t)x + b(t) \\ \hline 1 \end{pmatrix}$$

$$\frac{1}{2}[X,Y](x,t) = \begin{pmatrix} -A(t)\ b(t) + b'(t) \\ \hline 0 \end{pmatrix} = \begin{pmatrix} D_A\ b(t) \\ \hline 0 \end{pmatrix}$$

and

$$\frac{1}{2}[Y,X](x,t) = -\frac{1}{2}[X\ Y](x,t)\quad .$$

Compute :

$$\tfrac{1}{2}[[X,Y],X]\ (x,t)=\left(\begin{array}{c|c} 0 & (-A(t)b(t)+b'(t))' \\ \hline 0 & 0 \end{array}\right)\left(\begin{array}{c} A(t)x+b(t) \\ \hline 1 \end{array}\right)-$$

$$-\left(\begin{array}{c|c} A(t) & A'(t)x+b'(t) \\ \hline 0 & 0 \end{array}\right)\left(\begin{array}{c} -A(t)b(t)+b'(t)) \\ \hline 0 \end{array}\right)$$

$$\tfrac{1}{2}[[X,Y],X]\ (x,t)=\left(\begin{array}{c} -A(t)(-A(t)b(t)+b'(t))+(-A(t)b(t)+b'(t)) \\ \hline 0 \end{array}\right)=$$

$$\left(\begin{array}{c} D_A^2\ b(t) \\ \hline 0 \end{array}\right)$$

and in the same way :

$$\tfrac{1}{2}[[X,Y],\ Y]\ (x,t) = \begin{pmatrix} D_A^2\ b(t) \\ 0 \end{pmatrix} \quad ;$$

thus it is obvious now that we have :

$\mathrm{rank}(G(X,Y))$ at $t_o = \mathrm{rank}(b(t_o),\ D_A b(t_o),\dots,D_A^n b(t_o),\dots)+1$

so if the D.P. has rank $n+1$, the positive trajectory $G^+(X,Y)(0,t_o)$ has interior points in any neighbourhood of $(0,t_o)$; choose an interior point (Y,t_1) such that : $t_1-t_o \le \varepsilon$; the set :

$$G^+(X,Y)(0,t_o) \cap \{ (x,t)\ \ R^{n+1};\ \ t=t_1\}$$

is contained in the reachable set from $(0,t_o)$ at time t_1 for system III.1.1 (see I.2) so we have proved :

if rank of system $b(t),\ D_A b(t),\dots,D_A^n b(t)\ \dots$ is n for every $t \ge t_o$ then for any ε the reachable set at time $t_o \le t_1 \le t_o+\varepsilon$, by B.B controls has non empty interior ;

thus the reachable set at time $t_o+\varepsilon$ which contains reachable set at time t_1 by B.B. control, and is convex and symmetric (from linearity of the system), is a neighbourhood of 0 . The rank condition is necessary in analytical case, because if it is not satisfied the positive orbit $G^+(X,Y)(0,t_o)$ is a submanifold of R^{n+1} of codimension greater than 1 which is transverse to the hyperplane $\{ (x,t)\ ;\ t=t_o+\varepsilon\}$ so the reachable set at time $t_o+\varepsilon$ has an empty interior. This achieves the proof of theorem III.1.2.

As an other application let us get the classical :

III.1.3 - Theorem : (Kalman (1)) For the linear system :

(1) $$\frac{dx}{dt} = Ax + Bu \; ; \; x \in R^n \; ; \; u \in R^p$$

there exists a change of coordinate such that the system (1) is equivalent to system :

(2) $$\frac{dX_1}{dt} = A_{11}X_1 + A_{12}X_2 + B_1$$

$$\frac{dX_2}{dt} = A_{22}X_2 \;;$$

the first equation being completely controllable, moreover the dimension of X_1 is the rank of the matrix :

$$(B, AB, \ldots, A^{n-1}B) \; .$$

Proof of th.III.1.3 : Consider the family of vector fields in R^n :

$$D = \{ Ax \pm b_i \; ; \; i=1,2,\ldots,p\}$$

where b_i is the ith column of B. We are in the analytical case. The positive trajectory $G^+(D).0$ is (because of linearity) a linear subspace of R^n ; we know (II.3.1) that $G^+(D).0$ has a non empty interior in $G(D).0$, thus the dimension of $G^+(D).0$ is the dimension of $G(D).0$ i.e. the rank of the system at 0 . From the preceding computation (Th.III.1.2) and from Cayley Hamilton Theorem we know that this rank is precisely the rank of the matrix $(B,AB,\ldots,A^{n-1}B)$. Take a basis of R^n, $\varepsilon_1,\varepsilon_2,\ldots,\varepsilon_n$ such that ; $\varepsilon_1,\varepsilon_2,\ldots,\varepsilon_p$ is a basis of $G^+(D).0$ (which is the linear space generated by $(B,AB,\ldots,A^{n-1}B)$ because the columns of this matrix are tangent to $G(D).0$ at point 0) ; then denote by X_1 the first p-component of X and the last n-p by X_2 ; the system is :

$$\frac{dX_1}{dt} = A_{11}\,X_1 + A_{12}\,X_2 + B_1u$$

$$\frac{dX_2}{dt} = A_{21}\,X_1 + A_{22}\,X_2 + B_2u$$

any response to this system, issued from 0, satisfies $X_2(t)=0$ so one have :

$$0 = A_{21} X_1(t, \mathfrak{U}) + B_2 \quad (t)$$

for any $\mathfrak{U}(t)$; because $X_1(0, \mathfrak{U})$ is 0 , if B_2 is not zero we get a contradiction. Now, we have :

$$0 = A_{21} X_1(t, \mathfrak{U})$$

which implies $A_{21} = 0$ because the first equation is completely controllable. This proves the Th.III.1.3.

III.1.4 - Remark : This result extends to the non linear case (at least locally) if we suppose that the orbits G(D).y define a locally trivial foliation of M ; i.e. the orbits have the same dimension p and in suitable coordinate system they are defined by equations :

$$x_{p+1} = \alpha_1 \quad x_{p+2} = \alpha_2 \quad \dots \quad x_n = \alpha_n \quad ;$$

Actually the problem is to check it on the system.

§2. Controllability of non linear systems.

Let us define a symmetric system.

III.2.1 - Definition : A control system :

$$\frac{dx}{dt} = f(x,u) \; ; \; x \in R^n \quad ; \quad u \in U$$

is symetric if, for any u in U there exists $\bar{u}$ in U such that :

$$f(x,u) = -f(x,\bar{u}).$$

Notice that symmetric systems are very particular ; the symmetry assumption excludes non autonomous systems, and second order systems.

III.2.1 - Theorem : A symmetric control system :

$$\frac{dx}{dt} = f(x,u) \; ; \; x \in R^n \quad ; \quad u \in U$$

is completely controllable by piecewise constant controls if (and only if in analytical case), the rank of the dynamical polysystem generated by :

$$D = \{ f(x,u) \; ; \; u \in U \}$$

is n at every point.

The proof is trivial because, in this case, $G^{+}(D).x$ is equal to $G(D).x$; thus theorem follows from Th.II.1.7 and II.2.5.

We can remove this assumption of symmetry in the case of compact riemannian manifolds ; let us recall that a dynamical system is conservative if it preserves the natural measure of the manifold, by Carathe-Labry-Poincaré theorem one knows that Poisson stable points are dense for conservative systems.

III.2.2 - Definition : A conservative dynamical polysystem is a dynamical polysystem generated by conservative vector fields.

III.2.3 - Theorem : A conservative dynamical polysystem on a compact manifold M of dimension n is controllable (i.e. for any x the positive orbit is M itself) if (and only if in analytical case) the rank is n at every point.

Outline of the proof of th.III.2.3 (see Lobry (4) for complete proof).

One has to prove that $G^{+}(D).x = G(D).x$. First, we show $G^{+}(D).x$ is dense in $G(D).x$. This comes from the fact that if x is poisson-stable for some vector field X then the point $X_t(x)$ $(t>0)$ can be made arbitrarily close to $X_\tau(x)$ $(\tau<0)$ if one chooses t large enough ; so every backward motion can be replaced by a foreward close one. To conclude one has just to remark that $G^{-}(D).y$ has a non empty interior and y $G^{+}(D).z$ if z is an element of $G^{-}(D).y$.

It can be proved (see Lobry (2)) : in the class of conservative systems almost all systems satisfy rank condition ; thus, almost all conservative systems are controllable.

§3. Perturbation of a non linear system.

Consider a system of the form :

III.3.1 $$\frac{dx}{dt} = f(x,u) = f_o(x)+ \sum_{i=1}^{p} u_i f_i(x) \quad U=\{0 \leq u_i \leq 1\} ; \; x \in R^n \quad ;$$

and define a Bang Bang control as a control of the form : $(0,0,\ldots,1,0,\ldots,0)$. Denote by $\mathcal{R}(x_o,f)$ reachable set for this system by mesurable controls, by $\mathcal{R}_{B.B}(x_o,f)$ piecewise smooth B.B. controls. Let :

III.3.2 $$\frac{dx}{dt} = f(x,u) + g(x,u)$$

be a perturbed system, (g is the perturbation) and denote by $\mathcal{R}_{B.B}(x_o,f+g)$ the reachable set by piecewise B.B. controls for this system.

III.3.3 - Theorem : If the dynamical polysystem generated by the familly $D = \{ f(0,u) \; ; \; u \in U\}$ is in general position (see II.2.8), for any compact subset K contained in the interior of $\mathcal{R}(x_o,f)$ there exists an ε such that $\mathcal{R}_{B.B}(x_o,f+g)$ contains K for any g such that :

$$\| g(x,u)\| \leq \varepsilon \qquad x \in R^n \quad ; \quad u \in U \ .$$

Notice that if the perturbation is 0 this theorem asserts that the interior of the reachable set is accessible by B.B. control. (See in Krener (1) a more deep result giving conditions under which $\mathcal{R}(x_o,f) = \mathcal{R}_{BB}(x_o,f)$). A detailed proof of this theorem will be given in Brunovsky-Lobry (1).

III.3.4 - Corollary : A small perturbation of a completely controllable system on a compact manifold is completely controllable.

APPENDIX A : Some facts from geometry.

A.1 - Smooth mappings.

- Smooth means with infinitely many derivatives.
- A smooth mapping from R^p into R^q is actually a mapping defined on some open subset of R^p which is not precised.
- A local diffeomorphism is a one to one smooth mapping between two open subsets of R^n ; if x is contained in the source of the diffeomorphism one says that it is a local diffeomorphism at x .
- The rank of a smooth mapping is the rank of its jacobian matrix.
- The differential of a mapping f at point x is denoted by $Df(x)$.

Inverse function theorem : If f is a smooth mapping of rank n at point x from R^n into R^n then its restriction to a suitable neighbourhood of x is a local diffeomorphism at x.

Rank theorem : Let f be a smooth mapping from R^n into R^m whose rank is constant in a neighbourhood of x and equal to p ; then there exist two local diffeomorphisms, g at point x and h at point $f(x)$ such that, locally the mapping : $hofog^{-1}$ from R^n into R^m is defined by :

$$(x_1,x_2,\ldots,x_n) \to (x_1,x_2,\ldots,x_p,0,0,\ldots,0) \in R^m .$$

Weak Sard's theorem : The range of a smooth mapping from R^n into R^{n+k} $(k \geq 1)$ has a zero Lebesgue measure.

A.2 - Smooth manifolds.

The manifolds we will define are immersed in some euclidian space R^n. By a theorem of Whitney there is no loss of generality to assume that. In this case, a manifold is a subset of R^n and the definitions are more intuitive.

A.2.1 - Definition : A local parametrization of a subset M of R^n at point x is a smooth mapping $\varphi_x : \mathcal{V}_x \to M$ of an open neighbourhood $\mathcal{V}_x$ of 0 in R^p which satisfies the three conditions :

φ_x is one to one onto its image

φ_x is of maximum rank p on $\mathcal{V}_x$

$\varphi_x(0) = 0$.

The inverse of a local parametrization at x is called a system of local coordinate.

Roughly speaking a manifold is a set which is covered by local parametrizations ; but the subset $\{(x,y) ; xy = 0\}$ of R^2 which is covered by local parametrizations does not correspond to our idea of smoothness because it has corner ; so we introduce the :

A.2.2 - Definition :Two local parametrizations $\varphi_x : \mathcal{V}_x \to M$ and $\varphi_y : \mathcal{V}_y \to M$ are compatibles if the two subsets : $\varphi_x^{-1}(\varphi_y(\mathcal{V}_y))$ and $\varphi_y^{-1}(\varphi_x(\mathcal{V}_x))$ are open ; it follows from rank and weak Sard theorem that then $\mathcal{V}_x$ and $\mathcal{V}_y$ belong to euclidian spaces of same dimension and the two mappings :

$$\varphi_x^{-1} \circ \varphi_y \quad \text{and} \quad \varphi_y^{-1} \circ \varphi_x$$

are smooth mappings.

A.2.3 - Definition : An atlas on M is a family $(\varphi_x, \mathcal{V}_x)_{x \in M}$ of local parametrization which are compatible.

Two atlasses on M are equivalent if any local parametrization $(\varphi_x, \mathcal{V}_x)$ of the first one is compatible with any local parametrization $(\psi_y, \mathcal{U}_y)$ of the second one.

A.2.4 - Definition : An equivalence class of atlasses on M is smooth manifold structure on M .

A smooth manifold will be a subset M of some space R^n on which one is given a smooth manifold structure, so it would be better to denote it by $M, \mathcal{S}$ ($\mathcal{S}$ standing for the structure), but in general there is no possible confusion. Let $\mathcal{S}$ be a smooth manifold structure, then the strongest topology for which all the local parametrization of some atlas of the structure are continuous (this does not depend of the atlas) is referred as the topology of the manifold ; for each connected component the local parametrizations $(\varphi_x, \mathcal{V}_x)$ are defined on an euclidian space of the same dimension (see A.2.2) ; this dimension is the dimension of the connected component.

A.2.5 - Definition : A manifold M is a pair $(M, \mathcal{S})$ where M is a subset of some R^n, $\mathcal{S}$ is a smooth manifold structure on M , and it is assumed that each connected component of M , for its manifold topology, has the same dimension p . The number p is the dimension of the manifold.

A.3 - Smooth mappings on manifolds and tangent space.

Let M and N be two manifold and f be a mapping from some open subset of M into N .

A.3.1 - Definition : The mapping $f : M \to N$ is smooth if for any x in M , the composed mapping : $\psi_{f(x)}^{-1} \circ f \circ \varphi_x$, where φ_x is a local parametrization at x and $\psi_{f(x)}$ a local parametrization at $f(x)$, is a smooth mapping.

Because of the chain rule for computation of derivative of composed mappings this definition does not depend on the atlas choosed in the smooth manifold structure.

Now, we define the tangent space at x ; intuitively it represents the linear (affine) subspace of R^n which best approximate M at point x, that is to say the "tangent plane" to the surface.

A.3.2 - <u>Definition</u> : The tangent space TM_x of manifold M at point x is the linear subspace of R^n defined as the range of the linear mapping :

$$D\psi_x(0) : R^p \to R^n$$

where ψ_x is some local parametrization ; the definition does not depend of the atlas in the smooth manifold structure.

The whole tangent space TM is the "abstract union" of all the sets TM_x when x ranges over m ; it can be considered as the subset of $R^n \times R^n$ (where R^n is some euclidian space containing M) defined by $TM = \{(x,V) \; ; \; v \in TM_x\}$. One can give TM a natural structure of manifold by considering the local parametrizations $\varphi_{x,0}: (\mathcal{V}_x, R^p) \to TM$ defined by :

$$\varphi_{x,0}(y,\xi) = (\varphi_x(y) \; ; \; D\varphi_x(y).\xi)$$

where φ_x is some local parametrization of M at point x. We shall not go into further details on this subjet ; all we need is to know that the tangent space has a canonical manifold structure. The natural projection from TM onto M is the mapping which associate the point x in M to an element of TM_x.

A.3.3 - <u>Definition</u> : Let M and N be two manifolds ; f a smooth mapping from M into N. The differential $Df(x)$ of f at point x is the linear mapping :

$$Df(x) : TM_x \to TN_{f(x)}$$

defined by the formula (which does not depends on the atlasses):

$$Df(x).V = (D\psi_{f(x)}(0)) \circ D(\psi^{-1}_{f(x)} \circ f \circ \varphi_x)(0) \circ (D\varphi_x(0))^{-1} V .$$

A.4 - Submanifolds.

Let M and N be two manifolds and let i be a smooth one to one mapping defined on the whole manifold N into M ; this is an inclusion map, we denote it by :

$$i : N \hookrightarrow M .$$

A.4.1 - Definition : If the mapping i defined above has maximum rank at any point then N is an immersed submanifold of M ; if moreover the topology of N is induced by topology of M then N is a regularly imbedded (say imbedded) submanifold of M.

Example : The subset of R^2 :

$$N = \{ (x,y) \ ; \ x > 0 \ ; \ y = \sin \frac{1}{x} \}$$

is an imbedded submanifold of R^2.

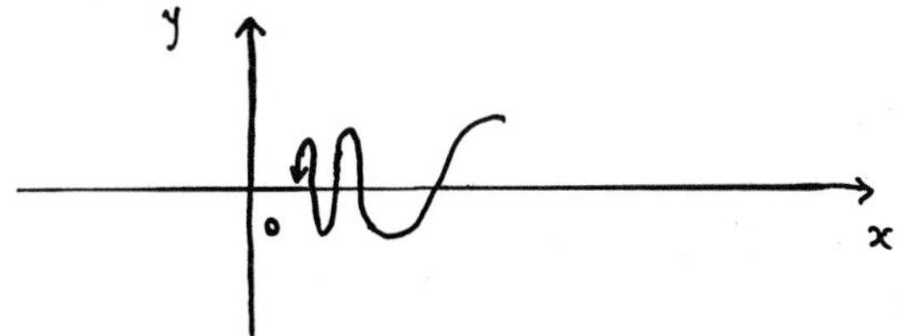

- An imbedded submanifold.

The subset of R^2 :

$$N = \{ (x,y) \ ; x > 0; y = \sin \frac{1}{x} , x=0 \quad |y| < 1\}$$

is an immersed submanifold of R^2 .

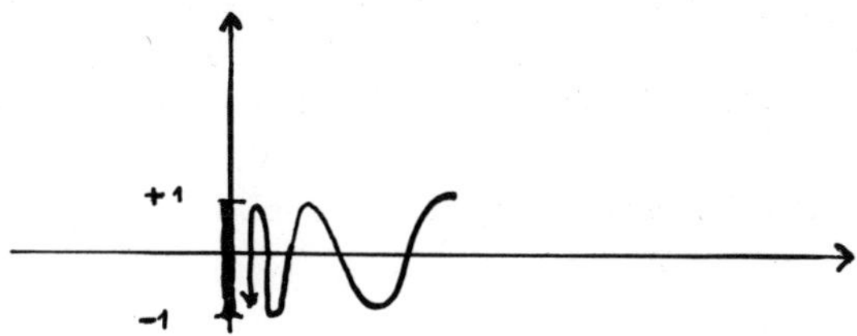

- An immersed submanifold.

The subset of R^2 :

$$N = \{ (x,y) \ ; \ x^2 + y^2 = 1\}$$

is an imbedded submanifold. Everywhere dense trajectories on the torus are immersed submanifold.

From rank theorem one can prove the :

A.4.2 - Proposition : Let N be an immersed submanifold of M. For any x in N there exists an open neighbourhood $\mathcal{U}_x$ of x, an open neighbourhood $\mathcal{U}_{i(x)}$ of $i(x)$, and a smooth mapping $\psi_{i(x)} = (\psi_x^1, \psi_x^2, \dots, \psi_x^{m-n})$ from $\mathcal{U}_{i(x)}$ into R^{m-n} of maximum rank such that :

$$i(\mathcal{U}_x) = \{y \in \mathcal{U}_{i(x)} \ ; \ \psi_{i(x)}(y) = 0\} .$$

Here m is the dimension of M, n the dimension of N ; $m-n$ is the codimension and $\psi_{i(x)}$ is called a system of local equations. Moreover if N is regularly imbedded one can choose $\mathcal{U}_{i(x)}$ such that :

$$i(N) \cap \mathcal{U}_{i(x)} = \{y \in \mathcal{U}_{i(x)} \ ; \ \psi_{i(x)}(y) = 0\} .$$

From now, when N is a submanifold, we omit to write the inclusion map and write only $N \subset M$.

A.5 - Vector fields.

A vector field on R^n is a smooth mapping $X : R^n \to R^n$ defined on the whole R^n.

A.5.1 - Definition : A smooth vector field on a manifold M is a smooth mapping :

$$X : M \to TM$$

such that $X(x)$ belongs to the tangent space TM_x at x.

A smooth curve on a manifold M is a smooth mapping $\alpha : I \to M$ from some interval I of R into M. The derivative $\alpha'(t)$ is by definition the vector of $TM_{\alpha(t)}$ defined by $\alpha'(t) = D\alpha(t).1$.

A.5.2 - Definition : A Dynamical system is a smooth mapping :

$$\Pi : M \times R \to M$$

such that :

$$\Pi(x,0) = x$$

$$\Pi(x, t_1+t_2) = \Pi(\Pi(x, t_1), t_2) .$$

Notice that for any x in M, $t \to \Pi(x,\alpha)$ is a smooth curve.

A.5.3 - Definition : A smooth vector field X is said to be complete if for any x in M there exists a smooth curve :

$$t \to \gamma_x(t) \qquad t \in R \ ,$$

such that :

$$\gamma_x(0) = x$$

$$\gamma_x'(t) = X(\gamma_x(t)) \ .$$

This curve is unique.

Notice that if we just ask to the curve to be defined on an open interval countaining 0 , from well known theorem on differential equations, such a curve always exists, and a maximal one is unique. In R^n vector fields which satisfy a growth condition are complete ; on a compact manifold every smooth vector field is complete. From theorem on existence, unicity, continuous dependance it follows the :

A.5.4 - Theorem : Let X be a complete vector field on M. The mapping :

$$\Pi : M \times R \to M$$

defined by :

$$\Pi(x,t) = \gamma_x(t)$$

where $\gamma_x(t)$ is the curve of def.A.5.3 defines a smooth dynamical system on M . This dynamical system is called the dynamical system generated by X (or the dynamical system whose infinitesimal generator is X) and is denoted by :

$$(x,t) \to X_t(x) \ .$$

A.5.5 - If N is a submanifold of M , one says that a vector field X is tangent to N if for every x in M the vector X(x) belongs to TN_x ; it follows from unicity of solutions of a differential equation that if a vector field X is tangent to a submanifold N then every trajectory of X crosses N according to an open set.

We turn now to define the Lie (or Jacobi) bracket of two vector fields.

A.5.6 - Definition : Let X and Y be two vector fields on M . Let us consider at the point x , the family of vectors in TM_x defined by :

$$V_x(t) = (DY_t(x))^{-1}. \ X(Y_t(x))$$

and denote by :

$$[X,Y](x) = (V'_x(t))_{t=0} \quad ;$$

we define by this way a new vector field $[X,Y]$ which is called the Lie bracket of X and Y.

One easily proves, in a local coordinate system on M, the formula :

A.5.7 $$[X,Y](x) = DX(x).Y(x) - DY(x).X(x) .$$

A.5.8 - Notice that the vector $V_x(t)$ of definition above is the derivative at 0 of the smooth curve :

$$s \to Y_{-t} \circ X_s \, Y_t(x) = \gamma(s) .$$

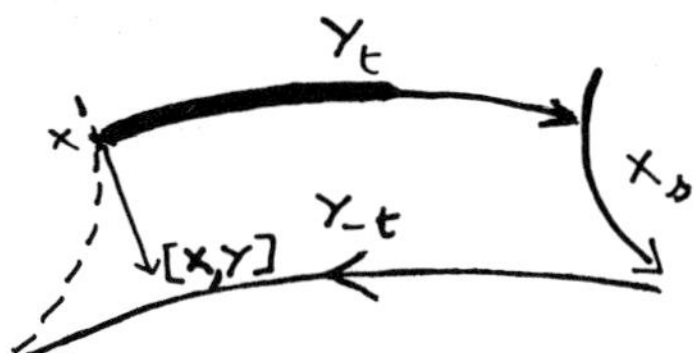

- Geometric interpretation of X,Y .

From this it follows that the bracket of two vector fields which are tangent to a submanifold is also tangent to the same submanifold.

A.6 - An example of manifold.

A.6.1 - Proposition : The subset $M_k(p,q)$ of the set $M(p,q)$ of $p \times q$ matrices composed of these matrices whose rank is exactly k is a regularly imbedded submanifold of $M(p,q)$ of codimension $(p-k)(q-k)$.

Proof : See Levine (1).

APPENDIX B : A density lemma.

We prove, in a particular simple case, Thom's density lemma (Thom (1)).

B.1 - k-Jets.

Let us call the set of k-jet at the origin of a smooth mapping from R^p into R^q, and denote it by $J^k_{p,q}$ the linear space $R^{\tau(k)}$ where $\tau(k)$ is the number :

$$\tau(k) = q(1 + C^1_{p+1-1} + \ldots + C^\ell_{p+\ell-1} + \ldots + C^k_{p+k-1}) .$$

The elements of this vector space have to be understood as the taylor coefficients a 0 of a mapping from R^p into R^q ; we adopt the following rules :

an element χ of $J^k_{p,q}$ is a sequence

$$\chi = (\chi_o, \ldots, \chi_\ell , \ldots, \chi_k)$$

where every χ_ℓ is a sequence :

$$\chi_o = \xi_o$$

$$\chi_\ell = (\xi_{\ell,1} , \xi_{\ell,2} , \ldots, \xi_{\ell j} , \ldots, \xi_{\ell, C^\ell_{p+\ell-1}})$$

of elements of R^q . For every integer ℓ one chooses a one to one mapping between the set $1,2,\ldots,C^\ell_{p+p-1}$ and set of sequences $\{ (i_1, i_2, \ldots, i_p) ; i_1+i_2+\ldots+i_p = \ell\}$ and we denote it by ρ_ℓ .

Let f be a smooth mapping from R^p into R^q , we denote by $J_k f$ the smooth mapping from R^p into $J^k_{p,q}$ defined by :

$$x \to J_k f(x) = (\chi_o(x), \chi_1(x), \ldots, \chi_\ell(x), \ldots, \chi_k(x))$$

$$\chi_o(x) = f(x)$$

$$\chi_\ell(x) = (\xi_{\ell,1}(x), \xi_{\ell,2}(x), \ldots, \xi_{\ell,j}(x), \ldots, \xi_{\ell, C^\ell_{p+\ell-1}})$$

with :

$$\xi_{\ell,j}(x) = \frac{\partial^\ell f(x)}{\partial x_p^{i_p} \ldots \partial x_2^{i_2} \partial x_1^{i_1}} \quad i_1+i_2+\ldots+i_p=\ell \quad \rho_\ell(j)=(i_1;i_2,\ldots,i_p).$$

Conversely to every k-jet χ one can associate a smooth mapping, namely a polynomial P_χ , whose k-jet at 0 is precisely χ :

$$P_\chi(x) = \sum_{\ell=0}^{p} \sum_{j=1}^{C^{\ell}_{\ell+p-1}} \frac{1}{i_1!} \times \frac{1}{i_2!} \times \dots \times \frac{1}{i_p!} \xi_{\ell,j}\, x_1^{i_1} x_2^{i_2} \dots x_p^{i_p} \quad ;$$
$$\rho_\ell(j) = (i_1, i_2, \dots, i_p) \; .$$

Let us remark that the linear mapping :

$$\chi \to (J_k\, P_\chi)(x)$$

is represented by a matrix of the following type :

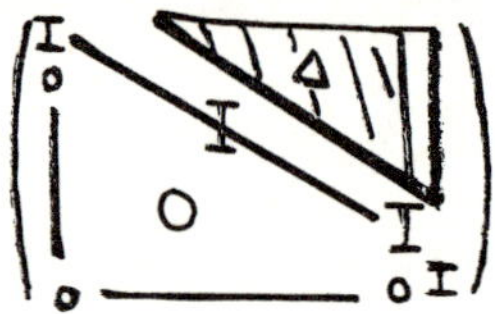

where I is actually an $q\times q$ identity matrix. Thus the mapping is non singular.

B.2 - Density lemma.

Let us consider an open subset U of R^p ; for every compact subset K of U let us denote by $\| \|_{k,K}$ the norm of uniform convergence up to order K for restrictions of smooth mappings to K ; it can be considered as a semi-norm on the set of smooth mappings on U .

We consider now on $U\times J^k_{p,q}$ a finite number $S_1, S_2, \dots, S_r$ of submanifolds, regularly imbeded. Denote by S the union $S_1\ S_2 \dots\ S_p$; we say that the codimension of S is the minimum codimension of the submanifolds S_i .

B.2.1 - Proposition (Thom's lemma) : Suppose codimension of S strictly greater than p . For every integer k and every compact K the subset of smooth mappings from U into R^p having the following property :

$$x \in U \qquad (x, J_k f(x)) \notin S$$

is a dense subset for the norm $\| \|_{k,K}$.

Proof of proposition B.2.1 : Let us choose a smooth mapping from U into R^p and let us consider the mapping :

$$T_f : U \times J^k_{p,q} \to U \times J^k_{p,q}$$

$$(x,\chi) \to T_f(x,\chi) = (x, J_k(P_\chi + f)(x)) \ ;$$

first we remark that this mapping is of maximum rank : with respect to x it is identity mapping and for x fixed $\chi \to T_f(x,\chi)$ is an affine mapping which linear part is of maximum rank as we remarked in B.1 . For this reason the set $T_f^{-1}(S) = T_f^{-1}(S_1) \quad T_f^{-1}(S_2) \ldots \quad T_f^{-1}(S_n)$ is the union of r regularly imbedded submanifolds of codimension strictly greater than p (the codimension of a manifold is the number of independant local equations ; this number is not changed by composition with a smooth mapping).

Let us denote by Π the canonical projection of $U \times J^k_{p,q}$ into $J^k_{p,q}$; and we look to $\Pi(T_\ell^{-1}(S_i))$. The set $T_f^{-1}(S_i)$ is a regularly imbedded submanifold of codimension strictly greater than p . Consider a denumerable set of open subsets θ_i of $U \times J^k_{p,q}$ covering $T_f^{-1}(S_i)$ for which we have a local parametrization $(\varphi_j , \mathcal{V}_j)$ such that :

$$\varphi_j(\mathcal{V}_j) \subseteq \theta_j \cap T_f^{-1}(S_i) \quad ;$$

this is obviously possible for an imbedded submanifold. Because the codimension of $T_f^{-1}(S_i)$ is strictly greater than p the set $\mathcal{V}_j$ belongs to some R^τ , where τ (the dimension of $T_f^{-1}(S_i)$) is strictly smaller than dimension of $J^k_{p,q}$. We have :

$$\Pi(T_f^{-1}(S_i)) \subset \bigcup_{j=1}^{\infty} \Pi(\varphi_j(\mathcal{V}_j)) \ .$$

By weack Sard's theorem each of the set $\Pi(\varphi_j(\mathcal{V}_j))$ has zero measure ; so the set $\Pi(T_f^{-1}(S)) = \cup \Pi(T_f^{-1}(S_i))$ has zero measure. For this reason one can choose a point χ in $J^k_{(p,q)} \setminus \Pi(T_f^{-1}(S))$ close enough to zero in order that the norm :

$$\| f - f + P_\chi \|_{k,K}$$

is less than ε , for any prescribed ε . By construction

$\chi \in \Pi(T_f^{-1}(S))$, that means :

$$x \in U \quad (x,\chi) \notin T_f^{-1}(S)$$
$$x \in U \quad T_f(x,\chi) \notin S$$
$$x \in U \quad (x,J^k(P_\chi + f)(x)) \notin S$$

thus the lemma is proved.

B.3 - Almost all linear control systems are "good".

As an illustration of Thom's lemma we prove that in a reasonable sense, almost all linear systems are "good".

Let us consider the system :

$$\text{(1)} \qquad \dot{X} = A(t)X + b(t)u \quad X \in R^n \ , \ u \in [-1,+1]$$

and assume that the mappings $t \to b(t)$ and $t \to A(t)$ are smooth. It is known that the following condition is the generalization of normality condition for autonomous systems.

Define for a mapping $t \to V(t) \in R^n$, the new mapping $t \to D_A V(t)$ by the formula :

$$D_A V(t) = -A(t)\ V(t) + V'(t) \ .$$

B.3.1 - Definition : One says that system (1) is normal if for every t in R the rank of :

$$(b(t),\ D_A b(t),\ldots,\ D_A^n b(t))$$

is equal to n .

B.3.2 - Proposition : For normal systems the followings are true :

- Controllable set is an open neighbourhood of 0
- Maximum principle is a sufficient condition for time optimality
- The optimal controller is unique, Bang Bang, and has a finite number of swichings.

Proof of Prop.B.3.2 : It follows by more or less standard arguments from the fact that the normal system is expanding, i.e : for every $t_2 > t_1 > t_o$ the controllable set on $[t_o, t_2]$ contains the controllable set on $[t_o, t_1]$ in its interior. (See Hermes-La Salle (1), Weiss (1) and Lobry (2)).

We prove now the following :

B.3.3 - <u>Proposition</u> : The set of mappings $t \to b(t)$ for which system (1) is normal is :

i) - Open in the topology defined by $\| \ \|_{n,[t_o,t_1]}$

ii)- Dense for every topology defined by $\| \ \|_{k,[t_o,t_1]}$; $k \in N$.

Proof of Prop.B.3.3 : Let us prove i). Consider an element :

$$(\chi_o, \chi_1, \ldots, \chi_n)$$

of $J^n_{1,n}$; (here every χ_ℓ is an n-vector). Denote by $A(t)\chi$ the n-jet :

$$A(t)\chi = (-A(t)\chi_o + \chi_1 , \chi_2 , \chi_3 , \ldots, \chi_n , 0)$$

and let G the mapping :

$$G : R \times J^n_{1,n} \to M(n,n+1)$$

defined by :

$$G(t,\chi) = (n{\times}n{+}1)\text{-matrix}(\chi_o,(A(t)\chi)_o,\ldots,(A^n(t)\chi)_o) .$$

Denote by S the subset of M(n,n+1) of matrices whose rank is less than n-1 . This set is closed, so $\complement S$ is open. From the definition it is clear that normality condition is equivalent to :

$$\forall t \quad G(t,(J_n b)(t)) \notin S$$

thus the set of $t \to b(t)$ which satisfy this condition is open in the topology defined by the norm $\| \ \|_{n, [t_o,t_1]}$.

Let us prove ii). From A.6.1, the set S is an union of submanifolds of codimension strictly greater than 1 . The mapping G is of maximum rank so $G^{-1}(S)$ in an union of submanifolds of codimension strictly greater than 1 so prop. B.2.1 applies.

BUSHAW D. (1) "Optimal Discontinuous Forcing Terms"
Contribution to the theory of nonlinear oscillations;
Vol IV (1958).

(2) "Dynamical Polysystems and Optimisation"
Contribution to differential equation 2, p. 351-365.

BRUNOVSKY-LOBRY (1) "Bang-Bang, Smooth controllability and perturbation of nonlinear systems" ; to appear.

BROCKETT R.W. (1) "System Theory on Groups Manifolds and Coset Spaces".
SIAM Journal on Control, vol 10, n° 2, (1972)
pp. 265-284.

ELLIOT D.L. (1) "A Consequence of Controllability"
J. of Diff. Equ. 10 (1971), pp. 364-370.

FRANCIS G.K. (1) "The Maximum Principle for Control Distributions".
University of Illinois, Urbana, Illinois ; to appear.

HAYNES-HERMES (1) "Non linear Controllability via Lie Theory".
SIAM Journal on control, vol. 8 (1970) pp. 450-460.

HERMANN R. (1) "On the Accessibility Problem in Control Theory."
Intermat. Symp. on Nonlinear Diff. Equ. and Nonlinear Mech. Academic Press. N.Y. (1963).

HERMES-LA SALLE (1) "Fuctionnal Analysis and Time Optimal Control".
Academic Press, (1969).

JURDJEVIC V. (1) "Certain Controllability Property of Analytic Control Systems".
SIAM Journal on Control vol. 10-2, p. 354-360 (1972)

JURDJEVIC-SUSSMANN (1) "Controllability of Nonlinear Systems"
J. Diff. Equations, 12 (1972) p. 470-476.

KALMAN R.E. (1) "Contribution to the Theory of Optimal Control".
Bol. Soc. Mat. Mex., 5, p. 102-119 (1960).

KRENER A. (1) "A Generalization of Chow's Theorem and the Bang Bang theorem to Nonlinear Control Problems".
to appear.

KUČERA. (1) "Solution in large of Control problem :
$\dot{x} = (A(1-u) + Bu)\ x$. "
Czech. Math. J. 16 (91), (1966), pp. 600-623.

KUČERA. (2) "Solution in large of Control Problem : $\dot{x} = (Au + Bv)\ x$. "
Czech. Math. J. 17 (92) (1967) pp. 91-96.

LEVINE H. (1) "Singularities of Differentiable Mapping."
Proceeding of Liverpool Singularities Symposium
Springer-Verlag Lecture Notes 192.

LOBRY C. (1) "Contrôlabilité des Systèmes Non linéaires".
SIAM Journal on Control vol. 8, n° 4 (1970) pp.573-605.

(2) "Quelques aspects qualitatifs de la théorie de la commande."
Thèse sciences math. Grenoble, mai 1972.

(3) "Une propriété générique des couples de champs de vecteurs".
Czech. Math, j. 22 (97 / 1972).

(4) "Controllability of Nonlinear Systems on Compact Manifolds".
to appear in SIAM J. Control.(1973.)

STEFAN (1) "Attainable Sets are Manifolds."
University of Wales. U.C.N.W. Bangaï ; to appear.

SUSSMANN H. (1) "Orbits of families of vector fields and integrability of distributions".
to appear in Trans. Amer. Math. Soc.

THOM (1) "Les singularités des applications différentielles".
Annales de l'Institut Fourier, VI, (1956) p. 43-87.

WEISS (1) "Controllability and Observability".
Lecture Notes, Summer School, CIME (1968).

General references

LEE-MARKUS (1) "Foundations of Optimal Control Theory"
John Wiley and Sons, Inc. New-York, London Sidney (1967).

CHEVALLEY C. (1) "Theory of Lie groups". Princeton University
Press, Princeton New Jersey, (1946).

BISHOP-CRITTENDEN (1) "Geometry of Manifolds".
Academic Press, New York, (1964).

LIE ALGEBRAS AND LIE GROUPS IN CONTROL THEORY

Roger W. Brockett

Harvard University

Cambridge, Massachusetts

---

*This work was supported in part by the U.S. Office of Naval Research under the Joint Services Electronics Program by Contract N00014-67-A-0298-0006 and by the National Aeronautics and Space Administration under Grant NGR 22-007-172.

PREFACE

The theory of differential equations and control have been linked very closely because most of the early applications of control theory were to engineering problems of the type which are most naturally described by ordinary differential equations. The questions of importance in control have helped to revitalize certain problem areas in differential equations and methods and tools from control have been useful in obtaining new results in differential equation theory. On the other hand, going back to the era of Lie himself, there has been close ties between Lie theory and differential equations. Thus it is not surprising that one finds that Lie theory and control are also closely connected. This "triangle" is the subject of this set of notes.

In control theory, Lie algebras make their appearance as Lie algebras of vector fields. Topological properties associated with Lie groups show up in the study of controllability and stability. Partial differential operators arise in the Fokker-Planck equations modeling the uncertainty of the environment and our uncertainty about the measurements we make of it. The problems which are of interest in control frequently require a generalization of the usual treatment of topics such as existence of geodesics, expressions for the spectrum of the Laplacian etc. The modification is, roughly speaking, to include the possibility of a metric which is "infinite" in certain directions, subject only to the condition that the directions along which it is finite can be combined in such a way as to make the distance between any two points finite. These notes contain a brief account of some of these topics, together with references where complete proofs can be found.

I have included a few exercises for the reader, both to indicate some results which do not exactly fit the format chosen here and to indicate some partial results and suggestions on additional problems of interest. Most of the examples are to be found in the exercises as well.

It is a pleasure to thank Prof. David Mayne for organizing such a stimulating forum for the exchange of ideas on system theory.

I. THE ALGEBRAIC THEORY OF LINEAR DIFFERENTIAL EQUATIONS

1.1 Lie Algebras and Linear Differential Equations

Clearly any linear differential equation of the form

$$\dot{x}(t) = A(t)x(t); \quad x(t) \in \mathbb{R}^n$$

can be expressed as

$$\dot{x}(t) = \left(\sum_{i=1}^{m} u_i(t)A_i\right)x(t)$$

with the A_i constant matrices and the $u_i(t)$ scalar functions of time. In view of the fact that the solution of the equation with a single A_i, i.e.

$$\dot{x}(t) = u(t)Ax(t)$$

is

$$x(t) = e^{A\int_0^t u(\sigma)d\sigma} x(0)$$

the question arises as to when the solution of the general problem can be written as the composition of a number of such solutions

$$x(t) = e^{A_1 g_1(t)} e^{A_2 g_2(t)} \ldots e^{A_m g_m(t)} x(0)$$

for a suitable choice of the $g_i(\cdot)$. Otherwise stated, we would like to know if the solutions of the matrix differential equation

$$\dot{X}(t) = (\sum_{i=1}^{m} u_i(t)A_i)X(t); \quad X(0) = I \quad \text{(identity)}$$

can be written as

$$X(t) = e^{A_1 g_1(t)} e^{A_2 g_2(t)} \ldots e^{A_m g_m(t)}$$

for a reasonably wide class of $u_i(t)$ and over some interval of time, say $|t| < \varepsilon$.

The above question is basically answered by a classical theorem of Frobenius [1]. However the theorem of Frobenius which applied here is a theorem in differential geometry. To use the insight of his result we need to look at the problem posed from a geometrical point of view. Consider the identity matrix as a point in the set of all nonsingular n by n matrices. Suppose that the one parameter curves $e^{A_i t}$ leave the identity as indicated in figure 1.

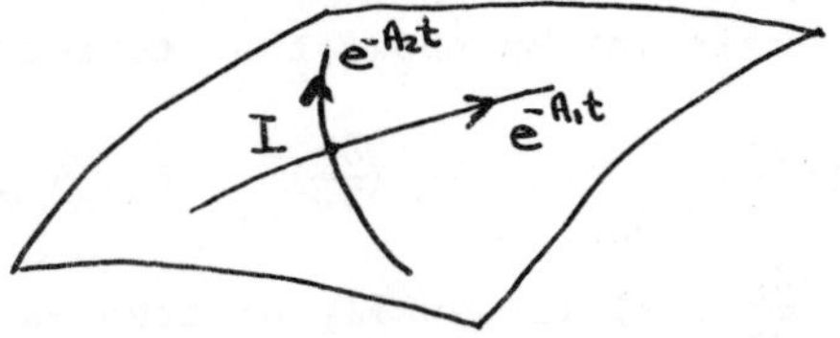

Figure 1: Neighborhood of I in the set of all n by n matrices

We regard the set of all points of the form

$$S = \{X : X = \prod_{i=1}^{m} e^{A_i \alpha_i}; \qquad \alpha_i \in \mathbb{R}\}$$

as a subset of the set of all nonsingular n by n matrices. Our question is, when do the integral curves of the given matrix differential equation corresponding to a wide class of $u_i(\cdot)$ lie in S? In order for this to be true for all piecewise continuous u's we require, for example, that

$$e^{A_1 t} e^{A_2 t} e^{-A_1 t} e^{-A_2 t}$$

be expressible as an element of S. To see why this is so we point out that the choice

$$u_1(\sigma) = \begin{cases} -1 & t \leqslant \sigma < 2t \\ 0 & 0 \leqslant \sigma < t; \quad 2t \leqslant \sigma < 3t \\ 1 & 3t \leqslant \sigma < 4t \end{cases}$$

$$u_2(\sigma) = \begin{cases} -1 & 0 \leqslant \sigma < t \\ 0 & t \leqslant \sigma < 2t; \quad 3t \leqslant \sigma < 4t \\ 1 & 2t \leqslant \sigma < 3t \end{cases}$$

$$u_i(\sigma) = 0 \quad i > 2$$

yields

$$X(4t) = e^{A_1 t} e^{A_2 t} e^{-A_1 t} e^{-A_2 t}$$

Geometrically, what we are asking is that in following the 4-sided path shown in figure 2 we should not be lead out of the set S.

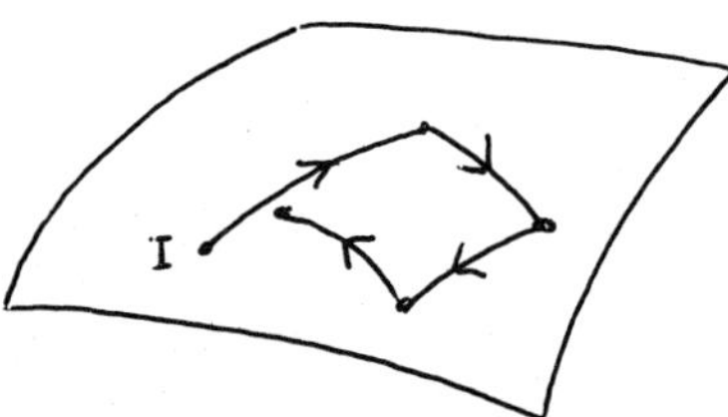

Figure 2: Illustrating the path leading to $e^{A_1 t} e^{A_2 t} e^{-A_1 t} e^{-A_2 t}$

More generally if f_1 and f_2 are smooth maps of $\mathbb{R}^n$ into $\mathbb{R}^n$ and if we apply the above choice of $u(\cdot)$ to the system

$$\dot{x}(t) = u_1(t) f[x(t)] + u_2(t) g[x(t)]; \quad x(0) = x_0$$

then a slightly messy calculation shows that to second order in t we have

$$x(4t) \simeq x_0 + \{(\tfrac{\partial f}{\partial x})_{x=x_0} g(x_0) - (\tfrac{\partial g}{\partial x})_{x=x_0} f(x_0)\} t^2$$

The quantity $\frac{\partial f}{\partial x} g(x) - \frac{\partial g}{\partial x} f(x)$ is usually written as $[f,g]$ and is called the Lie bracket of f and g. One calls a set of vectors $f_i : \mathbb{R}^n \to \mathbb{R}^n$ <u>involutive</u> if the Lie bracket of any two is a linear combination of the $\{f_i\}$. Frobenius showed that the set of points near x_0 which can be reached from x_0 along integral curves of

$$\dot{x}(t) = \sum_{i=1}^{m} u_i(t) f_i(x)$$

with $\{f_i\}$ involutive can be expressed as

$$\phi_m(t_m, \ldots, \phi_3(t_3, \phi_2(t_2, \phi_1, x))) \ldots)$$

where $\phi_i(t,x)$ are the solutions of

$$\dot{x}(t) = f_i[x(t)]$$

The reason the set $\{f_i\}$ must be involutive is that otherwise the special choice of $u(\cdot)$ outlined above will, for small t, surely lead out of set of points expressible as $\phi_m(t_m,\phi(t_{m-1}, \ldots\phi(t,x_o))\ldots))$.

Applying this type of thinking to the linear case, we see first of all that the Lie bracket of A_1x and A_2x is $[A_1x, A_2x]=(A_1A_2-A_2A_1)x$ That is, the Lie bracket of the vector fields is expressible as the commutator of the matrices. We write $[A_i,A_j]$ for $A_iA_j-A_jA_i$. Thus if the set of matrices $\{A_i\}$ have the property that

$$[A_i,A_j] = \sum_{k=1}^{m} \gamma_{ijk}A_k$$

then the theorem of Frobenius would imply that for small $|t|$ we can write

$$X(t)x_o = \prod_{i=1}^{m} e^{A_ig_i(t)} x_o$$

A linear space of square matrices which is closed under $[\cdot,\cdot]$ is a matrix Lie algebra. Of course if the original set $\{A_i\}$ does not form a basis for a Lie algebra we simply supplement it with additional A's until it does. If x is of dimension n then there are only n^2 linearly independent matrices so this process always results in a finite set.

Wei and Norman [2] have given a direct verification of the above representation based on the implicit function theorem and have developed a set of nonlinear differential equations for the $g_i(\cdot)$. The basis for their derivation is the Baker-Campbell-Hausdorff formula

$$e^ABe^{-A} = B+[A,B]+ \frac{1}{2} [A,[A,B]+ \frac{1}{3!} [A,[A,[A,B]]\ldots]$$

Thus if one assumes a solution of the form

$$X(t)X_o= e^{A_1g_1(t)} e^{A_2g_2(t)} \ldots e^{A_mg_m(t)}$$

and then differentiates, the result is

$$\begin{aligned}\dot{X}(t) &= A_1\dot{g}_1(t)e^{A_1g_1(t)} e^{A_2g_2(t)} \ldots e^{A_mg_m(t)}\\ &+ e^{A_1g_1(t)} A_2\dot{g}_2(t)e^{A_2g_2(t)} \ldots e^{A_mg_m(t)}\\ &\cdot\ \cdot\ \cdot\ \cdot\ \cdot\\ &+ e^{A_1g_1(t)} e^{A_2g_2(t)} \ldots A_m\dot{g}_m(t)e^{A_mg_m(t)}\end{aligned}$$

Now we must collect all the A's together at the left in order to compare this expression for $\dot{X}$ with that given by the differential equation. The Baker-Campbell-Hausdorff formula provides the means to do this. To see how this happens, observe that by inserting $e^{-A_ig_i(t)} e^{A_ig_i(t)}$ freely we can arrive at

$$\dot g_1A_1+\dot g_2 e^{A_1g_1(t)}A_2e^{-A_1g_1(t)}+\ldots\dot g_m e^{A_1g_1(t)}e^{A_2g_2(t)}\ldots A_m\ldots e^{A_1g_1(t)}$$

$$= A_1u_1(t)+A_2u_2(t) + \ldots + A_mu_m(t)$$

We apply the Baker-Campbell-Hausdorff expansion to each term on the left. If the set $\{A_i\}$ is a basis for a Lie algebra then we can express the result as a linear combination of the A_i. Since the A_i are linearly independent we can equate coefficients on each side and thereby get a set of differential equations for the g_i. It is important to note that the differential equations for the g_i only depend on the A_i through the commutation rules

$$[A_i,A_j] = \sum_{k=1}^{m} \gamma_{ijk}A_k$$

Thus when a differential equation is solved by this method a whole class of differential equations are solved at the same time -- one for each set of A's which satisfy the given commutation relation.

Exercises

1. Show that if the A_i in

$$\dot X(t) = \sum_{i=1}^{m} u_i(t)A_iX(t)$$

are all upper triangular then it is possible to express the solution of the differential equations for the $g_i(\cdot)$ explicitly in terms of integrals.

2. Show that the smallest Lie algebra of matrices which contains A_1 and A_2

$$A_1 = \begin{pmatrix}0 & 1\\ 0 & 1\end{pmatrix} ; \quad A_2 = \begin{pmatrix}0 & 0\\ 1 & 0\end{pmatrix}$$

is 4 dimensional.

3. Study the definition of Euler angles from the point of view of the Wei-Norman equations. In particular explain why it is generally not possible to obtain a Wei-Norman representation the entire half-line $[0,\infty)$ in terms of the degeneracy of the Euler angles.

4. Show that for any square matrix P the set of all solutions of $PA+A'P = 0$ from a Lie algebra.

1.2 The $x^{[p]}$ and $x^{(p)}$ Equations

Associated with each linear map of $\mathbb{R}^n$ into $\mathbb{R}^n$ are two families of linear maps which may be described as follows. Choose a basis in $\mathbb{R}^n$ and let the original map be represented by the matrix A. Then we easily see that

$$y_i = \sum_{j=1}^{n} a_{ij}x_j$$

implies that the $n(n+1)/2$ linearly independent terms of the form y_iy_j depend linearly on the $n(n+1)/2$ linearly independent terms of the form x_ix_j. More generally the set of all linearly independent p-degree terms $y_iy_j \ldots y_k$ depend linearly on the set of all linearly independent p-degree terms $x_ix_j \ldots x_k$. How many linearly independent terms of degree p are there in n variables? If we denote this integer by N_n^p then it is easy to see that

$$N_{n+1}^{p+1} = N_{n+1}^{p} + N_{n}^{p+1}$$

from which an induction gives $N_n^p = \binom{n+p-1}{p}$. Thus associated with each map of $\mathbb{R}^n$ into $\mathbb{R}^n$ is a sequence of maps, the pth one mapping $\mathbb{R}^{N_n^p}$ into $\mathbb{R}^{N_n^p}$.

In order to give this family of maps a matrix description we need to choose a basis in $\mathbb{R}^{N_n^p}$ which is in some way convenient. The principle which guides our choice of basis is this: let $<x,y>$ be the ordinary inner product

$$<x,y> = \sum_{i=1}^{n} x_iy_i$$

If the map of $\mathbb{R}^n$ into $\mathbb{R}^n$ defined by A preserves length, we would like the maps of $\mathbb{R}^{N_n^p}$ into $\mathbb{R}^{N_n^p}$ to preserve length as well. To achieve this we introduce the basis elements

$$\sqrt{\binom{p}{p_1}\binom{p-p_1}{p_2}\cdots\binom{p-p_1-\cdots p_{p-1}}{p_p}}\; x_1^{p_1}x_2^{p_2}\ldots x_n^{p_n};\quad \sum_{i=1}^{n} p_i=p;\ p_i \geq 0$$

For example if n=p=3 we have basis elements

$$x_1^3,\ \sqrt{3}x_1^2x_2,\ \sqrt{3}x_1^2x_3,\ \sqrt{3}x_1x_2^2,\ \sqrt{6}x_1x_2x_3,\ \sqrt{3}x_1x_3^2,\ x_2^3,\ \sqrt{3}x_2^2x_3, \sqrt{3}x_2x_3^2, x_3^3$$

If we denote this vector, ordered lexigraphically, by $x^{[p]}$ then the choice of basis is such that $(||x||=(<x,x>)^{1/2})$

$$||x^{[p]}|| = ||x||^p$$

More generally, we have

$$<x,y>^p = <x^{[p]},y^{[p]}>$$

We denote by $A^{[p]}$ the map, or matrix, which verifies

$$y = Ax \Rightarrow y^{[p]} = A^{[p]}x^{[p]}$$

The principle properties of $A^{[p]}$ are covered by the following theorem.

Theorem 1: Suppose we are given A and B. $A : \mathbb{R}^n \to \mathbb{R}^n$ and $B : \mathbb{R}^n \to \mathbb{R}^n$. Then $A^{[p]}$ and $B^{[p]}$ satisfy

i) $I_n^{[p]} = I_{N_n^p}$

ii) $(AB)^{[p]} = A^{[p]}B^{[p]}$

iii) $(A^q)^{[p]} = (A^{[p]})^q$; q integer; A^q defined

iv) $(A')^{[p]} = (A^{[p]})'$

Proof: i) Clear from definition. ii) Let z=Ay=ABx. Then $z^{[p]}=A^{[p]}y^{[p]}=A^{[p]}B^{[p]}x^{[p]}=[AB]^p x^{[p]}$. iii) This follows from ii) on letting B=A (or $B=A^{-1}$ if A is invertible) and using induction. iv) This follows from the identity $\langle x,y\rangle^p=\langle x^{[p]},y^{[p]}\rangle$ and $\langle x,Ay\rangle = \langle A'x,y\rangle$.

A second series of maps associated with A are the so called compounds of A which we write as $A^{(p)}$ and define in terms of matrices as

$$A^{(p)} = \begin{pmatrix} \text{matrix of all p by p minors} \\ \text{of A ordered lexographically} \end{pmatrix}$$

Since there are $\binom{n}{p}$ ways to select the rows and $\binom{n}{p}$ ways to select the columns in a p by p minor of an n by n matrix we see that $A^{(p)}$ is an $\binom{n}{p}$ by $\binom{n}{p}$ matrix. The following properties of $A^{(p)}$ are well known. See for example [2] or [3].

Theorem 2: Let A and B be given; $A: \mathbb{R}^n \to \mathbb{R}^n$ and $B: \mathbb{R}^n \to \mathbb{R}^n$. Then $A^{(p)}$ and $B^{(p)}$ for $0 \leq p \leq n$ maps $\mathbb{R}^{\binom{n}{p}}$ into $\mathbb{R}^{\binom{n}{p}}$ and

i) $I_n^{(p)} = I_{\binom{n}{p}}$

ii) $(AB)^{(p)} = A^{(p)}B^{(p)}$

iii) $(A^q)^{(p)} = (A^{(p)})^q$ q integer; A^q defined

iv) $(A')^{(p)} = (A^{(p)})'$

We have used two different points of view in defining $A^{[p]}$ and $A^{(p)}$. The construction of $A^{[p]}$ from A was described in terms of linear maps whereas in the definition of $A^{(p)}$ we used matrices exclusively. Alternative approaches are available which give $A^{(p)}$ a geometric meaning in terms of skew symmetric forms of degree p in n variables.

These two constructions are specializations of the tensor product in the following way. If $A: \mathbb{R}^n \to \mathbb{R}^n$ and $B: \mathbb{R}^n \to \mathbb{R}^n$ then we may identify the tensor product of $A\eta$ and $B\gamma$ with $A\eta(B\gamma)'$; i.e.

$$A\eta \otimes B\gamma = A\eta(B\gamma)' = A(\eta\gamma')B'$$

If we consider the linear map of the space of n by n matrices into itself defined by $L(Q)=AQB'$ then $L_A(Q)=AQA'$ when restricted to act on symmetric matrices has $A^{[2]}$ as a matrix representation and when restricted to the complementary space of skew symmetric matrices, it has $A^{(2)}$ as its matrix representation. Thus if we let $\simeq$ indicate "similar to" then we have

$$A \otimes A \simeq A(\cdot)A' \simeq \begin{bmatrix} A^{[2]} & 0 \\ 0 & A^{(2)} \end{bmatrix}$$

One can also see that $A \otimes A \otimes A$ "contains" $A^{[3]}$ and $A^{(3)}$ but there are more than 2 symmetry types for a 3 index tensor so that $A^{[3]} \oplus A^{(3)}$ is only part of $A \otimes A \otimes A$. (Check the dimensionality; $n(n+1)(n+2)/6$ and $n(n-1)(n-2)/6$ does not add up to n^3.)

Now consider a linear differential equation in $\mathbb{R}^n$

$$\dot{x}(t) = A(t)x(t)$$

Observe that

$$x^{[p]}(t+h)=(I+hA(t))^{[p]}x^{[p]}(t)+0(h^2)$$

so that

$$x^{[p]}(t+h)-x^{[p]}(t)=[(I-hA)(t))^{[p]}-I]x^{[p]}(t)+0(h^2)$$

Thus

$$\frac{d}{dt}x^{[p]}(t) = (\lim_{h\to 0}\frac{1}{h}[(I-hA(t))^{[p]}-I])x^{[p]}(t)$$

(Note that the dimensions of the identity matrices in these equations are n and N_n^p respectively.) We define $A_{[p]}$ to be the coefficient matrix in this differential equation.

$$\frac{d}{dt}x^{[p]}(t) = A_{[p]}(t)x^{[p]}(t);\quad p=1,2,3,\ldots$$

Thus the set of all p-degrees forms in $\{x_1,x_2,\ldots,x_n\}$ satisfies a <u>linear</u> differential equation with a coefficient matrix which is easily derived from A.

Starting with a matrix equation

$$\dot{X}(t) = A(t)X(t)$$

we can make an analogous construction using compound matrices (round brackets). The estimate

$$X^{(p)}(t+h) = (I+hA(t))^{(p)}X^{(p)}(t) + 0(h^2)$$

leads to

$$\frac{d}{dt}X^{(p)}(t) = (\lim_{h\to 0}\frac{1}{h}[(I+hA(t))^{(p)}-I])X^{(p)}(t)$$

which we write as

$$\frac{d}{dt}X^{(p)}(t) = A_{(p)}(t)X^{(p)}(t);\quad p=1,2,\ldots,n$$

The special case in which p=n is the basis for well known Able-Jacobi-Liouville formula obtained by integrating the scalar equation

$$\frac{d}{dt}(\det X) = (\operatorname{tr} A(t))\det X(t)$$

Thus we see that $A_{[p]}$ and $A_{(p)}$ are infinitesimal versions of

$A^{[p]}$ and $A^{(p)}$ respectively. As such, they depend linearly on the elements of A. This has some significant implications.

We also have the infinitesimal version of the tensor product reduction given above. It takes the form

$$A(\cdot)+(\cdot)A' \simeq I \otimes A + A \otimes I \simeq \begin{bmatrix} A_{[2]} & 0 \\ 0 & A_{(2)} \end{bmatrix}$$

There are important relationships between A, $A_{[p]}$ and $A_{(p)}$ which are more or less clear from derivation. First of all, if A has all distinct eigenvalues $\{\lambda_i\}$ then the solutions of $\dot{x}(t)=Ax(t)$ consists of a sum of terms of the form $\alpha_i e^{\lambda_i t}$. Thus $x^{[p]}$ consists of products, p at a time, of such terms

$$x^{[p]} = \Sigma\beta_{ij\ldots k} e^{(\lambda_i+\lambda_j+\ldots\lambda_k)t}$$

Thus the eigenvalues of the $\binom{n+p-1}{p}$ by $\binom{n+p-1}{p}$ matrix $A_{[p]}$ are the $\binom{n+p-1}{p}$ sums over distinct (unordered) index sets

$$\lambda_i+\lambda_j+\ldots\lambda_k; \quad \text{p terms}$$

The same is true for the case where A has eigenvalues of higher multiplicity. Similarly, the eigenvalues of $A_{(p)}$ consist of sums p at a time of the eigenvalues of A but in this case the indices i,j,..,k must all be distinct.

A second fact involves the transition matrix $\Phi_A(t)$ which satisfies

$$\dot{\Phi}(t) = A(t)\Phi(t); \quad \Phi(0) = I$$

By the above construction we see that

$$\Phi_{A_{[p]}}(t) = \Phi_A^{[p]}(t)$$

and

$$\Phi_{A_{(p)}}(t) = \Phi_A^{(p)}(t)$$

(Again, the last of these is the Able-Jacobi-Liouville formula if p=n.)

Finally, if $\{A_i\}$ is a basis for a Lie algebra and if

$$[A_i,A_j] = \sum_{k=1}^{m} \gamma_{ijk}A_k$$

then

$$[A_{i_{[p]}},A_{j_{[p]}}] = \sum_{i=1}^{m} \gamma_{ijk}A_{k_{[p]}}$$

That is, the $\{A_{i_{[p]}}\}$ form a Lie algebra with the same structural constants. To see this we need to show that

$$[A,B]_{[p]} = [A_{[p]},B_{[p]}]$$

but this can be seen from the approximations

$$\begin{aligned} e^{[A,B]_{[p]}t^2} &= (e^{[A,B]t^2})^{[p]} \\ &\approx e^{A_{[p]}t} e^{B_{[p]}t} e^{-A_{[p]}t} e^{-B_{[p]}t} \\ &\approx e^{[A_{[p]},B_{[p]}]t^2} \end{aligned}$$

where in all cases the approximations are valid up to and including terms of second order in t. Identical formulas hold with [p] replaced by (p).

This circle of ideas is of great importance in the theory of representations of Lie algebras; see [4] or [5]. However in control theory and differential equations there exist many problems where one can use these ideas, and other ideas from representation theory, to simplify calculations and to provide insight. A particular example is the study of the moment equations for stochastic differential equations. See, for example, reference [6].

Exercises

1. Show that

$$\begin{bmatrix} \dot{x}_1 \\ \dot{x}_2 \end{bmatrix} = \begin{bmatrix} 0 & 1 \\ -k(t) & -1 \end{bmatrix} \begin{bmatrix} x_1 \\ x_2 \end{bmatrix}$$

and

$$\begin{bmatrix} \dot{x}_1 \\ \dot{x}_2 \\ \dot{x}_3 \end{bmatrix} = \begin{bmatrix} 0 & 2 & 0 \\ -\sqrt{2}k(t) & -1 & \sqrt{2} \\ 0 & -\sqrt{2}k(t) & -1 \end{bmatrix} \begin{bmatrix} x_1 \\ x_2 \\ x_3 \end{bmatrix}$$

are an A, $A_{[2]}$ pair.

2. Show that $A^{[p]}$ is orthogonal if A is orthogonal. What about $A^{(p)}$?

3. Describe in full the decomposition of $A \otimes A \otimes A$.

4. Give a definition of $A^{[p]}$ for which $z = Ax$ implies $z^{[p]} = A^{[p]}x^{[p]}$ but which does not require A to be square.

1.3 Matrix Lie Algebras and the Matrix Exponential

In section 1 we saw that the solution of the differential equation

$$\dot{x}(t) = \left(\sum_{i=1}^{m} u_i(t)A_i\right)x(t); \quad x(0) = x_0$$

could be expressed for small $|t|$ as

$$x(t) = e^{A_1 g_1(t)} e^{A_2 g_2(t)} \cdots e^{A_m g_m(t)} x_0$$

provided the A_i form a basis for a Lie algebra. On the strength of the theorem of Frobenius, similar statements can be made for

$$\dot{x}(t) = \sum_{i=1}^{m} u_i(t) f_i[x(t)]; \quad x(0) = x_o$$

provided the set of vectors $\{f_i(\cdot)\}$ are involutive. There is a sort of converse question. If the set $\{A_i\}$ does not form the basis for a Lie algebra to what extent is it necessary to add elements to these sets in order to cover all possibilities? We know already that by adding enough elements to $\{A_i\}$ so as to obtain a basis for a Lie algebra we can be assured of a representation of the above form. However, it might happen that for

$$\dot{x}(t) = u_1(t)A_1x(t)+u_2(t)A_2x(t); \quad x(t) \in \mathbb{R}^n$$

the smallest Lie algebra which contains A_1 and A_2 is of dimension n^2. Are all of the n^2-2 elements which we add in order to get a Lie algebra really necessary?

In 1939 Chow [7] published a generalization of an earlier theorem of Caratheodory proving that if some regularity conditions hold, then along solution curves of

$$\dot{x}(t) = \sum_{i=1}^{m} u_i(t) f_i[x(t)]; \quad x_o = x(0)$$

one can reach the same points as one can along the solution curves of

$$\dot{x}(t) = \sum_{i=1}^{m} u_i(t) f_i[x(t)] + \sum_{i=1}^{\nu} v_i(t) g_i[x(t)]$$

where $g_i(x)$ are obtained as Lie brackets of the f_i, Lie brackets of these Lie brackets, etc. Thus on the basis of this "reachability" theorem of Chow we see that no matter how many elements we must add to get a basis for a Lie algebra, nothing short of the full set will suffice.

We formalize this discussion as follows. Let B denote any subspace of $g\ell(n)$. Let $\{B\}_A$ denote the smallest Lie algebra which contains B. Let C be any subset of $G\ell(n)$ and let $\{C\}_G$ denote the smallest group which contains C.

Theorem 1: With the above definitions

$$\{\exp B\}_G = \{\exp \{B\}_A\}_G$$

Perhaps the most elementary proof of this result appears in [8].

After sufficient insight is built up it is frequently possible to evaluate $\{\exp \{B\}_A\}_G$ by inspection. The insight comes from a handful of special cases and general formulas such as $\exp A_{[p]}$ $(\exp A)^{[p]}$. The notation for the principle special casesis this:

We take the field to be $\mathbb{R}$ and let $J = \begin{pmatrix} 0 & I \\ -I & 0 \end{pmatrix}$.

$$
\begin{aligned}
g\ell(n) &= \{X : X = n \text{ by } n \text{ matrices}\} \\
s\ell(n) &= \{X : X \in g\ell(n);\ \operatorname{tr} X = 0\} \\
so(n) &= \{X : X \in g\ell(n);\ X'+X = 0\} \\
sp(n) &= \{X : X \in g\ell(n);\ X'J+JX = 0\}
\end{aligned}
$$

Matrices satisfying the last condition are often called Hamiltonian because they take the form familiar in Hamiltonian mechanics

$$\begin{bmatrix} A & Q \\ R & -A' \end{bmatrix} \quad ; \quad Q = Q'; \quad R = R'$$

It is very important to keep in mind that $J^2 = -I$ so that $J^{-1} = -J$.

Associated with each of these algebras is a multiplicative group of matrices which are defined in a corresponding way

$$
\begin{aligned}
G\ell(n) &= \{X : X \text{ is } n \text{ by } n \text{ matrix; } \det X \neq 0\} \\
S\ell(n) &= \{X : X \in G\ell(n);\ \det X = 1\} \\
So(n) &= \{X : X \in G\ell(n);\ X'X = I\} \\
Sp(n) &= \{X : X \in G\ell(n);\ X'JX = J\}
\end{aligned}
$$

These groups are called the general linear group, the special linear group, the special orthogonal group and the symplectic group, respectively.

It is easy to verify that in any of these cases exp X belongs to a particular group if X belongs to the corresponding algebra. This corresponds to the following well known facts

i) exp M is nonsingular for all M
ii) $\det(\exp M) = \exp(\operatorname{tr} M) = 1$ if $\operatorname{tr} M = 0$
iii) exp A is orthogonal if A is skew symmetric since $(e^A)' = e^{A'} = e^{-A} = (e^A)^{-1}$ if $A = -A'$.
iv) exp A is symplectic if A is Hamiltonian since $e^{A'}Je^A = Je^{J'A'J}e^A = J$ if $A'J+JA = 0$.

Notice that the set of n by n symmetric matrices do not form a Lie algebra; alternatively, the nonsingular symmetric matrices do not form a group.

The implication for the study of differential equations is as follows. If X is an n by n matrix which satisfies the equation

$$\dot{X}(t) = A(t)X(t)$$

Then of course the fundamental solution $\Phi_n(t)$ is going to belong to the general linear group. But if A at all points in time belongs to one of the above subalgebras of $g\ell(n)$ then $\Phi_A(t)$ will belong to the corresponding subgroup of $G\ell(n)$. This group-algebra relationship provides qualitative information about the solution without actually solving the equations of motion.

To what extent are the above maps of the algebra into the group actually onto the group? It is well known that a real nonsingular matrix need not have a real logarithm. Thus as far as the real field is concerned, exp does not map $g\ell(n)$ onto $G\ell(n)$. However if the field is either the reals or the complexes, then every matrix sufficiently close to the identity does have a logarithm in the appropriate field and it is easy to see that exp maps a neighborhood of zero in the algebra onto a neighborhood of the identity in the group in a one to one way.

Exercises

1. Consider the set of n by n matrices whose column sums are zero. Show that they form a Lie algebra. If we denote this algebra by L then characterize $\{\exp L\}_G$.

2. Let so(p,q) denote the set of matrices satisfying

$$A'\Sigma(p,q) + \Sigma(p,q)A = 0$$

where $\Sigma(p,q)$ is defined by

$$\Sigma(p,q) = \begin{bmatrix} I_p & 0 \\ 0 & -I_q \end{bmatrix}$$

Show that this set of matrices forms a Lie algebra and show that for all matrices M in $\exp\{so(p,q)\}$ we have

$$\Sigma(p,q) = M'\Sigma(p,q)M$$

These are often called the pseudo orthogonal groups since they preserve the pseudo length $x'\Sigma(p,q)x$.

1.4 Cones and Semigroups

A semigroup of real n by n matrices is simply a subset of the n by n matrices which is closed under matrix multiplication. A cone in a real vector space is a subset closed under addition and multiplication by positive real numbers. Consider a real Lie algebra L in the set of n by n matrices. Let K be a conical subset of L. In general K will not be closed under Lie bracketing but it could be. Let $\{\exp K\}_{SG}$ indicate the smallest semigroup which contains exp K. As we will see, a number of problems in control lead to the question of characterizing $\{\exp K\}_{SG}$ in terms of K. The connection between a Lie algebra and its corresponding Lie group suggests analogous relationships between cones in the algebra and semigroups in the corresponding group. This kind of relationship is illustrated in the following example.

Example: Let K be the cone in $g\ell(n)$ consisting of all n by n matrices A such that A'+A is nonnegative definite. Then $\{\exp K\}_{SG}$ includes all orthogonal matrices since all skew symmetric matrices belong to K. Moreover, all symmetric matrices with eigenvalues greater than or equal to one belong to $\{\exp K\}_{SG}$ by well known properties of the exponential map. Thus by appealing to the fact that any matrix can be written in polar form $M = \theta R$ with θ orthogonal and R positive definite we see that if for all vectors x of unit length $||Mx||^2 = ||\theta Rx||^2 = ||Rx||^2 \geq 1$ then M belongs to $\{\exp K\}_{SG}$. It is easy to see that if $||Mx|| < 1$ for some x of unit length then we can not express M in the required way thus this condition is necessary and sufficient. We conclude that the semigroup of "expansive" matrices is the exponential of the nonnegative definite ones. Likewise, the semigroup of (nonsingular) "contractive" matrices is the exponential of the cone of nonpositive definite matrices.

This example can be generalized somewhat to give a theorem with broader scope.

Theorem 1: Let K be as above and let L_P be the Lie algebra of matrices satisfying $A'P+PA = 0$ with $P'P = I$. Then $\{\exp K \cap L_P\}_{SG} = \{\exp K\}_{SG} \cap \{\exp L_P\}_G$ i.e. the expansive matrices in $\{\exp L_P\}_G$.

Proof: Given any orthogonal matrix P, the group of matrices satisfying $M'PM = P$ has the property that the polar representations of each element has both its factors in the group. That is, if $M = e^{\Omega}e^{R}$ with e^{Ω} orthogonal and e^{R} positive definite and symmetric, then $e^{\Omega'}Pe^{\Omega} = P$, $e^{R}Pe^{R} = P$. To prove this we note that if $e^{R}e^{\Omega'}Pe^{\Omega}e^{R\Omega} = P$ then $e^{R}e^{\Omega'} = Pe^{-R}P'Pe^{-\Omega}P'$. However the term of the right is a polar decomposition since $Pe^{-R}P'$ is symmetric and positive definite and $Pe^{-\Omega}P'$ is orthogonal. Thus by uniqueness of the polar decomposition we see that $e^{R} = Pe^{-R}P'$ and $e^{\Omega} = Pe^{\Omega}P'$ which shows that each factor belongs to the given group.

Now if M has the polar form $M = e^{\Omega}e^{R}$ and if M belongs to $\{\exp K\}_{SG} \cap \{\exp L_P\}_G$ then $R \geq 0$ and Ω and R belong to L_P. Thus Ω belongs to $L_P \cap K$ and so does R.

Typically the relationship between a cone in the Lie algebra and the semigroup which the exponential maps it into is very difficult to describe. One problem of this type which has been investigated extensively arises in probability theory. Let $x_o \in \mathbb{R}^n$ have nonnegative components which sum to one. Suppose that $x(t)$ evolves in time according to

$$\dot{x}(t) = A(t)x(t); \quad x(0) = x_o$$

If $A(\cdot)$ has the two properties:

(i) the off-diagonal elements of $A(t)$ are nonnegative for all t
(ii) the sums of the columns of $A(t)$ are zero for all t,

then $x(t)$ will have nonnegative components which sum to one for all $t \geq 0$. This is equivalent to saying that subject to the above restrictions on $A(\cdot)$ the solution of the matrix equation

$$\dot{X}(t) = A(t)X(t); \quad X(0) = I \qquad (*)$$

is a stochastic matrix; i.e. a matrix with nonnegative entries whose columns sum to 1. The imbedding problem [9] is that of determining which stochastic matrices ϕ can be reached from the identity along solutions of (*) given only that $A(t)$ must satisfy (i) and (ii). Of course the set of matrices which satisfy (i) and (ii) form a cone and the set of reachable matrices form a semigroup. It is not true however that for $n > 2$ this semigroup consists of all stochastic matrices.

In control applications there is particular interest in the case of cones of the form

$$K = \{X : X = \alpha A + \Sigma\beta_i B_i;\quad \alpha \geq 0;\ \beta_i \text{ unrestricted}\}$$

i.e. cones which are half spaces.The first point to make is that by virtue of theorem 3.1 we may as well assume that the B_i form a basis for a Lie algebra since by adding elements to $\{B_i\}$ to make the basis of the Lie algebra generated by $\{B_i\}$ we do not enlarge the reachable set. Moreover, it is also clear from theorem 3.1 that

$$\{\exp\{A,B_i\}_A\}_G \supseteq \{\exp K\}_{SG} \supseteq \{\exp\{B_i\}_A\}_G$$

It is more or less clear that if e^{At} is periodic then

$$\{\exp\{A,B_i\}_A\}_G = \{\exp K\}_{SG}$$

and Jurdjevic and Sussmann [10] have shown that this is also true if e^{At} is almost periodic.

It is also true that $Ad_A^k B_i$ belongs to the Lie algebra generated by the B_i's then

$$\exp K = e^{\alpha A}\{\exp\{B_i\}_A\}_G$$

For a proof and some generalizations see the thesis of Hirschorn [11].

Exercises

1. Calculate $\{\exp N\}_{SG}$ where N is the cone

$$N = \{X : X = \begin{bmatrix} a & b \\ c & -a \end{bmatrix};\quad X+X' \leq 0\}$$

2. It is well known that the elements of $\Phi_A(t)$ are nonnegative for all $t \geq 0$ if $A(t)$ itself as elements which are nonnegative off the diagonal -- the diagonals may have any sign. Give an example which shows that $\{\exp K\}_{SG}$ is not the entire semigroup of square matrices with nonnegative entries if K is the cone of A's described above. (Find a matrix with positive entires and negative determinant.)

3. Explore the relationship between #2 and the imbedding problem.

II. INPUT-OUTPUT SYSTEMS

In this chapter we consider input/output systems which can be represented by a pair of equations of the form

$$\dot{X}(t) = (A+\sum_{i=1}^{m} u_i(t)B_i)X(t);\quad y(t) = C(X(t)) \tag{*}$$

Here X is an n by n matrix as are A and $B_1, B_2, \ldots, B_m$; the map C is subject to certain restrictions to be described later. The differential equation is said to be of the "right invariant type" because a multiplication on the right by a fixed element of $G\ell(n)$ gives an equation

$$\dot{X}(t)M = (A+\sum_{i=1}^{m} u_i(t)B_i)X(t)M$$

which is again of the same form and with the same coefficient matrices. This is to be contrasted with an equation such as

$$\dot{X}(t) = (A+\sum_{i=1}^{m} u_i(t)B_i)X(t)+X(t)(D+\sum_{i=1}^{m} u_i(t)E_i)$$

which does not have this invariance property. The basic idea is to understand as well as possible the properties of input-output maps which can be represented by equation (*). We will study controllability, observability and state space isomorphism theorems.

2.1 Controllability

If u_i is an m-dimensional piecewise continuous function of time and if t_i is a nonnegative number, then we given the pairs (u_i,t_i) a semigroup structure by defining

$$(u_1,t_1) \circ (u_2,t_2) = (u_1|u_2,t_1+t_2)$$

whereby $u_1|u_2 = u_3$ we mean

$$u_3(t) = \begin{cases} u_1(t); & 0 \leqslant t < t_1 \\ u_2(t-t_1); & t_1 \leqslant t < t_2 \end{cases}$$

This is the concatenation semigroup with due regard for the domain of definition of the functions being concatenated. We denote it by U^m.

Consider the time invariant control system

$$\dot{x}(t) = f[x(t),u(t)] \quad ; \quad x(t) \in \mathbb{R}^n \qquad (**)$$

with f well enough behaved so as to guarantee the existence of a unique solution for each starting point $x_o \in \mathbb{R}^n$ and each $(u,t) \in U^m$. Let T^n be the semigroup of one to one continuous maps of $\mathbb{R}^n$ into $\mathbb{R}^n$ with composition as the semigroup operation. Then the control system (**) defines a homomorphism of U^m into T^n. We denote this homomorphism by ϕ and, by analogy with automata theory, call the image of U^m under ϕ the Myhill semigroup of the system.

The main thing which is special about bilinear systems is that the Myhill semigroup is easily identified with a matrix semigroup. That is, if we have a system in $\mathbb{R}^n$

$$\dot{x}(t) = (A+\sum_{i=1}^{m} u_i(t)B_i)x(t)$$

then the matrix equation

$$\dot{X}(t) = (A+\sum_{i=1}^{m} u_i(t)B_i)X(t); \quad X(0) = I$$

describes the relationship between U^m and T^n -- each matrix being associated with an element of T^n in the standard way

$$M \mapsto \quad f(x) = Mx$$

If A is absent in the above equation then it is clear that the Myhill semigroup is actually a group since if $u(\cdot) \in U^m$ steers the system from I to M at time t_1 then $v(\cdot) \in U^m$ and defined by

$$v(t) = -u(t_1-t)$$

steers the system to M^{-1} at $t = t_1$.

Given an initial state x_o, the set of states reachable from x_o can be identified with the set of points which x_o is mapped into by the various elements of the Myhill semigroup. That is, the Myhill semigroup acts on the state space

$$S : \Sigma \to \Sigma$$

The reachable set from x_o is the "orbit" through x_o defined by this action.

We now give various examples of reachability theorems.

Theorem 1: There exists a control which steers the system

$$\dot{X}(t) = (\sum_{i=1}^{m} u_i(t)B_i)X(t)$$

from X_o to X_1 in time $t_1 > 0$ if and only if $X_1X_o^{-1}$ belongs to $\{\exp\{B_i\}_A\}_G$.

Proof: This is an immediate consequence of Theorem 1.3.1.

It is also easy to see that if A belongs to $\{B_i\}_A$ then the reachable set for

$$\dot{X}(t) = (A+\sum_{i=1}^{m} u_i(t)B_i)X(t)$$

is just the same as it would be if A were absent.

Notice that the reachable set does not depend on t_1 as long as t_1 is positive. If A is absent and if one restricts the controls to be bounded, say $|u_i(t)| \leq 1$ then all points of the above form are reachable after a suitably long time but the time required will depend on the point to be reached.

A second result which we want to use in a moment is this.

Theorem 2: The reachable set at time t for

$$\dot{X}(t) = (\tilde{A}+\sum_{i=1}^{m} u_i(t)\tilde{B}_i)X(t); \quad X(0) = I$$

and

$$\tilde{A} = \begin{bmatrix} A & 0 \\ 0 & 0 \end{bmatrix} \qquad \tilde{B}_i = \begin{bmatrix} 0 & B_i \\ 0 & 0 \end{bmatrix}$$

with A square is

$$R(t) = e^{At}\{\exp Ad_{\tilde{A}}, \tilde{B}_i\}_A\}_G$$

Here $\{Ad_{\tilde{A}},\tilde{B}_i\}_A$ indicates the smallest Lie algebra which contains $\{\tilde{B}_i\}_A$ and is closed under the action of $Ad_{\tilde{A}}$.

Proof: See reference [8], Theorem 7.

We can combine theorems 1 and 2 in an obvious way to get the following more general result.

Theorem 3: The reachable set at time t for

$$\dot{X}(t) = \tilde{A}x(t)+ \sum_{i=1}^{m} u_i(t)\tilde{B}_i X(t)+ \sum_{i=1}^{q} v_i(t)\tilde{C}_i X(t)$$

where

$$\tilde{A} = \begin{bmatrix} A & 0 \\ 0 & 0 \end{bmatrix}; \quad \tilde{B}_i = \begin{bmatrix} 0 & 0 \\ 0 & B_i \end{bmatrix} \quad ; \quad \tilde{C}_i = \begin{bmatrix} 0 & C_i \\ 0 & 0 \end{bmatrix}$$

with A and B_i square is

$$R(t) = \exp At\{\exp\{Ad_{\tilde{A}}, \tilde{B}_i, \tilde{C}_i\}_A\}_G$$

Finally, one can get additional results by using a nice lemma of Jurdjevic and Sussmann [10].

Theorem 4: The reachable set for the $\mathbb{R}^n$ system at time t starting from x=0 at t=0 and governed by

$$\dot{x}(t) = (A+ \sum_{i=1}^{m} u_i(t)B_i)x(t)+ \sum_{i=1}^{p} v_i(t)g_i; \quad x(t) \in \mathbb{R}^n$$

is the vector space generated by $\{L_i^k g_i\}$ where k indicates powers and L_i is a basis for the associative algebra generated by $\{A,B_i\}$.

Proof: To begin we observe that if x_1 is reached at $t=t_1$ starting from x=0 at t=0 using the control (u,v) then the control $(u,\alpha v)$ steers the system to αx_1 at $t=t_1$. Also, we know that if we write the system as

$$\frac{d}{dt}\begin{bmatrix} x(t) \\ 1 \end{bmatrix} = \left(\begin{bmatrix} A & 0 \\ 0 & 0 \end{bmatrix} + u_i(t) \begin{bmatrix} B_i & 0 \\ 0 & 0 \end{bmatrix} + v_i \begin{bmatrix} 0 & g_i \\ 0 & 0 \end{bmatrix}\right)\begin{bmatrix} x(t) \\ 1 \end{bmatrix}$$

then the reachable set has a nonempty interior in

$$R= \{\exp\{\tilde{A}, \tilde{B}, \tilde{G}\}_{LA}\}_G \begin{bmatrix} 0 \\ 1 \end{bmatrix}$$

where

$$\tilde{A} = \begin{bmatrix} A & 0 \\ 0 & 0 \end{bmatrix} ; \quad \tilde{B}_i = \begin{bmatrix} B_i & 0 \\ 0 & 0 \end{bmatrix} ; \quad \tilde{G}_i = \begin{bmatrix} 0 & g_i \\ 0 & 0 \end{bmatrix}$$

There exists a nonzero control of the form (0,v) which steers the system back to zero at time $t=t_1$ from 0 at t=0 -- use u=0 and invoke standard linear theory. According to lemma 6.1 of [10] we obtain on taking perturbations about this control an open set in R containing 0. Using the cone property mentioned in the first sentence we see that the reachable set is a vector space. Lie algebras tell us which one.

A particular problem in controllability theory which has received a good deal of attention is

$$\dot{x}(t) = Ax(t)+u(t)b\langle c,x(t)\rangle \quad ; \quad x(t) \in \mathbb{R}^n$$

where u(·) is a scalar, and b is a column vector. Of course the linear system

$$\dot{x}(t) = Ax(t)+bv(t)$$

is controllable in $\mathbb{R}^n$ if and only if $(b,Ab,..,A^{n-1}b)$ is of full rank. If the linear system is controllable it might be supposed that the bilinear one is also controllable since if v is a control which drives the state of the linear system from x_o to x_1 then the control

$$u(t) = v(t)/\langle c,x(t)\rangle$$

drives the bilinear system from x_o to x_1. This argument has the obvious fallacy that $\langle c,x(t)\rangle$ might vanish along the trajectory leaving u(t) undefined. In particular, if x(0) = 0 then of course x vanishes identically for all future time. Thus the most one could hope for is that any nonzero state could be steered to any nonzero state. It turns out that this is too much to hope for also. A simple pair of examples which illustrate that no amount of work can salvage this argument and which at the same time suggest the nature of the problem are these.

Consider the system

$$\begin{bmatrix}\dot{x}_1(t)\\ \dot{x}_2(t)\end{bmatrix} = \begin{bmatrix}0 & 1\\ 1 & 0\end{bmatrix}\begin{bmatrix}x_1(t)\\ x_2(t)\end{bmatrix} + u(t)\begin{bmatrix}0 & 0\\ 0 & 1\end{bmatrix}\begin{bmatrix}x_1(t)\\ x_2(t)\end{bmatrix}$$

which has the form

$$\dot{x}(t) = Ax(t) + u(t)b\langle c,x(t)\rangle$$

with [A,b,c] a minimal realization of $s/(s^2-1)$. However for any given x_o there exists x_1 such that x_1 is not reachable from x_o because regardless of k, the off-diagonal elements of $(A+k(t)bc')$ are always positive so that $\phi(t,t_o)$, the transition matrix, has all entries nonnegative for $t > t_o$. Thus if x(0) has nonnegative entries for all t > 0. This argument shows that the system is not controllable.

Consider the system

$$\begin{bmatrix}\dot{x}_1(t)\\ \dot{x}_2(t)\end{bmatrix} = \begin{bmatrix}0 & 1\\ -1 & 0\end{bmatrix}\begin{bmatrix}x_1(t)\\ x_2(t)\end{bmatrix} + k(t)\begin{bmatrix}0 & 0\\ 0 & 1\end{bmatrix}\begin{bmatrix}x_1(t)\\ x_2(t)\end{bmatrix}$$

which has the form $\dot{x}(t) = Ax(t)+k(t)bc'x(t)$ with [A,b,c] a minimal realization of $s/(s^2+1)$. In this case we see that the system is controllable on $R^2-\{0\}$. (See reference [12] for details.)

Exercises

1. Show that the Myhill semigroup for the linear system

$$\dot{x}(t) = Ax(t)+bu(t); \quad x(t) \in \mathbb{R}^n$$

can be identified with the multiplicative matrix semigroup

$$S = \{X : X = \begin{bmatrix} e^{At} & x \\ 0 & 1 \end{bmatrix} ; \quad t \geq 0; \; x \in \text{span}(b, Ab, \ldots A^{n-1}b)\}$$

2. Consider a bilinear system

$$\dot{x}(t) = Ax(t) + u(t)Bx(t)$$

on $\mathbb{R}^n - \{0\}$. Is it true that if there exists any state x_o such that all points in $\mathbb{R}^n - \{0\}$ are reachable from x_o then all states have this property?

3. Consider the linear system

$$\dot{X}(t) = A_\ell X(t) + X(t)A_r + \sum_{i=1}^{m} u_i(t)B_i$$

Here $X(t)$ is an n by q matrix and A_ℓ and A_r are n by n and q by q respectively; the B_i are n by q. Show that the Myhill semigroup equation can be identified with

$$\frac{d}{dt}\begin{bmatrix} S_1(t) & S_3(t) \\ 0 & S_2(t) \end{bmatrix} = \left(\begin{bmatrix} A_\ell & 0 \\ 0 & A_r \end{bmatrix} + \sum_{i=1}^{m} u_i(t)\begin{bmatrix} 0 & B_i \\ 0 & 0 \end{bmatrix}\right)\begin{bmatrix} S_1(t) & S_3(t) \\ 0 & S_2(t) \end{bmatrix}$$

Show that the reachable set at time t for the Myhill equation is

$$\exp At \cdot \exp\{Ad_A, \tilde{B}_i\}$$

2.2 Observability

We now consider systems with an output

$$\dot{X}(t) = (A(t) + \sum_{i=1}^{m} u_i(t)B_i(t))X(t); \quad y(t) = C(X(t)); \quad X(t) \in G\ell(n)$$

The exact nature of the output map is not essential. We give the output space no structure -- it is just a set. The critical assumption is that there should exist subgroups H_ℓ and H_r of $G\ell(n)$ such that $C(X_1) = C(X_2)$ if and only if

$$H_1X_1H_2 = X_2$$

for some H_1 in H_ℓ and some H_2 in H_r. Under this assumption $C(X)$ identifies X to within a multiplication on the left by an element of H_ℓ and a multiplication on the right with an element of H_r. We call systems of this form homogeneous.

In such a set up, the observation of y, even over a period of time, can at most determine X to within a right multiplication by an element of H_r. Thus we might as well regard the system as evolving on the coset space $G\ell(n)/H_r$. Whether or not the observation of y and the knowledge of u over the interval $[0,\infty)$ serves to identify uniquely an element of X/H_r as a starting state is then subject to investigation.

Theorem 1: Consider the above system with H_r and H_ℓ given. Let R denote the set of X's reachable from I. Suppose that R is a group.

Then two points X_1H_r and X_2H_r in $G\ell(n)/H_r$ give rise to the same input/output map if and only if for each R_1 in R there exists $H_1(R)$ in H_ℓ such that

$$R^{-1}H_1(R)RX_1H_r = X_2H_r$$

If we denote by P the subgroup

$$P = \{X : R^{-1}XR \in H_\ell; \quad \forall R \in S\}$$

then any two elements of the form X_1H_r and $P_1X_1H_r$ with P_1 in P are not distinguishable.

Proof: If X_1H_r and X_2H_r are to be indistinguishable as starting states we must have

$$H_\ell R_i X_1 H_r = H_\ell R_i X_2 H_r$$

for all R_i in R. Since H_ℓ and H_r are groups and since R is a subgroup of $G\ell(n)$, the above condition is equivalent to asking that for each R_i in R there exist $H_1(R)$ in H_ℓ such that

$$R_i^{-1}H_1(R_i)R_iX_1H_r = X_2H_r$$

The remainder of the conclusions are clear.

Exercises

1. Assuming that the evolution equations are of the form

$$\dot{x}(t) = Ax(t) + \sum_{i=1}^{m} u_i(t)B_ix(t); \quad y(t) = H_\ell x(t)H_r$$

with

$$H_\ell = \{\exp\{C_i\}_A\}_G; \quad H_r = \{\exp\{D_i\}_A\}_G$$

give an observability condition in terms of Lie algebras. (See ref. [8] for some results along this line.)

2. Apply the results of problem 1 to the bilinear problem

$$\dot{x}(t) = Ax(t) + \sum_{i=1}^{m} u_i(t)B_ix(t); \quad y(t) = c[x(t)]$$

by identifying $\mathbb{R}^n$ with the n dimensional affine group modulo $G\ell(n)$.

2.3 Isomorphic Systems

The two scalar realizations

$$\dot{x}(t) = x(t)+u(t)x(t); \quad y(t) = x^3(t); \quad x(0) = 1$$

and

$$\dot{z}(t) = 3z(t)+3u(t)z(t); \quad y(t) = z(t); \quad z(0) = 1$$

realize the same input-output map. They are each controllable on $(0,\infty)$ and any two reachable states are distinguishable. They are related by the automorphism of the multiplicative group $(0,\infty)$ defined by

$$z = x^3$$

Thus despite the apparent differences between these two realizations they are closely related. The following theorem describes a general result of this type.

Theorem 1: Consider the two homogeneous realizations of the same input-output map

$$\dot{X}(t) = (A+\sum_{i=1}^{m} u_i(t)B_i)X(t); \quad y(t) = c[X(t)]$$

$$\dot{Z}(t) = (F+\sum_{i=1}^{m} u_i(t)G_i)Z(t); \quad y(t) = h[Z(t)]$$

which evolve in $G\ell(n_1)$ and $G\ell(n_2)$ respectively and which have reachable sets from the identity, **R and** $\hat{R}$, which are groups. Suppose H_ℓ, H_r and $\hat{H}_\ell$, $\hat{H}_r$ are given subgroups of $G\ell(n_1)$ and $\hat{G}\hat{\ell}(n_2)$ respectively such that c and h are one to one on $H_\ell RH_r$ and $\hat{H}_\ell\hat{R}\hat{H}_r$ and such that the systems are observable on RH_r and $\hat{R}\hat{H}_r$. Finally, suppose that there is no normal subgroup of R which has a non-trivial intersection with $R \cap H_r$ and the same for $\hat{R}$ and $\hat{H}_r$. Then there exists an isomorphism $\phi : R \to \hat{R}$ such that

$$\phi(e^{At}) = e^{Ft}; \quad \phi(e^{B_i t}) = e^{G_i t}$$

Proof: Suppose that there exists a control (u,T) in U^m which takes the first system from I to $D_1 \neq I$ and takes the second system from I to I. Let D denote the set of all such points. By virtue of the observability hypothesis we see that D is a subset of H_r and, in fact, a subgroup of H_r. Moreover it is easily seen to be a normal subgroup of R and hence of $R \cap H_r$. By hypothesis D is trivial. This implies that there is a one to one correspondence between points in $R \cap H_r$ and $\hat{R} \cap \hat{H}_r$ which is, in fact, a homomorphism.

We see that R and $\hat{R}$ are both homomorphic images of U^m. If a pre-image of R in U^m is in U_R then what is the image under the action of the second system of U_R? It is clearly $\hat{R}$ or else a subgroup of $\hat{R}$. If it is a subgroup then the subgroup must contain $\hat{R} \cap \hat{H}_r$ but there is a one to one and onto correspondence between $\hat{R}/R \cap H_r$ and $\hat{R}/\hat{R} \cap \hat{H}_r$ and an isomorphism between $R \cap H_r$ and $\hat{R} \cap \hat{H}_r$ Using the properties of the system maps we see that the above map must be onto R and thus it establishes an isomorphism. The remaining claims then follow.

Exercises

1. Develop the Lie algebra analog of Theorem 1.
2. Apply the above results to bilinear systems of the form

$$\dot{x}(t) = Ax(t)+\sum_{i=1}^{m} u_i(t)B_i x(t); \quad y(t) = cx(t); \; x(0) = x_o$$

See P. d'Alessandro, A. Isidori and A. Ruberti [13] and Brockett [14].

III. OPTIMAL CONTROL

This chapter is quite brief due to the absence in the literature of results relating specifically to the Lie group case. We discuss only two problem areas -- the question of existence of optimal controls in the bang bang case and questions centering around minimum "energy" transfer.

3.1 Bang-Bang Theorems

It is well known that under very weak assumptions on the matrices $A(\cdot)$ and $B(\cdot)$ the linear system

$$\dot{x}(t) = A(t)x(t)+B(t)u(t);\ x(0) = \text{given}$$

with controls constrained by

$$|u_i(t)| = 1$$

has a set of reachable points at any time $t_1 > 0$ which is the same as the set of points reachable with the constraint relaxed to

$$|u_i(t)| \leq 1$$

This is called a "bang-bang theorem" because the controls u_i need only take on their extreme values and not intermediate ones. Some generalizations of this have been investigated by Krenner [15] and Sussmann [16]. We examine only an easy case here.

Theorem 1: Let X satisfy the differential equation in $G\ell(n)$

$$\dot{X}(t) = AX(t) + \left(\sum_{i=1}^{m} B_i u_i(t)\right)X(t)$$

Then if $[\mathrm{Ad}_A^k(B_i), B_j]$ is zero for all i and j and $k=0,1,\ldots n^2-1$ then the set of states reachable at time t for $|u_i(t)| = 1$ is the same as the set reachable for $|u_i(t)| \leq 1$.

Proof: In view of the commutativity condition we can express the solution of the given equation as

$$X(t) = e^{At} e^{\int_0^t \sum_{i=1}^{m} e^{-A\sigma} B_i e^{A\sigma} u_i(\sigma)\,d\sigma} X(0)$$

See [8] Theorem 7 for details. Now since the bang-bang theorem is valid for the linear system

$$\dot{F}(t) = \sum_{i=1}^{m} e^{-At} B_i e^{At} u_i(t)$$

and since $X(t) = e^{At}e^{F(t)}$ we see that it holds for the systems defined here as well.

Exercises

1. The solution of the scalar differential equation
$$\dot{x}(t) = u(t)x(t)+v(t)$$

is

$$x(t) = e^{\int_0^t u(\sigma)\,d\sigma} x(0) + \int_0^t e^{\int_\sigma^t u(\rho)\,d\rho} v(\sigma)\,d\sigma$$

Is the bang-bang theorem valid if we regard u and v as controls?

2. Is the bang-bang theorem valid for the pair of scalar equations

$$\dot{z}(t) = u(t)z(t)$$

$$\dot{x}(t) = (u(t)+v(t))x(t)$$

3. Show that the bang-bang theorem is valid for

$$\dot{x}(t) = u(t)x(t)$$

$$\dot{y}(t) = -y(t)+u(t)$$

Generalize this result.

3.2 Least Squares Theory

Under the assumption used in the previous section we can develop a satisfactory theory for minimizing

$$\eta = \int_0^t \sum_{i=1}^m u_i^2(t)\,dt$$

subject to the constraint that the system

$$\dot{X}(t) = (A + \sum_{i=1}^m u_i(t)B_i)X(t) \qquad (*)$$

should be transferred from the state X_o at t=0 to the state X_1 at $t=t_1$.

Theorem 1: Let X(t) satisfy the $G\ell(n)$ equation (*). Suppose that $[Ad_A^k B_i, B_j] = 0$ for all i and j and $k=0,1,2,\ldots n^2-1$. Suppose that

$$X_1X_0^{-1} \in e^{At_1}\{\exp\{Ad_A, B_i\}_A\}_G$$

Then there exists a control $u(\cdot)$ which steers the system from X_o at t=0 to X_1 at $t=t_1$ and minimizes η. This control is the same as the control which steers the linear system

$$\dot{F}(t) = \sum_{i=1}^m e^{-At}B_i e^{At} u_i(t)$$

from 0 at t=0 to $\ell n(e^{-At_1}X_1X_o^{-1})$ at $t=t_1$ and minimizes η where ℓn denotes the real solution of

$$e^M = e^{-At_1}X_1X_o^{-1}$$

which results in the smallest value of η. The optimal control is of the form

$$u_i(t) = tr(M_i e^{-At} B_i e^{At})$$

for some constant matrices M_i.

Proof: As in the proof of the bang-bang theorem we see that

$$X(t) = e^{At}e^{F(t)}$$

where F(t) satisfies

$$\dot{F}(t) = \sum_{i=1}^{m} e^{-At}B_i e^{At}u_i(t)$$

From this point on everything follows from standard linear theory. See [17], section 22.

Exercises

1. Consider the system

$$\dot{x}(t) = x(t)+u(t)$$
$$\dot{y}(t) = u(t)y(t)$$

Suppose we want to steer this system from (α,β) to (γ,δ) in t_1 units of time and to minimize

$$\eta = \int_0^{t_1} u^2(t)dt$$

If δ/β is positive this transfer is possible and the $u(\cdot)$ which achieves the optimal is of the form ae^t+b. Generalize Theorem 1 in such a way as to capture this example.

2. If B_1 and B_2 commute, describe the solutions of

$$\prod_{i=1}^{\nu} (e^{B_1u_i}e^{B_2v_i}) = N$$

IV. STOCHASTIC DIFFERENTIAL EQUATIONS

Stochastic processes on spheres has been of interest in physics for some time. Debye [18] in his book on statistical mechanics gives one application of S^2 stochastic processes. Nuclear magnetic resonance phenomena account for some more recent interest in diffusions on S^2. See Chapter 15 of the recent text [19]. The French mathematical physicist Perin wrote a classical paper [20] on diffusion on SO(3). Recent interest in physics regarding models of the type under study here is discussed in Fox [21]. Transmission of electromagnetic waves through random media leads to stochastic processes on the symplectic group -- distance playing the role usually assumed by time. Tutubalin [22] can be consulted for recent results and references. Carrier [23] has examined an equation of this general type in connection with a gravity wave propagation problem. One can think of this study as a stochastic process on the two dimensional symplectic group. An engineering problem for which the theory is potentially interesting is the randomly switched electrical circuit.

4.1 Bilinear Stochastic Equations

In this paper all stochastic differential equations are to be interpreted in the Ito sense. All Wiener processes are of unity variance and Wiener processes with distinct indices are assumed to be uncorrelated. The reader is encouraged to study Clark [24] for more details on stochastic calculus.

Under what circumstances does the Ito equation

$$dx(t) = Ax(t)dt + \sum_{i=1}^{m} dw_i(t)B_i x(t) \qquad (*)$$

evolve on the manifold defined by $x'Qx$ = constant? If we expand to second order keeping in mind that $dw_i dw_j = \frac{1}{2}\delta_{ij}dt$ we get

$$dx'Qx = x'(A'Q+QA)xdt + \sum_{i=1}^{m} x'(B_i'Q + QB_i)xdw + \frac{1}{2}\sum_{i=1}^{m} x'B_i'QB_i x\, dt$$

Thus in order for the derivative of $x'Qx$ to vanish we require

$$A'Q + QA + \sum_{i=1}^{m} B_i'QB_i = 0$$

and also we require

$$B_i'Q + QB_i = 0$$

We see that the drift term A needs to be "corrected" by a term coming from the white noise. For example, if we want equation (*) to evolve on a sphere then A is not skew symmetric as it would be in the deterministic case but rather it has a correction term whose size depends on the B_i. On the other hand, the B_i must be skew symmetric.

In order to evolve on the symplectic group it is a skew symmetric form which must be preserved. Repeating the above with Hamiltonian matrices gives rise to the conditions that B_i and $A + \frac{1}{2}\sum_i B_i^2$ should be Hamiltonian.

<u>Exercises</u>

1. Show that the Ito equation

$$\begin{bmatrix} dx_1 & dx_2 \\ dx_3 & dx_4 \end{bmatrix} = \begin{bmatrix} \alpha & \beta \\ \gamma & \delta \end{bmatrix}\begin{bmatrix} x_1 & x_2 \\ x_3 & x_4 \end{bmatrix} dt + \begin{bmatrix} x_1 & x_2 \\ x_3 & x_4 \end{bmatrix}\begin{bmatrix} dw_1 & dw_2 \\ dw_3 & -dw_1 \end{bmatrix}$$

evolves on the special linear group $S\ell(2)$ if suitable restrictions are placed on α, β, γ, δ.

2. Generalize the previous problem to $S\ell(n)$.

4.2 The Moment Equations

Associated with the stochastic equation

$$dx(t) = Ax(t)dt + \sum_{i=1}^{m} B_i x(t)dw_i(t) \qquad (*)$$

is a family of higher order equations analogous to those given in section 1.2. These are the equations for $x^{[p]}$. In order to display their form it is necessary to work out section 1.2 using the Ito calculus. As an alternative, suggested to me by Martin Clark, one can convert (*) into an analogous Stratonovich equation, use the ordinary calculus to get the $x^{[p]}$ equation, and then convert back to the Ito form. This idea is particularly attractive in the present setup since we have the deterministic results already.

The Stratonovich analog of (*) is simply

$$\mathrm{đ}x(t) = (A-\frac{1}{2}\sum_{i=1}^{m} B_i^2)x(t)dt + \sum_{i=1}^{m} B_i x(t)\mathrm{đ}w_i(t)$$

where đ indicates Stratonovich differentials. Applying ordinary calculus we get

$$\mathrm{đ}x^{[p]}(t) = (A-\frac{1}{2}\sum_{i=1}^{m} B_i^2)_{[p]}x^{[p]}dt+\sum_{i=1}^{m} B_{i[p]}x^{[p]}(t)\mathrm{đ}w_i(t)$$

Now if we want to convert this back to an Ito form we must correct the drift term to get

$$dx^{[p]}(t)=[(A-\frac{1}{2}\sum_{i=1}^{m} B_i^2)_{[p]}+\sum_{i=1}^{m}(B_{i[p]})^2]x^{[p]}(t)dt+\sum_{i=1}^{m} B_{i[p]}x^{[p]}(t)dw_i(t)$$

We can easily take expectations to get the moment equation

$$\frac{d}{dt}(\mathscr{E}x^{[p]}(t))=[(A-\frac{1}{2}\sum_{i=1}^{m} B_i^2)_{[p]}+\sum_{i=1}^{m}(B_{i[p]})^2]\mathscr{E}x^{[p]}(t)$$

Notice that the apparently more general equation

$$dx(t) = Ax(t)dt + \sum_{i=1}^{m} B_i x(t)dw_i(t)+\sum_{i=1}^{m} e_i dw_2(t) \qquad (**)$$

is covered by these equations as well. To see this we let

$$\tilde{x} = \begin{bmatrix} x \\ 1 \end{bmatrix}$$

then x satisfies an equation of the form

$$d\tilde{x}(t) = \tilde{A}\tilde{x}(t)dt + \sum_{i=1}^{m}(\tilde{B}_i+\tilde{C}_i)\tilde{x}(t)dw_i(t)$$

There are many papers in the literature which analyze the stability of of these equations under various assumptions -- particular emphasis being placed on the case p=2. See, e.g. [25]. In reference [6] it is shown that under a suitable hypothesis all the moment equations are stable.

Exercises

1. Show that in the scalar case the moment equations for

$$dx(t) = \alpha(t)x(t)dt + \beta(t)x(t)dw(t)$$

are $$\frac{d}{dt}\mathscr{E}x^p(t) = [p(\alpha(t)-\frac{1}{2}\beta^2(t))+\frac{1}{2}p^2\beta^2(t)]\mathscr{E}x^p(t)$$

Notice that if α and $\beta \neq 0$ are constant then it can never happen that all moment equations are stable.

2. A problem of interest in geophysics leads to the stochastic equation

$$\begin{bmatrix} dx_1(t) \\ dx_2(t) \end{bmatrix} = \begin{bmatrix} 0 & dt \\ -dt+\varepsilon dw(t) & 0 \end{bmatrix} \begin{bmatrix} dx_1(t) \\ dx_2(t) \end{bmatrix} ; \quad \begin{bmatrix} x_1(0) \\ x_2(0) \end{bmatrix} = \begin{bmatrix} 1 \\ 0 \end{bmatrix}$$

Show that the autocorrelation is, for small ε, approximated by

$$\mathscr{E} x_1(t)x_1(\tau) \simeq e^{(\varepsilon^2/4)(t+\tau)} e^{-(\varepsilon^2/4)|t-\tau|} \cos(t-\tau)$$

(See Carrier [23]).

4.3 Fokker-Planck Equations

Associated with the Ito equation

$$dx(t) = Ax(t)dt + \sum_{i=1}^{m} dw_i B_i x(t)$$

is the formal Fokker-Plank equation

$$\frac{\partial\rho}{\partial t} - \frac{1}{2}\,\mathrm{tr}\Big(\sum_{i=1}^{m} B_i xx' B_i', \frac{\partial}{\partial x_i}\frac{\partial}{\partial x_j}\Big)\rho - \nabla_x \rho A x = 0$$

However, if x evolves on a manifold then this equation will not be especially useful unless the redundant variables are eliminated. In order to carry out this reduction it is necessary to coordinatize the manifold in some natural way. This coordinatization necessarily proceeds in a case by case way. To illustrate we work out four cases on the two-sphere S^2.

Consider the stochastic equations (Compare with McKean [26] who considers case b, case a being classical.)

$$\begin{bmatrix} dx_1 \\ dx_2 \\ dx_3 \end{bmatrix} = \begin{bmatrix} -dt & -dw_3 & dw_2 \\ dw_3 & -dt & -dw_1 \\ -dw_2 & dw_1 & -dt \end{bmatrix} \begin{bmatrix} x_1 \\ x_2 \\ x_3 \end{bmatrix} \quad (a)$$

$$\begin{bmatrix} dx_1 \\ dx_2 \\ dx_3 \end{bmatrix} = \begin{bmatrix} -\frac{1}{2}dt & 0 & dw_2 \\ 0 & -\frac{1}{2}dt & -dw_1 \\ -dw_2 & dw_1 & -dt \end{bmatrix} \begin{bmatrix} x_1 \\ x_2 \\ x_3 \end{bmatrix} \quad (b)$$

$$\begin{bmatrix} dx_1 \\ dx_2 \\ dx_3 \end{bmatrix} = \begin{bmatrix} -\frac{1}{2}dt & -dt & dw_2 \\ +dt & -\frac{1}{2}dt & -dw_1 \\ -dw_2 & dw_1 & -dt \end{bmatrix} \begin{bmatrix} x_1 \\ x_2 \\ x_3 \end{bmatrix} \quad (c)$$

$$\begin{bmatrix} dx_1 \\ dx_2 \\ dx_3 \end{bmatrix} = \begin{bmatrix} 0 & -dt & 0 \\ dt & -\frac{1}{2}dt & -dw_1 \\ 0 & dw_1 & -\frac{1}{2}dt \end{bmatrix} \begin{bmatrix} x_1 \\ x_2 \\ x_3 \end{bmatrix} \quad (d)$$

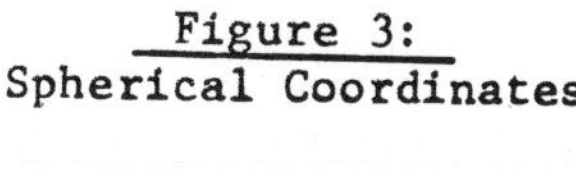

Figure 3:
Spherical Coordinates

We introduce polar coordinates according to figure 3. The Fokker-Planck equations corresponding to the above cases are then

$$\left[\frac{\partial}{\partial t} - \frac{1}{2}\left(\frac{1}{\sin\phi}\frac{\partial}{\partial\phi}\sin\phi\frac{\partial}{\partial\phi} + \frac{1}{\sin^2\phi}\frac{\partial^2}{\partial\theta^2}\right)\right]\rho(t,\phi,\theta) = 0 \qquad \text{(a)}$$

$$\left[\frac{\partial}{\partial t} - \frac{1}{2}\left(\frac{1}{\sin\phi}\frac{\partial}{\partial\phi}\sin\phi\frac{\partial}{\partial\phi} + \frac{1}{\tan^2\phi}\frac{\partial^2}{\partial\theta^2}\right)\right]\rho(t,\phi,\theta) = 0 \qquad \text{(b)}$$

$$\left[\frac{\partial}{\partial t} - \frac{1}{2}\left(\frac{1}{\sin\phi}\frac{\partial}{\partial\phi}\sin\phi\frac{\partial}{\partial\phi} + \frac{1}{\tan^2\phi}\frac{\partial^2}{\partial\theta^2}\right) + \frac{\partial}{\partial\theta}\right]\rho(t,\phi,\theta) = 0 \qquad \text{(c)}$$

$$\left[\frac{\partial}{\partial t} - \frac{1}{2}\left(\sin\theta\frac{\partial}{\partial\phi} + \cot\phi\cos\theta\frac{\partial}{\partial\theta}\right)^2 + \frac{\partial}{\partial\theta}\right]\rho(t,\phi,\theta) = 0 \qquad \text{(d)}$$

The idea behind the derivation of these equations is that each of the three generators

$$\begin{bmatrix} 0 & 1 & 0 \\ -1 & 0 & 0 \\ 0 & 0 & 0 \end{bmatrix}, \quad \begin{bmatrix} 0 & 0 & 1 \\ 0 & 0 & 0 \\ -1 & 0 & 0 \end{bmatrix}, \quad \begin{bmatrix} 0 & 0 & 0 \\ 0 & 0 & 1 \\ 0 & -1 & 0 \end{bmatrix}$$

can be associated with a first order partial differential operator which describes the effect of a drift around the corresponding axis of rotation and also with a second order partial differential operator which describes the effect of a diffusion around the corresponding axis of rotation. The derivation of these operators is an exercise in differential geometry, however the following insight is useful.

On a manifold with a Riemannian metric $(g_{ij}(x))$, the Laplace-Beltrami operator [27]

$$\nabla^2 = \frac{1}{\sqrt{\det(g_{ij}(x))}}\frac{\partial}{\partial x_i}(g_{ij}(x))^{-1}\sqrt{\det(g_{ij}(x))}\frac{\partial}{\partial x_j}$$

serves as the Laplacian, in that the basic heat equation, assuming constant conductivity, is

$$\left(\frac{\partial}{\partial t} - \frac{1}{2}\nabla^2\right)\phi(t,x) = 0$$

On S^2, in terms of the given coordinates, the usual metric is

$$(ds)^2 = [d\phi,\ d\theta]\begin{bmatrix} 1 & 0 \\ 0 & \sin^2\phi \end{bmatrix}\begin{bmatrix} d\phi \\ d\theta \end{bmatrix}$$

one sees easily that case a above corresponds to the heat equation.

As for case b, it is obtained from case a by removing one of the generators -- the one which corresponds to a diffusion about the x_3-axis. This is equivalent to subtracting $\frac{1}{2}(\partial^2/\partial\theta^2)$ from the operator appearing in case a.

Case c is obtained in an analogous way. We must add a drift term to the operator appearing in b corresponding to a rotation about the x_3-axis. Thus we add a $(\partial/\partial\theta)$ term to the operator appearing in b.

Case d is the most degenerate of all in that there is now only diffusion about one axis. There is a $(\partial/\partial\theta)$ drift term as in case c together with the operator which corresponds to diffusion about the x_1-axis.

It is of some interest to note that all these operators are studied in quantum theory. See Rose [28], appendix A.

Exercises

1. Consider the stochastic equation

$$\begin{bmatrix} dx_1 \\ dx_2 \\ dx_3 \end{bmatrix} \begin{bmatrix} -\frac{1}{2}\,dt & -dw & 0 \\ dw & -\frac{1}{2}\,dt & dt \\ 0 & dt & 0 \end{bmatrix} \begin{bmatrix} x_1 \\ x_2 \\ x_3 \end{bmatrix} \; ; \; x_1^2(0)+x_2^2(0)-x_3^2(0)=1$$

Show that it evolves on the manifold defined by $x_1^2+x_2^2-x_3^2=1$. Introduce coordinates in this manifold and work out the Fokker-Planck equation. Is there a limiting distribution?

2. Show that the moment equations associated with each of the four cases analyzed here are stable. (see [26])

4.4 Calculation of Diffusion Times

We continue with the analysis of the four cases of diffusions on spheres, now with a view toward determining, if possible, a complete solution to the Fokker-Plank equation. In cases where that proves too difficult we look for some measure of the relaxation time of the process.

To begin with, the standard S^2 diffusion (case a above) leads to the Fokker-Plank equation

$$\frac{\partial\rho(t,x)}{\partial t} - \frac{1}{2}\,\nabla^2\rho(t,x) = 0$$

Where ∇^2 is the usual Laplacian on the sphere. It is, of course, well known that the eigenvalues of the Laplacian on the sphere are $n(n+1)$, $n=0,1,2,\ldots$ with the nth being of multiplicity $2n+1$. Thus the general solution of the above equation starting from the singular distribution concentrated at $\theta = \phi = 0$ is

$$\rho(t,\theta,\phi) = \sum_{n=1}^{\infty} \sum_{k=-n}^{n} P_{nk}(\cos\phi)e^{ik\theta}e^{-n(n+1)t}$$

where P_{nk} are the spherical harmonics. We also see that the eigenvalues are a measure of the speed with which the density approaches steady state.

On the basis of this Green's function one can, of course, express the general solution of the Fokker-Plank equation in terms of its initial value. Thus we have, in terms of the spherical harmonics, a complete solution to the Fokker-Planck equation. This is classical.

On the other hand, it is possible to be almost as explicit in the other cases as well. This comes about because the 2n+1 equations for the coefficients of the spherical harmonics of the form $P_{nk}(\cos\ )e^{ik\theta}$ $k=0,\pm 1,\ldots \pm n$ are decoupled from those corresponding to $P_{n'k}(\cos\ )e^{ik\theta}$ for $n\neq n'$. Thus the solution of the Fokker-Planck equation reduces to a sequence of linear differential equations; the nth entry in the sequence being a coupled set of 2n+1 equations. It happens, however, that there is a simple connection between the moment equations of section 4.2 and the equations for the coefficients of the spherical harmonics. We describe this for the S^2 situation but similar results hold on spheres of any dimension.

For an S^2 equation x is a 3-vector and $x^{[p]}$ is of dimension (p+1)(p+2)/2. The equation for $x^{[p]}$ includes all linearly independent p-forms in x; thus it includes (p-1)(p)/2 terms of the form

$$(x_1^2+x_2^2+x_3^2)x^{[p-2]}$$

Hence we can partition $x^{[p]}$ into two parts of dimension (p-1)p/2 and (p+1)(p+2)/2 - (p-1)p/2 = 2p+1, respectively according to whether the components have a factor of $x_1^2+x_2^2+x_3^2$ or not. Now of course $x_1^2+x_2^2+x_3^2 = 1$ so that the components which do contain this factor can be thought of as moment equations of a lower order and hence they evolve independently of the second part of the equation. On the other hand, the 2p+1 components which do not contain $x_1^2+x_2^2+x_3^2$ as a factor evolve independently as well. Collecting these facts we see that the moment equations have the structure

$$\frac{d}{dt}\mathscr{E}x^{[p]}(t)=\begin{bmatrix} A_\delta & 0 & & \cdots & \\ 0 & A_{\delta+2} & & \cdots & \\ \cdots & \cdots & \ddots & \cdots & \\ \cdots & & & \tilde{A}_{p-2} & 0 \\ \cdots & & & 0 & \tilde{A}_p \end{bmatrix}\mathscr{E}x^{[p]}(t)$$

where δ is zero or one depending on whether p is even or odd. The dimension of A_p is (2p+1) by (2p+1) and the coefficients of the spherical harmonics of type P_{nk}, n fixed, $k=0,\pm 1,\pm 2,\ldots \pm n$ are governed by the differential equation

$$\dot{y}(t) = \tilde{A}_p y(t)$$

Thus the spectrum of the operators

$$(A - \frac{1}{2}\sum_{i=1}^{m} B_i^2)_{[p]} + \sum_{i=1}^{m} (B_{i[p]})^2$$

which were derived in section 4.2, governs the relaxation time of the process. In case a above we have already commented that the spectrum is $\frac{1}{2}(n(n+1))$ with the nth term being of multiplicity 2n+1. In case b there is less diffusion and one would expect the relaxation to be slower. This is the case; a calculation shows that the first few entries of the spectrum compares with case as follows.

$$\frac{1}{2}\begin{cases} \underbrace{0}_{I},\ \underbrace{2,\ 2}_{II},\ \underbrace{2\ \ 6,\ 6,\ 6,\ 6\ \ 6}_{III} \ldots \text{case a} \\ 0,\ 1,\ 1,\ 2\ \ 2,\ 2,\ 5,\ 5,\ 6 \ldots \text{case b} \end{cases}$$

Finally, we remark that examples b, c, and d are specific cases of the hypoelliptic operators of Hormander [29].

<u>Exercises</u>

1. Consider the linear stochastic equation

$$\begin{bmatrix} dx_1(t) \\ dx_2(t) \end{bmatrix} = \begin{bmatrix} -1/2 & 1 \\ -1 & -1/2 \end{bmatrix} dt + \begin{bmatrix} dw_1(t) \\ dw_2(t) \end{bmatrix}; \quad x(0) = 0$$

as an approximation to the first two components of the S^2 equation

$$\begin{bmatrix} dx_1(t) \\ dx_2(t) \\ dx_3(t) \end{bmatrix} = \begin{bmatrix} -\frac{1}{2}dt & dt & dw_1 \\ -dt & -\frac{1}{2}dt & dw_2 \\ -dw_1 & -dw_2 & -dt \end{bmatrix} \begin{bmatrix} x_1(t) \\ x_2(t) \\ x_3(t) \end{bmatrix}; \quad x(0) = \begin{bmatrix} 0 \\ 0 \\ 1 \end{bmatrix}$$

Compute the second moment in each case and compare.

2. Consider the stochastic equation on S^2 defined by

$$\begin{bmatrix} dx_1(t) \\ dx_2(t) \\ dx_3(t) \end{bmatrix} = \begin{bmatrix} -dt/2 & dw_1 & 0 \\ -dw_1 & -(1+\rho)dt/2 & \rho dw_2 \\ 0 & -\rho dw_2 & -\frac{\rho}{2} \end{bmatrix} \begin{bmatrix} x_1(t) \\ x_2(t) \\ x_3(t) \end{bmatrix}$$

Find the first few eigenvalues of corresponding Fokker-Planck operator as a function of ρ.

V. STABILITY THEORY

In the study of ordinary differential equations on Lie groups both linear and nonlinear problems are of interest, however in these notes we discuss linear problems only. Of course the most common stability problems encountered in control concern the general linear group. However in the study of specific applications other groups may occur. For example, in the case of problems arising in classical mechanics the symplectic group plays a major role. Moreover since tensoring will typically transform a system evolving in $G\ell(n)$ into one which evolves on some subgroup of $G\ell(q)$ is desirable to take a general point of view.

5.1 Stability of the $x^{[p]}$ Equations

The following theorem is an obvious consequence of the calculations in section 1.2.

Theorem 1: The null solution of the system

$$\dot{x}(t) = A(t)x(t)$$

is stable (asymptotically stable) if and only if the null solution of the equation

$$\dot{y}(t) = A_{[p]}(t)y(t)$$

is stable (asymptotically stable). Moreover if all solutions of the first equation are bounded by $|x(t)| \leqslant Me^{-\lambda t}$ then all solutions of the second are bounded by $|y(t)| \leqslant M_1e^{-p\lambda t}$.

When combined with standard estimates this theorem can give very precise information about high order systems which are either in the form of $\dot{y}(t) = A_{[p]}(t)y(t)$ or else in the form

$$\dot{y}(t) = A_{[p]}(t)y(t) + D(t)y(t)$$

with $D(t)$ small in some sense.

Example: We know from Liapunov [see e.g. [30]] that all solutions of the $Sp(2)$ equation

$$\begin{bmatrix} \dot{x}_1(t) \\ \dot{x}_2(t) \end{bmatrix} = \begin{bmatrix} 0 & 1 \\ -p(t) & 0 \end{bmatrix} \begin{bmatrix} x_1(t) \\ x_2(t) \end{bmatrix}$$

are bounded if $p(\cdot)$ is pointwise nonnegative, periodic of period T with positive average value and with

$$\int_0^T p(t)dt \leqslant 4/T$$

Thus we see that all solutions of the $x^{[2]}$ equation

$$\begin{bmatrix} \dot{y}_1(t) \\ \dot{y}_2(t) \\ \dot{y}_3(t) \end{bmatrix} = \begin{bmatrix} 0 & 1 & 0 \\ -2p(t) & 0 & 2 \\ 0 & -p(t) & 0 \end{bmatrix} \begin{bmatrix} y_1(t) \\ y_2(t) \\ y_3(t) \end{bmatrix}$$

are also bounded under the same hypothesis. (Here we have taken $y_2=2x_1x_2$ instead of $\sqrt{2}\, x_1x_2$). A change of basis puts this equation in a more symmetric form

$$\begin{bmatrix} \dot{z}_1(t) \\ \dot{z}_2(t) \\ \dot{z}_3(t) \end{bmatrix} = \begin{bmatrix} 0 & \frac{1}{2}(1-p(t)) & 0 \\ -\frac{1}{2}(1-p(t)) & 0 & \frac{1}{2}(1+p(t)) \\ 0 & \frac{1}{2}(1+p(t)) & 0 \end{bmatrix} \begin{bmatrix} z_1(t) \\ z_2(t) \\ z_3(t) \end{bmatrix}$$

This equation evolves on the pseudo-orthogonal group SO(2,1).

One particular fact which should be mentioned here is that systems with a single time varying parameter, say

$$\dot{x}(t) = Ax(t)+k(t)Bx(t) \qquad (*)$$

go into systems with a single time varying parameter e.g.

$$\dot{x}^{[p]}(t) = (A_{[p]}+k(t)B_{[p]})x^{[p]}(t)$$

Thus the many useful results about (*) (circle criterion, [17], etc.) can be extended in a nontrivial way.

Exercises

1. It is known that all solutions of the differential equation

$$\ddot{x} + \dot{x} + k(t)x(t) = 0$$

remain bounded if $0 \leq k(t) \leq \sim 3.9$ (see [17]). On the other hand, if one picks a positive definite quadratic form in x and $\dot{x}$ say $v(x,\dot{x})$ and computes its derivative along solutions of the given differential equation then there exists one quadratic form which implies stability via Liapunov theory, for $0 \leq k(t) \leq 1$ but the constant 1 cannot be improved on using a quadratic Liapunov function. However, if we look at the $x^{[p]}$ version of the differential equation then a quadratic Liapunov function for $x^{[p]}$ is a 2p-degree Liapunov function for the original equation and a more suitable Liapunov function can be found. Work out the details.

2. Consider a differential equation in $\mathbb{R}^n$

$$\dot{x}(t) = Ax(t)+k(t)Bx(t)$$

Suppose that A and B generate a four dimensional Lie algebra which is isomorphic with $g\ell(2)$. Use the theory of the representations of $g\ell(2)$ (see, e.g. Samelson [4] page 114) and the circle criterion (see, e.g. [17]) to derive stability criteria for the given system.

5.2 Periodic Self-Contragradient Systems

A matrix Lie algebra is said to be self-contragradient if there exists a matrix P such that

$$PLP^{-1} = -L'$$

for all L in the Lie algebra. For example, so(n) is self-contragradient with P=I and sp(n) is self-contragradient with P=J. As far as the stability of periodic systems is concerned, the important consequence of this assumption is that if A(t) satisfies $PA(t)P^{-1} = -A'(t)$ then the transition matrix for

$$\dot{x}(t) = A(t)x(t)$$

satisfies

$$\Phi_A'(t)P\Phi_A(t) = P$$

since

$$\frac{d}{dt}(\Phi_A'(t)P\Phi_A(t)) = \Phi_A'(t)(A'(t)P+PA(t))\Phi_A(t) = 0$$

Thus $\Phi_A(t)$ similar to $(\Phi_A^{-1})'$. As an immediate consequence of this fact we see that the eigenvalues of $\Phi_A(t)$ occur in reciprocal pairs -- if λ is an eigenvalue then so is $1/\lambda$. If we assume we are dealing with real systems then of course the eigenvalues occur in complex conjugate pairs as well.

If $A(t) = A(t+T)$ then the well known Floquet theory insures that the transition matrix for

$$\dot{x}(t) = A(t)x(t)$$

can be expressed as

$$\Phi_A(t) = Q(t)e^{Rt}; \quad Q(0) = I$$

with $Q(t+T) = Q(t)$ and R constant, though not necessarily real. The value of $\Phi_A(T)$ is decisive as far as the stability of a periodic system is concerned since $\Phi_A(nT) = [\Phi_A(T)]^n$.

If $A(t)$ is given by

$$A(t) = \sum_{i=1}^{m} a_i(t)A_i$$

with the A_i being a basis for a self-contragradient representation of a Lie algebra, then of course

$$\Phi_A'(t)P\Phi_A(t) = P$$

for all t. If $(\Phi_A(T))^n$ is bounded for n=1,2,... then we call $\Phi_A(T)$ stable. We call it P-strongly stable if it happens that for all sufficiently small R such that $R'P+PR = 0$, the matrix $e^R\Phi_A(T)$ is also stable. (Compare with [31].) In view of the fact that the eigenvalues of a matrix depend continuously on the elements of the matrix and in view of the fact that the eigenvalues of Φ_A must occur in reciprocal pairs, we see that if the eigenvalues of $\Phi_A(T)$ are distinct and if $\Phi_A(T)$ is stable, then it is P-strongly stable. However it can happen that $\Phi_A(T)$ is P-strongly stable even if the eigenvalues of $\Phi_A(T)$ are not distinct.

Theorem 1: If $\{A_i\}$ is the basis for a self-contragradient matrix Lie algebra, $A_i'P+PA_i = 0$, and if

$$\dot{x}(t) = (\sum_{i=1}^{m} a_i(t)A_i)x(t)$$

is periodic and if $\Phi_A(T)$ is P-strongly stable, then there exists $\varepsilon > 0$ such that for $|b_i(t)-a_i(t)| < \varepsilon$ and $b_i(t)$ periodic of period T the system

$$\dot{x}(t) = (\sum_{i=1}^{m} b_i(t)A_i)x(t)$$

is stable.

Exercises

1. Determine if for P = J the matrix

$$M = \begin{bmatrix} \cos\theta & 0 & \sin\theta & 0 \\ 0 & \cos & 0 & \sin\theta \\ -\sin\theta & 0 & \cos\theta & 0 \\ 0 & -\sin & 0 & \cos\theta \end{bmatrix} \quad ; \; 0 < \theta < \pi$$

is P-strongly stable or not. See [30], theorem 8.

2. Show that if p(t) is periodic of period T with average value zero and if

$$\begin{bmatrix} \dot{x}_1(t) \\ \dot{x}_2(t) \end{bmatrix} = \begin{bmatrix} 0 & 1 \\ -1 & -p(t) \end{bmatrix} \begin{bmatrix} x_1(t) \\ x_2(t) \end{bmatrix}$$

then $\Phi_A(T)$ is symplectic although $\Phi_A(t)$ for $t \neq T$ need not be. The corresponding $x^{[2]}$ equation is expressible as

$$\frac{d}{dt} \begin{bmatrix} x_1^2 \\ x_1 x_2 \\ x_2^2 \end{bmatrix} = \begin{bmatrix} 0 & 2 & 0 \\ -1 & -p(t) & -1 \\ 0 & -2 & -2p(t) \end{bmatrix} \begin{bmatrix} x_1^2 \\ x_1 x_2 \\ x_2^2 \end{bmatrix}$$

Use the idea of strong stability to investigate the stability of these systems.

3. If D is diagonal then D+H is similar to a diagonal matrix if H is any symmetric matrix. However if D is diagonal there may exist an n(n-1)/2 dimensional set, the upper triangular matrices, such that D+T is not diagonalizable; consider the identity. Relate this to strong stability.

5.3 The Symplectic Case

In the special case of the symplectic group Krein [30] has given an elegant theorem on how large the perturbation in Theorem 1 of the previous section can be. We give an application of this theorem and some facts about realizations of feedback systems as well.

Notice that the second order system with Q(t) symmetric

$$\ddot{x}(t) + Q(t)x(t) = 0; \quad x(t) \in \mathbb{R}^n$$

is equivalent to the symplectic system

$$\begin{bmatrix} \dot{x}_1(t) \\ \dot{x}_2(t) \end{bmatrix} = \begin{bmatrix} 0 & I \\ -Q(t) & 0 \end{bmatrix} \begin{bmatrix} x_1(t) \\ x_2(t) \end{bmatrix}$$

Krein has investigated this set of equations and more general ones. One of his results reads as follows.

Theorem 1: Let $P(t) = P(t+T) = P'(t)$, then all solutions of the equation

$$\ddot{x}(t)+P(t)x(t) = 0$$

are bounded if

i) $P(t) \geq 0$ all t

ii) $\int_0^T P(t)dt > 0$ (positive definite)

iii) $(4/T)I-\int_0^T P(t)dt > 0$ (positive definite)

Proof: See Krein [30], page 165.

As an example of an application of this result to problems of the type which arise frequently in system theory we prove the following theorem. (Compare with [32])

Theorem 2: Suppose that q(s) and p(s) are polynomials without common factors. Suppose further that q(s)/p(s) is an even function of s with all its poles and zeros on the imaginary axis and assume that the poles and zeros of sq(s)/p(s) interlace. Let D = d/dt and let k( ) be periodic of period T. Then all solutions of the nth order differential equation

$$p(D)x(t) + k(t)q(D)x(t) = 0$$

are bounded provided

$$0 < \int |\lambda(t)|^2 dt < 4/T$$

where $\lambda(t)$ denotes the zero of $p(s)+k(t)q(s) = 0$ which has the largest magnitude.

Proof: Write q(s)/p(s) as $r(s^2)/m(s^2)$ with r and m polynomials. This is possible because q(s)/p(s) is even. Write r(s)/m(s) as $b'(Is-A)^{-1}b$ with $A = A'$. This is possible because the poles and zeros of r(s)/m(s) interlace, (See [25]). Thus

$$q(s)/p(s) = b'(Is^2-A)^{-1}b$$

and the differential equation in the theorem statement is equivalent to the system

$$\ddot{x} + (A+k(t)bb')x(t) = 0$$

Krein's result implies stability if

$$I(T/4) - \int_0^T (A+k(t)bb')dt > 0$$

But since the largest eigenvalue of the sum of two positive definite symmetric matrices is less than or equal to the sum of the largest eigenvalues of the respective matrices there is a corresponding inequality for integrals and we see that

$$\lambda_{max}\int_0^T (A+k(t)bb')dt \leq \int_0^1 |\lambda(t)|^2 dt$$

The result then follows.

It is interesting to compare this result with the analogous facts about completely symmetric systems investigated in [25]. Also notice that this theorem captures Liapunov's original theorem as a special case, as does the basic theorem of Krein.

Exercises

1. Use these results to investigate the stability of the scalar equation

$$x^{(4)}+4x^{(2)}+3x+k(t)(x^{(2)}+x) = 0$$

with k(t) periodic.

2. Derive a matrix version of Theorem 2.

VI. REFERENCES

1. H. Flanders, Differential Forms, Academic Press, 1963, New York.
2. J. Wei and E. Norman, "On the Global Representation of the Solutions of Linear Differential Equations as a Product of Exponentials," Proc. Am. Math. Soc., April 1964.
3. R. F. Gantmacher, Theory of Matrices, Chelsea, New York, 1959.
4. H. Samelson, Notes on Lie Algebras, Van Nostrand, New York, 1969.
5. M. Hausner and J. T. Schwartz, Lie Groups ; Lie Algebras, Gordon and Breach, London, 1968.
6. R. W. Brockett, "Lie Theory and Control Systems Defined on Spheres," SIAM J. on Applied Math., Vol. 24, No. 5, Sept. 1973.
7. W. L. Chow, "Über Systeme Von linearen partiellen Differentialgleichungen erster Ordinung," Math. Ann. Vol. 117, (1939), pp. 98-105.
8. R. W. Brockett, "System Theory on Group Manifolds and Coset Spaces," SIAM Journal on Control, Vol. 10, No. 2, May 1972, pp. 265-284.
9. S. Jonhansen (This Volume).
10. V. Jurdjevic and H. J. Sussmann, "Control Systems on Lie Groups," J. of Differential Equations, Vol. 12, No. 2, (1972) pp. 313-329.
11. R. Hirschorn, Ph.D. Thesis, Applied Mathematics, Harvard University, 1973.
12. R. E. Rink and R. R. Mohler, "Completely Controllable Bilinear Systems," SIAM J. on Control, Vol. 6, No. 3, 1968.
13. P. d'Alessandro, A. Isidori and A. Ruberti, "Realization and Structure Theory of Bilinear Dynamical Systems," (to appear, SIAM J. on Control).
14. R. W. Brockett, "On the Algebraic Structure of Bilinear Systems," in Variable Structure Control Systems, (R. Mohler and A. Ruberti, eds.) Academic Press, 1972, pp. 153-168.

15. A. J. Krener, A Generalization of the Pontryagin Maximal Principle and the Bang-Bang Principle, Ph.D. Thesis, Dept. of Mathematics, University of California, Berkeley, 1971.
16. H. J. Sussmann, "The Bang-Bang Problem for Linear Control Systems in $G\ell(n)$," SIAM J. on Control, Vol. 10, No. 3, Aug. 1972
17. R. W. Brockett, Finite Dimensional Linear Systems, John Wiley and Sons, New York, 1970.
18. R. Debye, "Polar Molecules," The Chemical Catalogue Co., New York, 1929.
19. C. P. Poole and H. A. Farach, Relaxation in Magnetic Resonance, Academic Press, 1971.
20. F. Perrin, "Etude Mathematique du Mouvement Brownien de Rotation," Ann. Ecole Norm. Sup., Vol. 45, (1928), pp. 1-51.
21. R. F. Fox, Physical Applications of Multiplicative Stochastic Processes, J. of Mathematical Physics, Vol. 14, No. 1, Jan. 1973, pp. 20-25.
22. V. A. Tutubalin, "Multimode Waveguides and Probability Distributions on a Symplectic Group," Theory of Probability and Its Applications, Vol. 16, No. 4, pp. 631-642, 1971.
23. G. F. Carrier, "Stochastically Driven Dynamical Systems," J. of Fluid Mechanics, Vol. 44, Part 2 (1970) pp. 249-264.
24. Martin Clark, (This Volume).
25. R. W. Brockett and J. C. Willems, "Average Value Criteria for Stochastic Stability," Stability of Stochastic Dynamical Systems, (ed. Ruth Curtain), Springer-Verlag, 1972.
26. H. P. McKean, Stochastic Integrals, Academic Press, New York, 1969.
27. S. Helgason, Differential Geometry and Symmetric Spaces, Academic Press, New York, 1960.
28. M. E. Rose, Elementary Theory of Angular Momentum, J. Wiley, New York, 1957.
29. L. Hormander, "Hypoelliptic Second Order Differential Equations," Acta. Math., 119: 147-171, 1967.
30. M. G. Krein, "Introduction to the Geometry of Indefinite J-Spaces and to the Theory of Operators in Those Spaces," Am. Math. Soc. Translations, (Series 2) Vol. 93, 1970, pp. 103-176.
31. I. M. Gel'fand and V. B. Lidskii, "On the Structure of the Regions of Stability of Linear Canonical Systems of Differential Equations with Periodic Coefficients," Am. Math. Soc. Transl. Series 2, Vol. 8, (1958) pp. 143-182.
32. R. W. Brockett and A. Rahimi, "Lie Algebras and Linear Differential Equations," in Ordinary Differential Equations, (L. Weiss ed.) Academic Press, New York, 1972, pp. 379.

REALIZATION THEORY OF BILINEAR SYSTEMS

Alberto Isidori Antonio Ruberti

Istituto di Automatica

Università di Roma

PREFACE

Interest in bilinear systems theory and applications has grown in the recent years. The basic motivation of this growth is twofold: on the one hand, the feasibility to be a satisfactory model for large classes of systems (physical, biological, socio-economical, etc.); on the other hand the relative ease with which their theory can be set-up. The first characteristic is substantially related to the presence of a multiplicative control action (on the state motion) besides the additive one, which is the only one present in a linear system. The second characteristic lies in the fact that, for any fixed input, the equations are linear in the state.

Bilinear systems have been widely studied, both from the theoretical and the practical point of view (see, for example, [1] [2] and relative bibliography). Theoretical studies deal with controllability and observability properties, optimization and identification techniques, realization problems. These notes are substantially concerned with the latter, i.e. with the problem of associating a bilinear state space description to a given input/output map.

This work was supported in part by the Consiglio Nazionale delle Ricerche (Centro di Studio dei Sistemi di Controllo e Calcolo Automatici, Rome) and by the Fondazione Ugo Bordoni of Rome.

The approach that we shall follow was firstly presented in paper [3], and is mainly based in the search for the least linear ambient in which bilinear equations can be embedded. To this end a key role is played by the structure analysis of the state space, based on reachability and observability properties. Such analysis yields a state space decomposition theory, that we present at the beginning of these notes. Then we exploit the connection between the results of the structure theory and minimality of bilinear realizations and, finally, we conclude the notes with the statement of necessary and sufficient conditions for the existence of bilinear realizations of a prescribed input/output map, together with the description of some constructive procedures.

It is a pleasure to thank Professor David Mayne for having given us the opportunity to partecipate in this important meeting.

<u>List of symbols and abbreviations</u>

$<=>$	implies and is implied by
$\forall$	for all
r.h.s	right hand side
l.h.s	left hand side
$\sim$	is equivalent to
ε	is an element of
$\subseteq$	is a subset or coincides
$\supseteq$	contains or coincides
$\mathcal{R}(.)$	range space
$\mathcal{N}(.)$	null space
R^n	n dimensional euclidean space
$\perp$	orthogonal complement
$\oplus$	direct sum
$\otimes$	tensor product

<u>Note for the reader</u> - Each item (definition, theorem, formula, etc.) is given a pair numbers in the left hand margin, the first one to denote the section and the second one to number consecutively within each section. For cross references between different chapters a third number is added.

I. STRUCTURE ANALYSIS OF BILINEAR SYSTEMS

1. Introduction

The role played by the structure analysis of the state space in the theory of linear dynamical systems is well known. This analysis, based on the properties of reachability and unobservability, is the source of a number of basic results in the understanding of the relationships between state space description and input/output behaviour of the system. It provides a mean for proving the existence of canonical forms of the state variables equations (i.e. canonical forms of the coefficient matrices), for decomposing the system into four sub-systems of which only one influences the input/output behavior, for characterizing the minimality in the realization theory (in terms of structure properties), for reducing a given realization to its minimal form. In this chapter we shall develop a structure analysis of the state space of bilinear dynamical systems, with the purpose of extending, as far as possible, the above interesting results to this class of systems.

The chapter is organized as follows. We first introduce the equations for describing bilinear systems (section 2) and we analyze certain subspaces of R^n which play an essential role in the subsequent theory (section 3). Then, we present the theory for canonical decomposition of the state space of bilinear systems. This is still based (i.e. as in linear system theory) on the properties of reachability and unobservability; to be precise, as concerns the first-one of these, we consider reachability from an equilibrium state, in order to give a wider domain of validity to the theory (section 4). Based on this property and on unobservability (section 5) we show the existence of a decomposition of the state space that satisfies the requirements outlined above (section 6) (*).

(*) Section 3, 5 and 6 are derived, with minor modifications, from a previous joint paper of the authors and P. d'Alessandro [4]. Section 4 is substantially new. The analysis of the structure properties of bilinear systems (with a view to deducing a state-space decomposition) has been also studied by R.W. Brockett [5]; in fact, he has independently shown the existence of canonical forms for bilinear systems homogeneous in the state. In these notes we consider only continuous-time bilinear systems; however, a similar theory can also be developed for discrete-time systems [4].

2. Bilinear dynamical systems

We shall consider systems described by the following input-state-output equations

$$\begin{aligned} \dot{x}(t) &= Ax(t) + N[x(t) \otimes u(t)] + Bu(t) \\ y(t) &= Cx(t) \end{aligned} \tag{2.1}$$

where $x(t) \varepsilon R^n$, $u(t) \varepsilon R^p$, $y(t) \varepsilon R^q$ denote , respectively, state,input and output at time t. The tensor product $\otimes$ is defined in such a way that $[x(t) \otimes u(t)] \varepsilon R^{np}$ and A,N,B,C are constant matrices of proper dimensions.

Note that the term $N[x(t) \otimes u(t)]$ represents the most general bilinear map $R^p \times R^n \to R^n$. This map is specified by the matrix N and by the rule for performing the tensor product $\otimes$; this, in turn, specifies the order in which the columns of N must be considered. Among various rules for performing this tensor product the most interesting is the one specified by

$$a \otimes b \triangleq \begin{pmatrix} ab_1 \\ \cdot \\ \cdot \\ \cdot \\ ab_r \end{pmatrix} \qquad a \varepsilon R^s, \; b \varepsilon R^r \tag{2.2}$$

under which the bilinear term of (2.1) is represented as

$$N[x(t) \otimes u(t)] = [\sum_{i=1}^{p} N_i u_i(t)] x(t) \tag{2.3}$$

where the n×n matrices N_i are defined by the partition

$$N = (N_1 \; \ldots \; N_p) \tag{2.4}$$

In other words, the bilinear term appears as the result of a linear map $R^n \to R^n$, parametrized with u(t). This way of representing the bilinear term of (2.1) will be very useful in the structure analysis of the state space.

Assuming,from now on, the rule (2.2) for the tensor product, we might consider the set of matrices

$$\sigma = (A,N_1,\ldots,N_p,B,C) \tag{2.5}$$

as a model of the bilinear system (2.1).

The system will be called homogeneous in the state if $B=0$, homogeneous in the input if $A=0$, strictly bilinear if $A=0$ and $B=0$. Any inhomogeneous bilinear system can be reduced to a form homogeneous in the state by only adding an extra component to the state vector [5]. To this end, define the matrices

$$F = \begin{pmatrix} A & 0 \\ 0 & 0 \end{pmatrix}, \quad M_i = \begin{pmatrix} N_i & b_i \\ 0 & 0 \end{pmatrix}, \quad H = (C \quad 0) \tag{2.6}$$

where the $n\times 1$ vectors b_i are specified by the partition

$$B = (b_1 \; \ldots \; b_p) \; ; \tag{2.7}$$

then, one readily recognizes that each solution (for the output) of (2.1) from the initial state $x(0) = x_o$ is also a solution of

$$\begin{aligned} \dot{z}(t) &= Fz(t) + \Big[\sum_{i=1}^{p} M_i u_i(t)\Big] z(t) \\ y(t) &= Hz(t) \end{aligned} \tag{2.8}$$

from the initial state $z(0) = \binom{x_o}{1}$, and viceversa.

3. Some invariant subspaces

In this section we introduce two suitable subspaces of R^n which are important from the point of view of the properties of reachability and unobservability.

Consider a family of $n\times n$ matrices $F_1,F_2,\ldots,F_r$ and an $n\times s$ matrix G, and let $\text{gen}_{F_1,\ldots,F_r}(G)$ denote the least subspace of R^n invariant under $F_1,\ldots,F_r$ containing $\mathcal{R}(G)$. For the purpose of constructing this subspace, we introduce the matrix sequences

$$\begin{aligned} \bar{U}_1 &= G \\ \bar{U}_i &= (F_1\bar{U}_{i-1} \; \ldots \; F_r\bar{U}_{i-1}) \qquad (i=2,3,\ldots) \\ U_i &= (\bar{U}_1 \; \ldots \; \bar{U}_i) \qquad (i=1,2,\ldots) \end{aligned} \tag{3.1}$$

and we have the following

(3.2) Lemma - The least subspace of R^n invariant under $F_1,\ldots,F_r$ and containing $\mathcal{R}(G)$ can be expressed as

$$\text{gen}_{F_1,\ldots,F_r}(G) = \mathcal{R}(U_n) \tag{3.3}$$

Proof - If we define the sequence of subspaces

$$\begin{aligned} \mathcal{U}_1 &= \mathcal{R}(G) \\ \mathcal{U}_i &= \mathcal{U}_{i-1} + F_1\mathcal{U}_{i-1} + \ldots + F_r\mathcal{U}_{i-1} \end{aligned} \tag{3.4}$$

we easily verify that

$$\mathcal{U}_i = \mathcal{R}(U_i) \qquad \forall i \tag{3.5}$$

If, $\mathcal{U}_k = \mathcal{U}_{k-1}$, for some value k, then from (3.4) $\mathcal{U}_{k-1}$ is invariant under $F_1,\ldots,F_r$; moreover $\mathcal{U}_i = \mathcal{U}_{k-1}$ for all $i \geq k$. From this and from

$$\mathcal{U}_{i-1} \subseteq \mathcal{U}_i \subseteq R^n \tag{3.6}$$

which is an immediate consequence of (3.1), it follows that there exists an integer $k_o \leq n$ such that

$$\begin{aligned} \mathcal{U}_{i-1} &\subset \mathcal{U}_i \qquad \forall i \leq k_o \\ \mathcal{U}_{i-1} &= \mathcal{U}_i \qquad \forall i > k_o \end{aligned} \tag{3.7}$$

Therefore $\mathcal{U}_n$ is invariant under $F_1,\ldots,F_r$ and, by virtue of (3.4), it contains $\mathcal{R}(G) = \mathcal{U}_1$.

To prove that $\mathcal{U}_n$ is the least subspace with these properties, observe that any subspace $\mathcal{X}$ of this type must satisfy

$$\mathcal{X} \supseteq \mathcal{R}(G) = \mathcal{U}_1 \tag{3.8}$$

$$\mathcal{X} \supseteq F_1\mathcal{X} \supseteq F_1\mathcal{U}_1,\ldots,\mathcal{X} \supseteq F_r\mathcal{X} \supseteq F_r\mathcal{U}_1 \tag{3.9}$$

and, then,

(3.10) $\mathcal{X} \supseteq \mathcal{U}_2$

Again, from (3.10)

(3.11) $\mathcal{X} \supseteq F_1\mathcal{X} \supseteq F_1\mathcal{U}_2, \ldots, \mathcal{X} \supseteq F_r\mathcal{X} \supseteq F_r\mathcal{U}_2$

and, then

(3.12) $\mathcal{X} \supseteq \mathcal{U}_3$

Proceeding in this way we have that

(3.13) $\mathcal{X} \supseteq \mathcal{U}_i \qquad \forall i$

Therefore any subspace of R^n invariant under $F_1,\ldots,F_r$ and containing $\mathcal{R}(G)$ contains $\mathcal{U}_n$, which consequently is the least of them.

With reference to the same family of matrices $F_1,\ldots,F_r$ and G we consider also the largest subspace of R^n invariant under $F_1,\ldots,F_r$ and contained in $\mathcal{N}(G^*)$. This will be denoted by $\overline{\text{gen}}_{F_1,\ldots,F_r}(G^*)$.

In order to get the expression of this subspace it is convenient to establish a general result about a sort of duality between the two kind of invariant subspaces here considered. In fact we can prove the following

(3.14) Lemma - The largest subspace of R^n invariant under the matrices $F_1,\ldots,F_r$ and contained in a given subspace $\mathcal{N}(G^*)$ is equal to the orthogonal complement of the least subspace of R^n invariant under $F_1^*,\ldots,F_r^*$ and containing $\mathcal{R}(G)$, that is

(3.15) $$\overline{\text{gen}}_{F_1,\ldots,F_r}(G^*) = \left[\text{gen}_{F_1^*,\ldots,F_r^*}(G)\right]^{\perp}$$

Proof - We start from the property that, for any given subspace $\mathcal{Z}$ of R^n and any n×n matrix T,

(3.16) $$T\mathcal{Z} \subseteq \mathcal{Z} \Leftrightarrow T^*\mathcal{Z}^{\perp} \subseteq \mathcal{Z}^{\perp}$$

Therefore the r.h.s. of (3.15) is invariant under $F_1,\ldots,F_r$. Moreover, since

(3.17) $$\mathrm{gen}_{F_1^*,\dots,F_r^*}(G) \supseteq R(G) = N^{\perp}(G^*)$$

it follows that r.h.s. of (3.15) is contained in $N(G^*)$. Finally, we can prove by contradiction that the r.h.s. of (3.15) is the largest subspace of R^n with these properties. In fact, if there exists a subspace X of higher dimension, applying the above arguments, it would result that X is an invariant under $F_1^*,\dots F_r^*$ containing $R(G)$, of lower dimension than $\mathrm{gen}_{F_1^*,\dots,F_r^*}(G)$, and this is a contradiction.

As a consequence of the above Lemmas, introducing the sequences of matrices

$$\bar{V}_1 = G^*$$

(3.18) $$\bar{V}_i = \begin{pmatrix} \bar{V}_{i-1}F_1 \\ \vdots \\ \bar{V}_{i-1}F_r \end{pmatrix} \qquad (i=2,3,\dots)$$

$$V_i = \begin{pmatrix} \bar{V}_1 \\ \vdots \\ \bar{V}_i \end{pmatrix} \qquad (i=1,2,\dots)$$

we have the following

(3.19) Lemma - The largest subspace of R^n invariant under $F_1,\dots,F_r$ contained in $N(G^*)$ can be expressed as

(3.20) $$\overline{\mathrm{gen}}_{F_1,\dots,F_r}(G^*) = N(V_n)$$

Proof - By (3.15) we have that

(3.21) $$\overline{\mathrm{gen}}_{F_1,\dots,F_r}(G^*) = \left[\mathrm{gen}_{F_1^*,\dots,F_r^*}(G)\right]^{\perp}$$

The latter, in turn, by Lemma (3.2) is equal to the subspace $N(V_n)$ constructed from the sequence (3.18).

4. Reachability from an equilibrium state

The analysis of the structure of the state space of bilinear systems can be based on the concepts of reachability and unobservability. Generalizing the linear theory in the most direct way, one would be naturally brought to classify the states according to the property of reachability from the origin of the state space, and then to decompose the state space itself on the basis of such property. This is no longer possible when the bilinear system is homogeneous in the state (due to the absence of nontrivial reachable states). Unlike the linear case, however, there might exist regions of the state space that are still reachable, whenever the initial state is nonzero. In order to include both homogeneous and inhomogeneous (in the state) bilinear systems,one is therefore brought to consider reachability from a non (necessarily) zero initial state. A simplifying assumption, that however is consistent with the needs outlined above, will be the one of considering only equilibrium initial states, i.e. those states that satisfy the equation

(4.1) $$Ax_e = 0$$

We thus assume as a basis of our structure analysis the following

(4.2) Definition - A state x of the bilinear system (2.1) is said to be reachable from an equilibrium state x_e if there exists an admissible input function that transfers the state x_e into the state x in a finite interval of time.

Proceeding now with the problem of defining a canonical decomposition of the state space, we must observe that the set of states of system (2.1) reachable from a state x_e is not, in general a linear subspace; this is due to the arbitrariness of x_e as well as to the nonlinear nature of the system. The canonical decomposition, however, can still be achieved if the states of interest are included into a suitable subspace.

The latter is introduced on the basis of the following assumptions:

(a) we consider the least linear manifold that includes all the states of system (2.1) reachable from x_e, and

(b) we define the state space decomposition only in those cases in which this manifold includes the origin of the state space (i.e. when this manifold is itself a subspace).

The validity of the assumption (b) will be justified by the subsequent developments and, particularly, by the realization theory. For a first insight into its motivations, consider the case of a linear system evolving from a nonzero initial equilibrium state x_e. Then, if the system must be decomposed into four subsystems only one of which is relevant for the input/output behaviour, one naturally is brought to distinguish the case in which x_e is reachable (from the origin of the state space) from the one in which is not. This distinction corresponds to the one assumed in (b), since in the first case the set of the states reachable from x_e is a subspace, while in the second is a linear manifold.

At this point we can prove the following

(4.3) Theorem - The least linear manifold $\mathcal{M}(x_e)$ that includes all the states of system (2.1) reachable from a given equilibrium state x_e can be expressed as

$$(4.4) \qquad \mathcal{M}(x_e) = \{x_e + z : z \varepsilon \operatorname{gen}_{A,N_1,\dots,N_p} [(N_1 x_e \dots N_p x_e) + B]\}$$

Proof - Since $x_e \varepsilon \mathcal{M}(x_e)$, all that we have to do is to define the new variable

$$(4.5) \qquad z = x - x_e$$

and find the subspace

$$(4.6) \qquad \mathcal{Z} = \operatorname{span}\{z : z + x_e \text{ is reachable from } x_e \text{ in system (2.1)}\}$$

The differential equation for the variable z, obtained by (2.1) through condition (4.1), is

$$(4.7) \qquad \dot{z}(t) = Az(t) + \sum_{i=1}^{p} N_i u_i(t) z(t) + [(N_1 x_e \dots N_p x_e) + B] u(t)$$

with the initial condition

$$(4.8) \qquad z(0) = 0$$

Therefore we can rewrite the subspace (4.6) as

$$(4.9) \qquad Z = \text{span}\{z : z \text{ reachable from 0 in system (4.7)}\}$$

In fact, we have introduced a new bilinear system, described by the matrices $A, N_1, \ldots, N_p$ and

$$(4.10) \qquad B_{eq} = (N_1 x_e \ \ldots \ N_p x_e) + B$$

which is initially in the zero state.

The expression of the subspace Z can be obtained as follows. Consider that, for any differential equation $\dot{\xi}(t) = f[\xi(t), t]$ (with $\xi(t)$ defined in R^n) and for any subspace Ξ of R^n, the following is true

$$(4.11) \qquad \xi(t) \in \Xi \quad \forall t \in [0,T] \Rightarrow \dot{\xi}(t) \in \Xi \quad , \forall t \in [0,T]$$

By definition, at least a basis $\{z_1, z_2, \ldots, z_r\}$ of Z is reachable from the origin. Therefore, applying (4.11) to (4.7), we get

$$(4.12) \qquad \left[(A + \sum_{i=1}^{p} N_i u_i) z_j + B_{eq} u\right] \in Z \qquad \forall u \in R^p, \quad j = 1, 2, \ldots, r$$

But, again for (4.11), also

$$(4.13) \qquad B_{eq} u \in Z \qquad \forall u \in R^p$$

and, hence, from this and (4.12)

$$(4.14) \qquad (A + \sum_{i=1}^{p} N_i u_i) Z \subseteq Z \qquad \forall u \in R^p$$

From this and (4.13) we conclude that Z contains $R(B_{eq})$ and is invariant under $A, N_1, \ldots, N_p$. Therefore $Z \supseteq \text{gen}_{A, N_1, \ldots, N_p}(B_{eq})$, because the latter, by definition, is the least subspace with these properties. To prove the converse, observe that, by virtue of (4.13) and (4.14), any subspace of R^n invariant $A, N_1, \ldots, N_p$ and containing $R(B_{eq})$ is such that in any of its points it is possible to assign only velocities belonging to the subspace itself. Therefore any trajectory starting from the origin cannot leave each of such subspaces, and this implies that $\text{gen}_{A, N_1, \ldots, N_p}(B_{eq})$, the least of them, contains all the trajectories starting from origin, i.e. contains $\mathcal{Z}$.

We thus have

$$\mathcal{Z} = \text{gen}_{A,N_1,\ldots,N_p}(B_{eq}) \tag{4.15}$$

or, according to (4.5),

$$\mathcal{M}(x_e) = \{x_e + z : z \varepsilon \text{gen}_{A,N_1,\ldots,N_p}(B_{eq})\} \tag{4.16}$$

(4.17) Remark - The least linear subspace of R^n that includes all the states reachable from x_e can be expressed as

$$\text{span}\{x_e,\ \text{gen}_{A,N_1,\ldots,N_p}[(N_1 x_e \ldots N_p x_e)+B]\} \tag{4.18}$$

When $x_e = 0$, this obviously reduces to

$$\text{gen}_{A,N_1,\ldots,N_p}(B) \tag{4.19}$$

as previously proved in [3], [4]. When the system is homogeneous in the state, the subspace (4.18) reduces to

$$\text{gen}_{A,N_1,\ldots,N_p}(x_e) \tag{4.20}$$

In order to obtain the state space decomposition, under the assumption (b) of the previous discussion, it is useful to prove the following

(4.21) Corollary - If the least linear manifold including all the states of system (2.1) reachable from a given equilibrium state x_e is a linear subspace, then it can be expressed as

$$\mathcal{X}_p = \text{gen}_{A,N_1,\ldots,N_p}(B \quad x_e) \tag{4.22}$$

Proof - The basic hypothesis that $0 \varepsilon \mathcal{M}(x_e)$, by (4.16) is equivalent to

$$-x_e \varepsilon \text{gen}_{A,N_1,\ldots,N_p}(B_{eq}) \tag{4.23}$$

Therefore, recalling (4.10) and the invariance of the r.h.s. of (4.23) under $N_1,\ldots,N_p$, we can easily conclude that, if $0 \in M(x_e)$, then

$$\text{gen}_{A,N_1,\ldots,N_p}(B_{eq}) = \text{gen}_{A,N_1,\ldots,N_p}(B\ x_e) \tag{4.24}$$

which, in turn, according to (4.16), gives (4.22).

This concludes the proof.

5. Unobservability

The other structure property on which we base the state space decomposition is expressed by the following

(5.1) Definition - A state x of system (2.1) is said to be unobservable if it is indistinguishable from the origin of the state space.

As concerns the unobservable states we have the following

(5.2) Theorem - The subset of all the unobservable states of system (2.1) is a subspace X_q of R^n which can be expressed as

$$X_q = \overline{\text{gen}}_{A,N_1,\ldots,N_p}(C) \tag{5.3}$$

Proof - A state x is indistinguishable from the origin if and only if the difference between the response from x and the response from 0 is identically zero for every input function. This difference can be formally considered as a solution of (2.1) with $B = 0$. It follows that the set of all the unobservable states is a subspace. Moreover, applying (4.11) to this equation, we get

$$(A + \sum_{i=1}^{p} N_i u_i)\ X_q \subseteq X_q \qquad \forall u \in R^p \tag{5.4}$$

Moreover, from the observation equation of system (2.1), we have also

$$X_q \subseteq N(C) \tag{5.5}$$

Therefore X_q is invariant under $A, N_1, \ldots, N_p$ and is contained in $N(C)$, i.e. $X_q \subseteq \mathrm{gen}_{A,N_1,\ldots,N_p}(C)$ which is the largest subspace with this property. The converse, i.e. the inequality $X_q \supseteq \overline{\mathrm{gen}}_{A,N_1,\ldots,N_p}(C)$, can be proved as in the proof of Theorem (4.3).

We conclude this section by a comparison between (4.22) and (5.2)

(5.6) Remark - In the case that $x_e = 0$, Lemma (3.14) establishes a duality relationship between the subspaces X_p and X_q, that is between reachability from the origin and observability. This provides an extension to bilinear systems of the well -known result of the linear theory.

6. State space decomposition

Starting from the concepts of reachability from an equilibrium state x_e and unobservability or, more precisely, from the subspaces X_p and X_q defined in the previous sections, it is possible to decompose the state space X of the system (2.1) into the direct sum of four subspaces A, B, C, D, following the procedure proposed by R.E. Kalman in the case of linear constant systems, i.e.

$$
\begin{aligned}
A &= X_p \cap X_q \\
X_p &= A \oplus B \\
X_q &= A \oplus C \\
X &= A \oplus B \oplus C \oplus D
\end{aligned}
\qquad (6.1)
$$

On the basis of a decomposition of this type, which will be called a canonical decomposition, it is possible to prove the following

(6.2) Theorem - If the least linear manifold that includes all the states of system (2.1) reachable from a given equilibrium state x_e is a linear subspace, then, assuming as a basis in the state space the union of bases on the four subspaces A, B, C, D of a canonical decomposition, the set of matrices of the corresponding description (2.5) and the initial state x_e assume the form (canonical form)

$$\begin{pmatrix} A_{aa} & A_{ab} & A_{ac} & A_{ad} \\ 0 & A_{bb} & 0 & A_{bd} \\ 0 & 0 & A_{cc} & A_{cd} \\ 0 & 0 & 0 & A_{dd} \end{pmatrix}, \ldots, \begin{pmatrix} N^{(i)}_{aa} & N^{(i)}_{ab} & N^{(i)}_{ac} & N^{(i)}_{ad} \\ 0 & N^{(i)}_{bb} & 0 & N^{(i)}_{bd} \\ 0 & 0 & N^{(i)}_{cc} & N^{(i)}_{cd} \\ 0 & 0 & 0 & N^{(i)}_{dd} \end{pmatrix}, \ldots,$$

(6.3)

$$\begin{pmatrix} B_a \\ B_b \\ 0 \\ 0 \end{pmatrix}, \quad (0,\ C_b,\ 0,\ C_d) \quad , \begin{pmatrix} x_{e,a} \\ x_{e,b} \\ 0 \\ 0 \end{pmatrix}$$

Proof - The pattern of the zero submatrices in (6.3) is justified as follows. $X_p = A \oplus B$ is invariant under $A, N_1, \ldots, N_p$: this implies the zero blocks in the positions (3/1), (3/2), (4/1) and (4/2) of $A, N_1, \ldots, N_p$. $X_q = A \oplus C$ is invariant under $A, N_1, \ldots, N_p$, and this implies the remaining zero blocks. X_p contains both $R(B)$ and x_e and hence the blocks in the positions (3) and (4) of B and x_e are zero. Similarly X_q is contained in $N(C)$ and this implies the zero blocks of C.

(6.4) Theorem - The set of descriptions of system (2.1) which assume the canonical form (6.3) is an equivalence class modulo a coordinate transformation defined by a nonsingular matric such as

(6.5)
$$T = \begin{pmatrix} T_{aa} & T_{ab} & T_{ac} & T_{ad} \\ 0 & T_{bb} & 0 & T_{bd} \\ 0 & 0 & T_{cc} & T_{cd} \\ 0 & 0 & 0 & T_{dd} \end{pmatrix}$$

with the partitions consistent with the dimensions of the subspace A, B, C, D respectively.

Proof - It is easy to verify that the transformation defined by (6.5) maintains the pattern of zero submatrices in all matrices of the set (6.3). Consider now any two descriptions in canonical form; they correspond to two different choices of subspaces, B', C', D'

and $\mathcal{B}''$, $\mathcal{C}''$, $\mathcal{D}''$, in the decomposition (6.1) of the state space. The matrix representing the coordinate trasformation between the subspaces corresponding to these choices must have the pattern of zeros shown by (6.5). In fact, since $\mathcal{A} \oplus \mathcal{B}' = \mathcal{A} \oplus \mathcal{B}''$, the submatrices in the positions (3/1), (3/2), (4/1) and (4/2) must be zero matrices; similarly, the equality $\mathcal{A} \oplus \mathcal{C}' = \mathcal{A} \oplus \mathcal{C}''$ implies the other zero submatrices.

To conclude the section we note that

(6.6) Remark - If the system is initially in the state x_e and if the least linear manifold $M(x_e)$ that includes all the states reachable from x_e is a subspace, then the output of the bilinear system (2.1) depends only on the subsystem characterized by the matrices A_{bb}, $N_{bb}^{(1)},\ldots,N_{bb}^{(p)}$, B_b, C_b, in the initial state $x_{e,b}$.

II. THE PROPERTIES OF MINIMAL REALIZATIONS

1. Introduction

In the preceding Chapter we developed a structure analysis of the state space, based on the properties of reachability and unobservability. The main motivation of this analysis was the purpose of establishing a connection between state space decomposition and the theory of minimal realization. In the present Chapter we exploit this connection and we find:

(a) the relation between minimality, reachability and observability, and
(b) the description, where possible, of all minimal realizations.

The analysis is carried out quite independently of the realizability theory (i.e. the problem of finding conditions under which a given input/output map admits bilinear realizations and the means of constructing them). This will be considered in the third Chapter.

As observed previously, in order to give a wider domain of validity to the theory, we must include bilinear realizations being in a nonzero initial state (precisely, an equilibrium state). In this set up, we shall find a strong connection between theory of minimal realization and state space decomposition, only when the given input/output map admits realizations such that the initial state satisfies the assumptions under which the structure theory was based (i.e. when the least linear manifold that includes all the states reachable from the initial state is a subspace). These assumptions are not always satisfied and, therefore, this leads to distinguish between two classes of problems; in both cases, however, the theory can be developed (sections 3 and 5) and, also a test allowing the distinction may be given (section 4) (*).

2. A property of all bilinear realizations

We shall assume that the space $\mathcal{U}$ of input functions be a space of p-valued functions continuous on $[0,T]$; consequently, the space $\mathcal{Y}$ of output functions will be a space of q-valued functions continuous on $[0,T]$. Let x_e denote a fixed equilibrium state of system (I.2.1), i.e. a state such that

$$Ax_e = 0 \tag{2.1}$$

Equation (I.2.1), with $x(0) = x_e$, defines a map $\mathcal{U} \to \mathcal{Y}$, which will be called the input/output map (referred to the equilibrium state x_e). This map is associated to the set of matrices

$$\sigma = (A, N_1, \ldots, N_p, B, C) \tag{2.2}$$

and to the initial state x_e; we shall therefore use the symbol

$$y = f_{\sigma, x_e}(u) \qquad (u \in \mathcal{U},\ y \in \mathcal{Y}) \tag{2.3}$$

We thus give the following

(2.4) Definition - Let f be a prescribed map $\mathcal{U} \to \mathcal{Y}$. If there exist a set of matrices $\sigma = (A, N_1, \ldots, N_p, B, C)$ and an equilibrium state x_e such that

$$f_{\sigma, x_e}(u) = f(u) \qquad \text{for all} \quad u \in \mathcal{U} \tag{2.5}$$

(*) The analysis here developed is derived from the one presented in [4], but is suitably generalized in order to include the cases of nonzero initial states. In the present results one can also include those proved by R.W. Brockett [5], which refer to the case of state-homogeneous minimal realizations.

then f is said bilinearly realizable and the set $(A,N_1,\dots$ $\dots,N_p,B,C;x_e)$ is called a (finite dimensional) bilinear realization. A realization of the form $(A,N_1,\dots,N_p,B,C;0)$ is called a zero-state realization.

Before starting with the analysis of the properties of minimal realizations, we shall give a property which characterizes all bilinear realizations of a given input/output map. Recalling the formulae (I.3.1) and (I.3.18), define the sequences of matrices

(2.6)
$$\begin{aligned} \overline{P}_1 &= (N_1x_e \; \dots \; N_px_e) + B \\ \overline{P}_i &= (A\overline{P}_{i-1} \; N_1\overline{P}_{i-1} \; \dots \; N_p\overline{P}_{i-1}) \qquad (i=2,3,\dots) \\ P_i &= (\overline{P}_1 \; \overline{P}_2 \; \dots \; \overline{P}_i) \qquad (i=1,2,\dots) \end{aligned}$$

and

(2.7)
$$\overline{Q}_1 = C$$
$$\overline{Q}_i = \begin{pmatrix} \overline{Q}_{i-1}A \\ \overline{Q}_{i-1}N_1 \\ \vdots \\ \overline{Q}_{i-1}N_p \end{pmatrix} \qquad (i=2,3,\dots)$$
$$Q_i = \begin{pmatrix} \overline{Q}_1 \\ \overline{Q}_2 \\ \vdots \\ \overline{Q}_i \end{pmatrix} \qquad (i=1,2,\dots)$$

Then, we have

(2.8) <u>Lemma</u> - For every fixed r and s, the product

(2.9)
$$Q_r \cdot P_s$$

is invariant on the set of all bilinear realizations of a given input/output map.

Proof - We shall show that the entries of (2.9) are uniquely associated to the input/output map. Bearing in mind eqs. (I.4.5) and (I.4.7), we see that, for any fixed $u \in \mathcal{U}$, the input-state bilinear equation may be considered as equivalent to a linear equation with time-varying coefficients

$$\dot{z}(t) = \left[A + \sum_{i=1}^{p} N_i u_i(t)\right] z(t) + \left[(N_1 x_e \ldots N_p x_e) + B\right] u(t) \tag{2.10}$$

with initial condition $z(0) = 0$. The output response is characterized also by the observation equation which, in terms of z, is rewritten as

$$y(t) = C\, z(t) + C\, x_e \tag{2.11}$$

Apart from the constant term Cx_e (which obviously is invariant over all the realizations of the same input/output map), the overall response of eqs. (2.10)-(2.11) is uniquely characterized by the weighting-pattern

$$W(t,\tau) = C\,\Phi(t,\tau)\left[(N_1 x_e \ldots N_p x_e) + B\right] \tag{2.12}$$

where $\Phi(t,\tau)$ is the transition matrix of (2.10). If $\Phi(t,\tau)$ is expressed in terms of $\left[A + \sum_{i=1}^{p} N_i u_i(t)\right]$ with the Peano-Baker formula, then the weighting-pattern (2.12) is expanded as

$$W(t,\tau) = \overline{Q}_1 \overline{P}_1 + \int_\tau^t \overline{Q}_1 \left[A + \sum_{i=1}^{p} N_i u_i(\theta)\right] \overline{P}_1 \, d\theta \tag{2.13}$$

$$+ \int_\tau^t \int_\tau^{\theta_1} \overline{Q}_1 \left[A + \sum_{i=1}^{p} N_i u_i(\theta_1)\right]\left[A + \sum_{i=1}^{p} N_i u_i(\theta_2)\right] \overline{P}_1 \, d\theta_1 d\theta_2 + \ldots$$

with $\overline{P}_1, \overline{Q}_1$ defined by (2.6) and (2.7). The recursive structure of these latter, as well as that of (2.13), suggest a rearrangement of this expansion characterized in terms of parameters belonging to the infinite sequence $\{\overline{Q}_1 \overline{P}_i\}_1^\infty$.

We have thus proved (2.9) for $r=1$. This, however, is sufficient to prove (2.9) for all r, because the structure of (2.6) and (2.7) is such that each element of $\overline{Q}_i\overline{P}_j$ is also an element of $\overline{Q}_1\overline{P}_{i+j-1}$, and viceversa.

3. Minimality, reachability and observability

In this section we analyze the properties of the minimal bilinear realizations of a given input/output map. We start with the following

(3.1) Definition - A bilinear realization is minimal if the dimension of the state space is minimal over the set of all bilinear realizations of the given input/output map.

In order to exploit the connection between minimality and state space decomposition, our first concern is to identify the cases in which the structure analysis of Chapter I may be applied. To this end it is convenient to give the following:

(3.2) Definition - A bilinear realization $(A,N_1,\ldots,N_p,B,C;x_e)$ is said to be structurally connected if the least linear manifold that includes all the states reachable from x_e is a subspace.

In the present section we shall assume that the given map admits at least one realization with the property characterized by Definition (3.2) (*) and we develop the corresponding analysis. The analysis of the other case, as well as a means for testing whether a map admits structurally connected realizations or not, will be given in the following sections.

We start by giving a characterization of minimality, expressed by the following

(3.3) Theorem - Assume that the given input/output map admits at least one structurally connected bilinear realization. Then a given realization of it is minimal if and only if the state space coincides with the least linear manifold that includes all the states reachable from x_e and is observable.

(*) Note, for instance, that every zero state realization is structurally connected.

Proof - Let Σ denote the set of all structurally connected realizations of the given map and let $(\sigma;x_e)$ be any realization in Σ of least dimension. By contradiction we show that its state space coincides with $M(x_e)$ (i.e. the least linear manifold, that includes all the states reachable from x_e) and is observable. In fact, if a realization did not have these properties, then it would be possible to find in Σ a realization of lower dimension. This could be obtained by means of the state space decomposition (I.6.1) (see also Remark (I.6.6)).

Conversely, let $(\sigma;x_e)$, $(\hat{\sigma};\hat{x}_e)$ be any two realizations in Σ such that their own state spaces coincide respectively with $M(x_e)$, $M(\hat{x}_e)$ and are observable. Let $n,\hat{n}$ denote their dimension. On the basis of Lemma (2.8) and denoting $\max(n,\hat{n})$ by ν, we have

$$Q_\nu \cdot P_\nu = \hat{Q}_\nu \cdot \hat{P}_\nu \tag{3.4}$$

By theorems (I.4.3) and (I.5.2), the realization $(\sigma;x_e)$ is such that

$$\begin{aligned} R^n &= \mathrm{gen}_{A,N_1,\ldots,N_p}\left[(N_1x_e \ldots N_px_e) + B\right] \\ \{0\} &= \overline{\mathrm{gen}}_{A,N_1,\ldots,N_p}(C) \end{aligned} \tag{3.5}$$

Then, on the basis of Lemmas (I.3.2), (I.3.19) and of the formulae (2.6), (2.7), we conclude that the n rows of P_ν and the n columns of Q_ν are linearly independent. The same can be said about the $\hat{n}$ rows of $\hat{P}_\nu$ and the $\hat{n}$ columns of $\hat{Q}_\nu$. Consequently, the equality (3.4) implies $n=\hat{n}$. Together with the result proved at the beginning of the proof, this implies that any realization whose state space coincides with $M(x_e)$ and is observable has least dimension on Σ.

To complete the proof we shall show that all minimal realizations of the given map are included in Σ. Let n denote the least dimension of the realizations in Σ and assume that there exist a realization $(\sigma;x_e)\notin\Sigma$ of dimension $m\leq n$. Clearly $\dim M(x_e)\leq n-1$ and, therefore, by Theorem (I.4.3), Lemma (I.3.2) and formulae (2.6), we have that

$$\mathrm{rank}\ P_n \leq n-1 \tag{3.6}$$

This contradicts the property that $Q_n \cdot P_n$ has rank n (proved before for any realization belonging to Σ but valid everywhere, by Lemma (2.8)), thus completing the proof.

(3.7) Remark - As stated in the Theorem, and also clarified in its proof, the mutual relationships between minimal realizations and realizations are as follows

(3.8) $\{\text{realizations}\} \supseteq \{\text{structurally connected realizations}\} \supseteq$

$\supseteq \{\text{minimal realizations}\}$

(3.9) Remark - Since, by construction, the set $(A_{bb}, N_{bb}^{(1)}, \ldots$ $\ldots, N_{bb}^{(p)}, B_b, C_b; x_{e,b})$ identified by the canonical decomposition satisfies the conditions of Theorem (3.3), it can be considered as a minimal bilinear realization of the given input/output map. From this follows the possibility of developing algorithms for reducing any bilinear realization to a minimal one.

We shall now consider the uniqueness of minimal realizations.

(3.10) Theorem - Assume that the given input/output map admits at least one structurally connected bilinear realization. Then all its minimal bilinear realizations are a single equivalence class. Precisely,

$$(A, N_1, \ldots, N_p, B, C; x_e) \sim (\hat{A}, \hat{N}_1, \ldots, \hat{N}_p, \hat{B}, \hat{C}; \hat{x}_e)$$

if and only if

$$A = T\hat{A}T^{-1}$$

$$N_i = T\hat{N}_i T^{-1} \qquad (i=1,\ldots,p)$$

$$(N_1 x_e \; \ldots \; N_p x_e) + B = T\left[(\hat{N}_1 \hat{x}_e \; \ldots \; \hat{N}_p \hat{x}_e) + \hat{B}\right] \tag{3.11}$$

$$C = \hat{C}T^{-1}$$

$$C x_e = \hat{C}\hat{x}_e$$

Proof - The "if" part is trivial. To prove the converse, let σ, $\bar{\sigma}$ be any two minimal realizations, and let n denote their dimension.

Minimality and Theorem (3.3) imply that $Q_n^* \cdot Q_n$ is nonsingular; therefore we can define the n×n matrix

$$T = (Q_n^* \cdot Q_n)^{-1} \, Q_n^* \cdot \hat{Q}_n \tag{3.12}$$

Moreover, by Lemma (2.8), we have

$$Q_n \cdot P_n = \hat{Q}_n \cdot \hat{P}_n \tag{3.13}$$

and therefore

$$T^{-1} = \hat{P}_n \cdot P_n^* \, (P_n \cdot P_n^*)^{-1} \tag{3.14}$$

Now consider that the columns of AP_n are columns of P_{n+1}, the columns of N_iP_n $(i=1,\ldots,p)$ are again columns of P_{n+1}, $P_1 = [(N_1x_e \ldots \ldots N_px_e)+B]$ and $Q_1 = C$. By Lemma (2.8) we can write

$$\begin{aligned} Q_n AP_n &= \hat{Q}_n \hat{A} \hat{P}_n \\ Q_n N_i P_n &= \hat{Q}_n \hat{N}_i \hat{P}_n \qquad (i=1,2,\ldots,p) \\ Q_n[(N_1x_e \ldots N_px_e)+B] &= \hat{Q}_n[(\hat{N}_1\hat{x}_e \ldots \hat{N}_p\hat{x}_e)+\hat{B}] \\ CP_n &= \hat{C}\hat{P}_n \end{aligned} \tag{3.15}$$

from which, bearing in mind the definition (3.12) of T and the expression (3.14) of T^{-1}, we obtain the equivalence relation shown in (3.11).

(3.16) Remark - This theorem includes some previous results concerning the isomorphysm of minimal bilinear realizations. If we consider only zero-state minimal bilinear realizations, (3.15) shows that such realizations are equivalent modulo a coordinate transformation in the state space. The same is true if we consider only state-homogeneous bilinear realizations. Note, however, that a given map may admit both minimal zero-state and state-homogeneous realizations, simultaneously.

4. A test for the existence of structurally connected realizations

We start by proving the

(4.1) Proposition - If the map $y = f(u)$ is bilinearly realizable, then the map $y = f(u) - f(0)$ admits a zero-state bilinear realization.

Proof - Let $(A,N_1,\ldots,N_p,B,C;x_e)$ denote a realization of $f(u)$. Introducing the variable

$$z(t) = x(t) - x_e \tag{4.2}$$

we obtain equations (2.10) and (2.11). Since the substitution (4.2) does not affect the input/output map, denoting by σ^* the set $(A,N_1,\ldots,N_p,[(N_1x_e \ldots N_px_e) + B],C)$, we get

$$f(u) = f_{\sigma^*,0}(u) + f(0) \qquad \forall u \in \mathcal{U} \tag{4.3}$$

from which we conclude that $(\sigma^*;0)$ realizes $f(u)-f(0)$.

Therefore we have the following

(4.4) Proposition - Let $y = f(u)$ be a bilinearly realizable input/output map and $(A,N_1,\ldots,N_p,B,C;0)$ any zero-state minimal bilinear realization of the map $y = f(u)-f(0)$. Then the given map admits a structurally connected bilinear realization if and only if there exist a vector x_e such that

$$A\,x_e = 0 \tag{4.5}$$

$$C\,x_e = f(0) \tag{4.6}$$

Proof - The condition is sufficient. Equation (I.2.1), with $x(0)=0$ realizes, by hypothesis the map $y = f(u)-f(0)$. If (4.6) is satisfied, the equation

$$\begin{aligned} \dot{x}(t) &= A\,x(t) + \sum_{i=1}^{p} N_i u_i(t)\,x(t) + Bu(t) \\ y(t) &= C\,x(t) + C\,x_e \end{aligned} \tag{4.7}$$

with $x(0) = 0$ realizes the map $y = f(u)$. If also (4.5) is satisfied, then, by the change of variable

(4.8) $$z(t) = x(t) + x_e$$

we obtain the set $(A,N_1,\dots,N_p,B-(N_1x_e \dots N_px_e),C;x_e)$ which is a structurally connected realization of the given map.

The condition is necessary. If the given map admits structurally connected realizations, then all minimal realizations are structurally connected (see Remark (3.7)). Every minimal realization of the map $y=f(u)$ induces a zero-state minimal of the map $y=f(u)-f(0)$, satisfying the conditions (4.5) and (4.6) (see proof of Proposition (4.1)). Since all zero-state minimal realizations of a given map are unique modulo a coordinate transformation (see Remark (3.16)), if follows that (4.6) and (4.5) must be satisfied on each zero-state minimal realization of the map $y=f(u)-f(0)$.

(4.9) Remark - Note that this test, when satisfied, provides also a means for constructing a minimal bilinear realization of the map $y=f(u)$. The latter is simply derived from a zero-state minimal bilinear realization of the map $y=f(u)-f(0)$. It is worth noting that, even if the test is not satisfied, this realization can still be used to construct a minimal bilinear realization of the map $y=f(u)$; this point will be clarified after.

5. Further remarks on minimal realizations

A result to some extent similar to that shown by Theorem (3.3) can still be proved if the input/output map does not admit structurally connected realizations. In fact, we have the following

(5.1) Theorem - Assume that the given input/output map does not admit a structurally connected realization. Then, a given realization of it is minimal if and only if the state space is spanned by the states reachable from x_e and is observable.

Proof - The condition is necessary. Let $(A,N_1,\dots,N_p,B,C;x_e)$ be a minimal realization of the given map $y=f(u)$ and let n denote its dimension. To prove that R^n is spanned by the states reachable from x_e, we have to prove that $\dim M(x_e)=n-1$ (because, by hypothesis, $M(x_e)$ is not a subspace). The latter is proved by showing that the assumption $\dim M(x_e)=m<n-1$ would contradict the mini-

mality of the given realization. In this case, in fact, it would be possible to construct an m-dimensional zero-state realization of the map $y = f(u) - f(0)$ (Hint: consider the realization $(\sigma^*;0)$ described in the proof of Proposition (4.1); this can be reduced to dimension m on the basis of the results expressed by Theorem (3.3) and subsequent Remark (3.9)). Let $(\hat{A},\hat{N}_1,\ldots,\hat{N}_p,\hat{B},\hat{C};0)$ denote this realization. Then one easily sees that the set

$$(5.2)\qquad \begin{pmatrix}\hat{A} & 0\\ 0 & 0\end{pmatrix}, \begin{pmatrix}\hat{N}_1 & 0\\ 0 & 0\end{pmatrix}, \ldots, \begin{pmatrix}\hat{N}_p & 0\\ 0 & 0\end{pmatrix}, \begin{pmatrix}\hat{B}\\ 0\end{pmatrix}, (C\ y_e)\ ; \begin{pmatrix}0\\ 1\end{pmatrix}$$

with $y_e = f(0)$ realizes the map $y = f(u)$. But (5.2) has a dimension $m+1<n$ and this is a contradiction. The proof that every state in $R^n-\{0\}$ is observable is identical to that used in Theorem (3.3).

The condition is sufficient. It is possible to use the same technique as in Theorem (3.3), based on equality (3.4). Note that, in the present case, both P_ν and $\hat{P}_\nu$ have, respectively, only n-1 and $\hat{n}-1$ linearly independent rows. The proof, however, still leads to the equality $n=\hat{n}$.

(5.3) <u>Remark</u> - From the previous proof it is clear that a zero-state minimal bilinear realization of the map $y = f(u)-f(0)$ can still be used (see Remark (4.9)) to construct a minimal bilinear realization of the map $y = f(u)$, even if this does not admit structurally connected realizations.

Before concluding this section it is worth discussing also a different way of approaching the minimal realization problem. This approach is based on the result expressed by Proposition (4.1) and its proof; in fact, this suggests the possibility of realizing the map $y = f(u)$ by means of the set of equations

$$(5.4)\qquad \begin{aligned} \dot{x}(t) &= A\,x(t) + \sum_{i=1}^{p} N_i u_i(t)\,x(t) + B\,u(t) \qquad x(0) = 0\\ y(t) &= C\,x(t) + d \end{aligned}$$

in which the set $(A,N_1,\ldots,N_p,B,C)$ is derived by a zero-state bilinear realization of the map $y = f(u)-f(0)$ and $d = f(0)$.

By means of equations (5.4), which are bilinear in the state but inhomogeneous in the output, the realization problem is always reduced to a zero-state problem. On the basis of the results stated in section 3, specified for $x_e = 0$, we can also assert that

(a) a given realization (5.4) is minimal if and only if the state space is spanned by the states reachable from the origin and is observable,

(b) all minimal realizations (5.4) are a single equivalence class modulo a coordinate transformation in the state space.

This approach could seem more direct and simple, mainly because it avoids the distinction based on the existence of structurally connected realizations. However, we wish to stress that just on the basis of this distinction a better understanding of the state space structure and of the role played by the initial state was possible. This is the reason that led us to set up the theory as we did.

From a constructional point of view, the two approaches are substantially equivalent. In fact, if the input/output map admits structurally connected realizations, then the minimal bilinear realizations considered in section 3 have the same dimension as the ones of the form (5.4) and the same dynamical matrices $A, N_1, \ldots, N_p$.

If the map does not admit such realizations, the minimal realizations considered in Theorem (5.1) are essentially the same as the ones provided by (5.4). The effect of extra component of the state vector exhibited by (5.2) is substituted in (5.4) by the constant term on the output.

III. REALIZABILITY CONDITIONS AND REALIZATION PROCEDURES

1. Introduction

In the present Chapter we shall consider the problem of deducing conditions under which a prescribed input/output map admits bilinear realizations and, in the positive case, the problem of constructing a minimal one. The starting point is that of finding a suitable description of the input/output map of bilinear systems. Thanks to results proved in Chapter II (Proposition (4.1), Remark (4.9) and (5.3)), we may restrict the analysis to the case of a system in the initial state $x(0) = 0$. For this case we shall give two kinds of descriptions; the first by means of the sequence of kernels of a Volterra series expansion, the second by means of an infinite sequence of constant parameters which identifies uniquely such kernels. Correspondingly, realization problem can be stated as the one of matching a sequence of kernels or, alternatively, a sequence of parameters (section 2). Then we describe, in both cases, a necessary and sufficient condition for realizability and a constructive procedure (section 3 and 4). The analysis is completed by some considerations on the comparison between these two approaches and by two Appendices on complementary subjects (*).

2. The Volterra series expansion of the input/output map

In this section we give a representation for the input/output map of the bilinear system defined by (I.2.1). We assume, for sake of notational simplicity, that p=q=1 (one-dimensional input and output) and, as done at the beginning of Chapter II, that the input u belongs to a class of continuous functions on [0,T] (**). It is easy to show [8] that the input/output map of system (I.2.1), from the initial state x(0) = 0, can be expanded in series according to

$$(2.1) \qquad y_u(t) = \sum_{i=1}^{\infty} \frac{1}{i!} \int_0^t \cdots \int_0^t w_i(t_1,\dots,t_i) \left[\prod_{k=1}^{i} u(t-t_k)\right] dt_1 \dots dt_i$$

with symmetrical kernels given by

$$(2.2) \qquad w_1(t_1) = C\, e^{At_1} B$$

(*) The approach and the results described in section 3 are derived from the paper [3]; their generalization to multidimensional systems may be found in [4]. The approach and the results described in section 3 are derived from the paper [6]. Appendix I is a concise presentation of results discussed in [3]; Appendix II is deduced from [7].

(**) The analysis of the multidimensional case does not present any particular difficulty and, anyhow, may be found in [4].

(2.2) $w_i(t_1,\dots,t_i) =$

$$= \sum_{per} C e^{At_i} N e^{A(t_{i-1}-t_i)} \dots N e^{A(t_1-t_2)} B \prod_{k=0}^{i-2} \delta_{-1}(t_{k+1}-t_{k+2})$$

In the formulae $\delta_{-1}(\cdot)$ is the unit step function and the summation $\sum_{per}$ is carried out over all the i! permutations of $t_1,\dots,t_i$.

The expansion (2.1) is uniquely characterized by the infinite sequence of kernels $\{w_i(t_1,\dots,t_i)\}_1^\infty$; therefore the problem of finding zero-state bilinear realizations is equivalent to that of matching an infinite sequence of Volterra kernels. More precisely, according to (2.2) we can give the

(2.3) Definition - An infinite sequence $\{w_i(t_1,\dots,t_i)\}_1^\infty$ of symmetrical kernels of a Volterra series expansion admits bilinear realizations if there exist four constant matrices A,N,B,C, respectively n×n, n×n, n×1, 1×n, such that

(2.4)
$$C e^{At_1} B = w_1(t_1) \qquad \text{for all } t_1$$

$$\sum_{per} C e^{At_i} N e^{A(t_{i-1}-t_i)} \dots N e^{A(t_1-t_2)} B \prod_{k=0}^{i-2} \delta_{-1}(t_{k+1}-t_{k+2}) =$$

$$= w_i(t_1,\dots,t_i) \qquad \text{for all } t_1,\dots,t_i \qquad (i=2,3,\dots)$$

It is well known that, in linear system theory, the problem of matching a given kernel w(t) by

(2.5) $$C e^{At} B = w(t) \qquad \text{for all } t$$

is equivalent to that of matching an infinite sequence of parameters $\{S_j\}_1^\infty$ by

(2.6) $$C A^j B = S_j \qquad \text{for all } j$$

Parameters S_j are related to w(t) by

(2.7) $$S_j = \left(\frac{d^j w(t)}{dt^j}\right)_{t=0}$$

The same circumstance holds in the theory of bilinear systems. This is based on suitable Taylor series expansions of the kernels $\{w_i(t_1,\ldots,t_i)\}_1^\infty$; more precisely, defining

$$S^{(i)}_{k_1,\ldots,k_i} = \left[\frac{\partial^{k_1}}{\partial\theta_1^{k_1}}\cdots\frac{\partial^{k_i}}{\partial\theta_i^{k_i}} w_i(\theta_1+\theta_2+\ldots+\theta_i,\ldots,\theta_{i-1}+\theta_i,\theta_i)\right]_{\substack{\theta_1=0\\ \vdots\\ \theta_i=0}} \tag{2.8}$$

we have from (2.2) that

$$\begin{aligned} S^{(1)}_{k_1} &= C A^{k_1} B \\ S^{(i)}_{k_1,\ldots,k_i} &= C A^{k_i} N \ldots N A^{k_1} B \end{aligned} \tag{2.9}$$

The comparison between (2.9) and (2.6) shows that, in the case of bilinear systems, the problem of matching an infinite sequence of kernels can be reduced to that of matching an infinite sequence of constant parameters. For practical purposes, it is convenient to rearrange this sequence, to obtain a single-indexed ordering. It will be shown later (section 4) that the best way of re-indexing the sequence of parameters (2.9) is the one defined as follows:

(a) let w_j be a $1\times 2^{j-1}$ row vector whose r-th element, say, be $S^{(i)}_{k_1,k_2,\ldots,k_i}$;

(b) define w_{j+1} as a 1×2^j row vector whose r-th element is $S^{(i)}_{k_1+1,k_2,\ldots,k_i}$ and whose $(r+2^{j-1})$-th element is $S^{(i+1)}_{0,k_1,\ldots,k_i}$;

(c) inizialize the sequence $\{w_j\}_1^\infty$ by $w_1 = S^{(1)}_0$

It is easy to see that all parameters $S^{(i)}_{k_1,\dots,k_i}$ enter in the sequence $\{w_j\}_1^\infty$. For a clearer understanding, one may observe that, on the basis of (2.9), the w_j are such that

$$
\begin{aligned}
w_1 &= CB \\
w_2 &= (CAB \quad CNB) \\
w_3 &= (CA^2B \quad CANB \quad CNAB \quad CN^2B)
\end{aligned}
\tag{2.10}
$$

Parameters (2.10), to some extent, have the same structure as (2.6), provided that the j-th power of A is replaced by all 2^j different products, of length j, of matrices A and N.

At this point we can state alternatively the realization problem in the following way

(2.11) <u>Definition</u> - An infinite sequence $\{w_j\}_1^\infty$ of $1\times 2^{j-1}$ row vectors admits bilinear realizations if there exist four constant matrices A,N,B,C, respectively n×n, n×n, n×1, 1×n, such that

$$
C\overline{P}_j = w_j \qquad \text{for all } j \tag{2.12}
$$

where $\overline{P}_j$ are defined recursively by

$$
\begin{aligned}
\overline{P}_1 &= B \\
\overline{P}_j &= (A\overline{P}_{j-1} \quad N\overline{P}_{j-1})
\end{aligned}
\tag{2.13}
$$

(2.14) <u>Remark</u> - Note that the sequence (2.10) coincides with the sequence $\{\overline{Q}_1\overline{P}_j\}_1^\infty$ evidenced in the proof of Theorem (II.2.8) by means of the Peano-Baker formula.

(2.15) <u>Remark</u> - It is worth noting that the input/output map of system (I.2.1) from a given nonzero equilibrium state $x(0)=x_e$, can still be described by a Volterra series expansion, plus a constant term $y_e = C\,x_e$. The kernels of this series differ from the ones of the series (2.1) in the fact that vector B is replaced by vector $B + N\,x_e$. This is a confirm of the fact that, from the input/output point of view, there is no difference between the case in which the system is state-homogeneous (with nonzero initial state) and that in which the system is state-inhomogeneous.

3. A first realizability condition and a realization procedure

We can prove the following:

(3.1) Theorem - An infinite sequence $\{w_i(t_1,\ldots,t_i)\}_1^\infty$ of symmetrical kernels of a Volterra series expansion admits a bilinear realization if and only if

(a) $w_1(t_1)$ has a proper rational Laplace transform

(b) there exist three matrices $F(t)$, $G(t)$, $H(t)$, respectively $m\times m$, $m\times 1$, $1\times m$, of functions with proper rational Laplace transforms, such that the following relations are satisfied for all i

$$w_i(t_1,\ldots,t_i)=H(t_i)F(t_{i-1}-t_i)\ldots F(t_2-t_3)G(t_1-t_2) \tag{3.2}$$

on

$$S_i = \{(t_1,\ldots,t_i):t_1 > t_2 > \ldots > t_i\}$$

(3.3) Remark - The above condition holds also if the variables $t_1,\ldots,t_i$ in (3.2) are permutated in any way.

Proof - Necessity. If the sequence $\{w_i(t_1,\ldots,t_i)\}_1^\infty$ is realizable, then, by hypothesis, there exist four matrices A,N,B,C such that equations (2.4) are satisfied. The first of these implies the condition (a). The second, considered over the sets S_i defined above, reduces to

$$Ce^{At_i}Ne^{A(t_{i-1}-t_i)}\ldots Ne^{A(t_1-t_2)}B = w_i(t_1,\ldots,t_i) \tag{3.4}$$

If the matrix N is factored in the form

$$N = N'N'' \tag{3.5}$$

where N' is $n\times m$ and N'' is $m\times n$, one immediately sees that (3.4) can be reduced to the form (3.2), with the substitutions

$$Ce^{At}N' = H(t) \quad ; \quad N''e^{At}N' = F(t) \quad ; \quad N''e^{At}B = G(t) \tag{3.6}$$

Since all these functions have proper rational Laplace transforms, it follows that condition (b) is also satisfied.

Sufficiency. Suppose that (a) and (b) are both satisfied, and consider the matrix

$$L(t) = \begin{pmatrix} w_1(t) & H(t) \\ G(t) & F(t) \end{pmatrix} \tag{3.7}$$

Since all the elements of L(t) have proper rational Laplace transforms, this may be interpreted as the weighting pattern of a constant linear system of finite order with (m+1) outputs and (m+1) inputs. Consequently, there must exist three matrices A, R, S, respectively n×n, n×(m+1), (m+1)×n, such that

$$S\, e^{At} R = L(t) \tag{3.8}$$

By partitioning S and R in the form

$$S = \begin{pmatrix} S_1 \\ S_2 \end{pmatrix} \qquad R = (R_1 \quad R_2) \tag{3.9}$$

where S_1 is 1×n and R_1 is n×1, we get

$$\begin{aligned} w_1(t) &= S_1 e^{At} R_1 \quad ; \quad H(t) = S_1 e^{At} R_2 \\ G(t) &= S_2 e^{At} R_1 \quad ; \quad F(t) = S_2 e^{At} R_2 \end{aligned} \tag{3.10}$$

Now replace (3.10) in (3.2) and define

$$B = R_1 \quad ; \quad C = S_1 \quad ; \quad N = R_2 S_2 \tag{3.11}$$

It follows that the bilinear system characterized by the set A, N,B,C verifies the conditions (2.4) over the sets S_i.

On the other hand, since $w_i(t_1,\dots,t_i)$ is symmetrical by definition, (2.4) must be satisfied for all values of the variables $t_1,\dots,t_i$. This completes the proof.

Based on the proof of the above theorem and on the results of the analysis developed in the preceding chapters, it is possible to give immediately a procedure for constructing a minimal bilinear realization of the sequence $\{w_i(t_1,\ldots,t_i)\}_1^\infty$, once a factorization $\{F(t), G(t), H(t)\}$ of the sequence $\{w_i(t_1,\ldots,t_i)\}_2^\infty$ is available. The corresponding steps are the following:

(a) arrange $w_1(t)$, $F(t)$, $G(t)$, $H(t)$ as in (3.7)

(b) find a linear realization A, R, S of (3.7)

(c) define the bilinear realization A, N, B, C by the equations (3.8), (3.9), (3.11)

(d) from this latter derive a minimal realization, on the basis of the results outlined in Remark (II.3.9) (i.e., by means of the state-space decomposition).

It is convenient to stress that this realization procedure is substantially based on the use of techniques for linear realization and, also, that the reduction procedure can be carried out by means of linear algebraic tools (see Appendix II for a reduction algorithm).

4. A second realizability condition and another realization procedure

In this section we give a condition for the case in which the map is characterized by the infinite sequence of parameters $\{w_j\}_1^\infty$. The first step consists in defining a suitable infinite matrix that plays, in the present problem, a role similar to that of the Hankel matrix in the theory of linear systems. From the infinite sequence $\{w_j\}_1^\infty$ we define a doubly infinite sequence of matrices as follows

$$S_{1j} = w_j \qquad j=1,2,\ldots \tag{4.1}$$

S_{ij} $(i=2,3,\ldots;\ j=1,2,\ldots)$ is constructed from $S_{i-1,j+1}$ with this rule: form the partition

$$S_{i-1,j+1} = (S^1_{i-1,j+1} \quad S^2_{i-1,j+1}) \tag{4.2}$$

assigning the same number of columns to both blocks on the right-hand-side and put

$$S_{ij} = \begin{pmatrix} S^1_{i-1,j+1} \\ \\ S^2_{i-1,j+1} \end{pmatrix} \tag{4.3}$$

These operations are well-defined because S_{1j} has 2^{j-1} columns. As a result we obtain the desired sequence of matrices, in which S_{ij} is $2^{i-1}\times 2^{j-1}$. Note also that the matrices for which $i+j$ is constant are formed with the same elements.

We define the infinite Hankel matrix as the matrix

$$\Sigma = \begin{pmatrix} S_{11} & S_{12} & \cdots \\ S_{21} & S_{22} & \cdots \\ \cdot & \cdot & \cdots \\ \cdot & \cdot & \cdots \end{pmatrix} \tag{4.4}$$

and the finite Hankel matrix $\Sigma_{M'M}$ as the blockwise $M'\times M$ submatrix in the upper left-hand corner of the infinite matrix Σ. Note that $\Sigma_{M'M}$ is $(2^{M'}-1)\times(2^{M}-1)$.

Based upon the definition (4.4), we can prove the following realizability condition

(4.5) Theorem - An infinite sequence $\{w_j\}_1^\infty$ of $1\times 2^{j-1}$ row vectors admits a bilinear realization if and only if the infinite matrix Σ has finite rank, i.e. if and only if there exist two integers M and M' such that

$$\text{rank}\ \Sigma_{M'+i,M+j} = \text{rank}\ \Sigma_{M'M} \tag{4.6}$$

for all positive i and j.

Proof - Necessity. For conveniency we rewrite the formulae (II.2.6) and (II.2.7) (recall that we have assumed $p=1$ and $x_e=0$)

$$\begin{aligned} \overline{P}_1 &= B \\ \overline{P}_j &= (A\,\overline{P}_{j-1} \quad N\,\overline{P}_{j-1}) \qquad (j=2,3,\ldots) \\ P_j &= (\overline{P}_1 \quad \overline{P}_2 \ \ldots\ \overline{P}_j) \qquad (j=1,2,\ldots) \end{aligned} \tag{4.7}$$

$$\bar{Q}_1 = C$$

(4.8) $$\bar{Q}_i = \begin{pmatrix} \bar{Q}_{i-1} A \\ \\ \bar{Q}_{i-1} N \end{pmatrix} \qquad (i=2,3,\ldots)$$

$$Q_i = \begin{pmatrix} \bar{Q}_1 \\ \bar{Q}_2 \\ \vdots \\ \bar{Q}_i \end{pmatrix} \qquad (i=1,2,\ldots)$$

Then note that, by hypothesis, according to Definition (2.11) and formulae (4.1), (4.2), (4.3), we may write subsequently

(4.9) $$(S_{11} \;\; S_{12} \; \ldots \; S_{1k}) = (w_1 \;\; w_2 \; \ldots \; w_k) =$$

$$= (C\bar{P}_1 \;\; C(A\bar{P}_1 \;\; N\bar{P}_1) \; \ldots \; C(A\bar{P}_{k-1} \;\; N\bar{P}_{k-1})) = \bar{Q}_1 P_k$$

(4.10) $$(S_{21} \;\; S_{22} \; \ldots \; S_{2k}) = \begin{pmatrix} CA\bar{P}_1 & CA\bar{P}_2 & \ldots & CA\bar{P}_k \\ \\ CN\bar{P}_1 & CN\bar{P}_2 & \ldots & CN\bar{P}_k \end{pmatrix} =$$

$$= (\bar{Q}_2\bar{P}_1 \;\; \bar{Q}_2\bar{P}_2 \; \ldots \; \bar{Q}_2\bar{P}_k) = \bar{Q}_2 P_k$$

etc., that is

(4.11) $$\Sigma_{hk} = Q_h P_k$$

for all h and k. This enables us to assert that the infinite matrix Σ has finite rank.

Sufficiency. Consider firstly two special submatrices of $\Sigma_{M',M+1}$. The first one is formed by taking the dotted columns of

$$(4.12)\quad \Sigma_{M',M+1} = \begin{pmatrix} s_{11} & s^1_{12} & s^2_{12} & s^1_{13} & s^2_{13} & \cdots & s^1_{1,M+1} & s^2_{1,M+1} \\ s_{21} & s^1_{22} & s^2_{22} & s^1_{23} & s^2_{23} & \cdots & s^1_{2,M+1} & s^2_{2,M+1} \\ \vdots & \vdots & \vdots & \vdots & \vdots & & \vdots & \vdots \\ s_{M'1} & s^1_{M'2} & s^2_{M'2} & s^1_{M'3} & s^2_{M'3} & \cdots & s^1_{M',M+1} & s^2_{M',M+1} \end{pmatrix}$$

(see also (4.2)) and will be denoted with $\Sigma^1_{M'M}$. It is an easy matter to verify that $\Sigma^1_{M'M}$ coincides also with a suitable submatrix of $\Sigma_{M'+1,M}$; in fact, from the definition of $\{s_{ij}\}$ and Σ it is immediately seen that $\Sigma^1_{M'M}$ is formed by taking the dotted rows of

$$(4.13)\quad \Sigma_{M'+1,M} = \begin{pmatrix} s_{11} & s_{12} & \cdots & s_{1M} \\ s^1_{12} & s^1_{13} & \cdots & s^1_{1,M+1} \\ s^2_{12} & s^2_{13} & \cdots & s^2_{1,M+1} \\ s^1_{22} & s^1_{23} & \cdots & s^1_{2,M+1} \\ s^2_{22} & s^2_{23} & \cdots & s^2_{2,M+1} \\ \vdots & \vdots & & \vdots \\ s^1_{M'2} & s^1_{M'3} & \cdots & s^1_{M',M+1} \\ s^2_{M'2} & s^2_{M'3} & \cdots & s^2_{M',M+1} \end{pmatrix}$$

The submatrix constructed with the remaining columns of $\Sigma_{M',M+1}$, discarding the first one, is again of particular interest, and will be denoted with $\Sigma^2_{M'M}$. Similarly, this coincides with the submatrix of $\Sigma_{M'+1,M}$ constructed with the remaining rows, again discarding the first one.

Define now the linear spaces

$$U_M \triangleq R^{2^M - 1} \tag{4.14}$$

$$Y_{M'} \triangleq R^{2^{M'} - 1} \tag{4.15}$$

As a consequence of the special structure of matrices $\Sigma_{M',M}$, $\Sigma^1_{M',M}$ and $\Sigma^2_{M',M}$ we have the two commutative diagrams

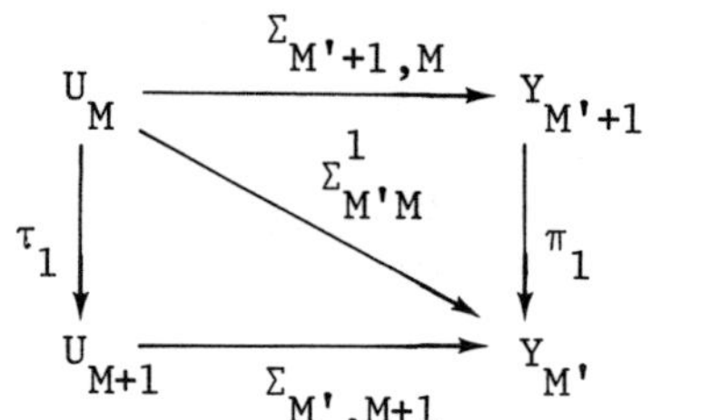

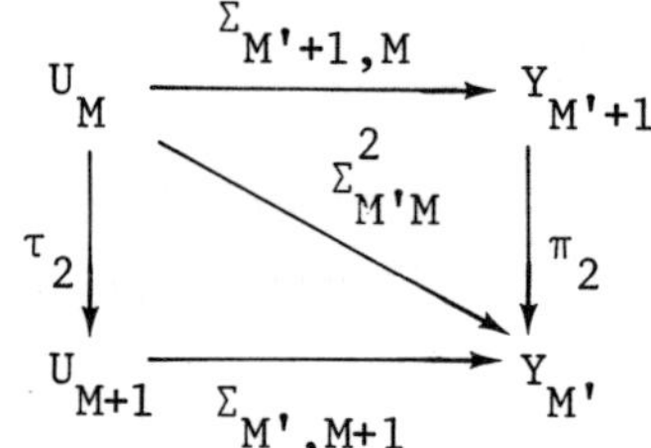

The existence of the matrices τ_1, π_1 (and τ_2, π_2) is implied by the relationships existing between $\Sigma^1_{M'M}$ (and $\Sigma^2_{M'M}$) and $\Sigma_{M',M+1}$, $\Sigma_{M'+1,M}$.

Moreover, if condition (4.6) is satisfied, there exist a matrix $\rho_{M'}$ and a matrix μ_M such that

$$\Sigma_{M'+1,M} = \rho_{M'} \Sigma_{M'M} \tag{4.16}$$

$$\Sigma_{M',M+1} = \Sigma_{M'M} \mu_M \tag{4.17}$$

This enables us to factor further the upper and lower edges of both diagrams. $\Sigma_{M'M}$ can, in turn, be factored "minimally" into the product

$$\Sigma_{M'M} = Q \cdot P \tag{4.18}$$

with the inner dimension equal to the rank of $\Sigma_{M'M}$. Let n denote this number and put

$$X \triangleq R^n \tag{4.19}$$

Therefore, from the previous commutative diagrams and from the factorizations (4.16), (4.17), (4.18), we get finally

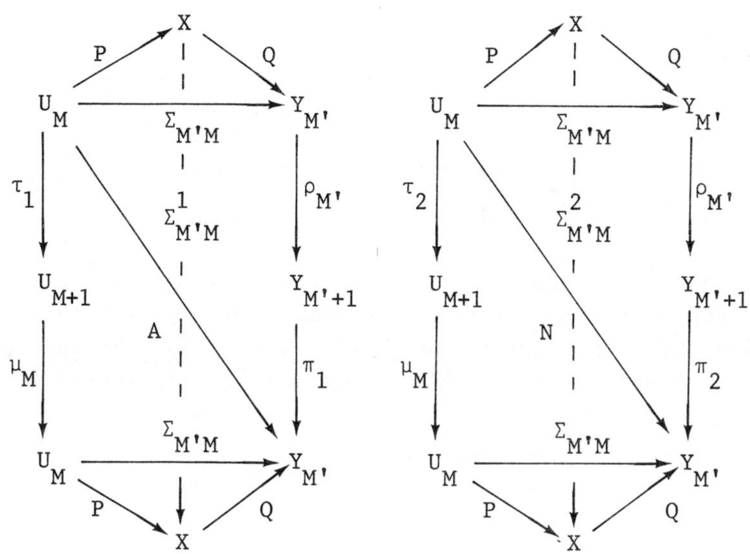

Since P (in the upper region) is onto and Q (in the lower region) in one-to-one, the application of Zeiger's fill-in Lemma [9] shows the existence of a unique matrix A (dashed arrow) such that the left-hand diagrm commutes and a unique matrix N such that the right-hand diagram commutes. Beside to matrices A and N thus defined, consider also the matrices B = first column of P, and C = first row of Q. Thus, P coincides with the matrix P_M constructed from A,N,B by means of (4.7) and Q coincides with the matrix $Q_{M'}$, constructed from A,N,C by means of (4.8). In fact, observe that the first column of P is B (or, what is the same, $\overline{P}_1$) by definition. From the left-hand portion of both diagrams we also obtain

(4.20) $A\overline{P}_1$ = first column of P $\mu_M\tau_1$ = second column of P

(4.21) $N\overline{P}_1$ = first column of P $\mu_M\tau_2$ = third column of P

where the second equalities are due to the special form of μ_M and τ_1,τ_2; therefore the matrix formed from the second and third columns of P coincides with $\overline{P}_2$. Again, on the basis of this result, we have

$$(4.22) \qquad A\overline{P}_2 = \text{second and third columns of } P\,\mu_M\tau_1 = \text{fourth and fifth columns of } P$$

$$(4.23) \qquad N\overline{P}_2 = \text{second and third columns of } P\,\mu_M\tau_2 = \text{sixth and seventh columns of } P$$

which proves that the matrix formed from the columns of P, from the fourth up to the seventh included, coincides with $\overline{P}_3$. By iterating this construction we obtain that $P = P_M$. The same procedure can be applied to prove that $Q = Q_{M'}$. In conclusion we have

$$(4.24) \qquad \Sigma_{M'M} = Q_{M'} \cdot P_M$$

and, also,

$$(4.25) \qquad \Sigma^1_{M'M} = Q_{M'}\,A\,P_M \quad \text{and} \quad \Sigma^2_{M'M} = Q_{M'}\,N P_M$$

These, by reversing the arguments used in the proof of the necessity, are sufficient to assert that the quadruplet (A,N,B,C) realizes all elements of $\Sigma_{M',M+1}$, i.e. the parameters $w_1,\dots,w_{M'+M}$.

But this,thanks to condition (4.6), is sufficient to prove that the quadruplet (A,N,B,C) realizes the whole sequence $\{w_j\}_1^\infty$. To this end one may apply A.J. Tether's continuation Lemma [10].

In this case, too, from the proof of the preceding Theorem,we can give a procedure for constructing a minimal bilinear realization of the sequence $\{w_j\}_1^\infty$. The corresponding steps are the following:

(a) choose M and M' such that (4.6) holds, and let $n = \text{rank}\ \Sigma_{M'M}$.

(b) find an $n\times(2^M-1)$ matrix P and a $(2^{M'}-1)\times n$ matrix Q such that

$$(4.26) \qquad Q\cdot P = \Sigma_{M'M}$$

(c) define the bilinear realization by

$$(4.27) \qquad A = (Q'Q)^{-1}\,Q'\,\Sigma^1_{M'M}\,P'(P\,P')^{-1}$$

(4.28) $$N = (Q'Q)^{-1} Q' \Sigma^2_{M'M} P' (P P')^{-1}$$

(4.29) $$B = \text{first column of } P$$

(4.30) $$C = \text{first row of } Q$$

where ' denotes the transpose.

5. Some concluding remarks about the realization problem

In the previous analysis we considered two different descriptions of the input/output map and, in both cases, we gave existence conditions and realization procedures. The first of these procedures presupposed the knowledge of a triplet of functions $\{F(t),G(t),H(t)\}$ that factorize the sequence $\{w_i(t_1,\ldots,t_i)\}_2^\infty$. From this we derived a bilinear realization and, then, we reduced it to minimal form on the basis of state space decomposition. The reduction to minimal form may also be preliminary accomplished on the triplet $\{F(t),G(t),H(t)\}$, thus obtaining a procedure which is in some sense complementary to the one described before. The discussion of this procedure requires a rather detailed analysis of the properties of such triplets and, for this reason, it will be given in Appendix I. It is also important to observe that the actual use of these procedures for computing a bilinear realization is directly related to the problems of finding algorithms for checking whether a given sequence $\{w_i(t_1,\ldots,t_i)\}_2^\infty$ satisfies condition (3.2) or not and, in the positive case, for computing a triplet $\{F(t),G(t),H(t)\}$.

The second realization procedure makes direct use of the parameters describing the input/output map and is closely connected with the realizability test. For these reasons it does not raises the aforementioned problems.

Before concluding we wish to point out that the input/output map, although formally described by an infinite sequence of kernels $\{w_i(t_1,\ldots,t_i)\}_1^\infty$ or parameters $\{w_j\}_1^\infty$, when is bilinearly realizable then is uniquely specified by a finite amount of data. In fact, as far as the kernels are concerned, Theorem (3.1) shows that they are identified by four matrices of functions with proper rational Laplace transform and, therefore, by a finite number of parameters (the coefficient of these rational functions). In the other case, Theorem (4.5) shows that the sequence $\{w_j\}_1^\infty$ is uniquely specified by the sub-sequence $\{w_j\}_1^{M'+M}$. This property allows an extension of the theory of partial realization also in the field of bilinear systems [6].

APPENDIX I - Factorizable sequences of kernels

The realizability condition expressed by Theorem (III.3.1) draws the attention upon the sequences of kernels that satisfy condition (III.3.2). An analysis of some general properties of such sequences, developed in [3], has revealed some interesting aspects that may be related to the theory of bilinear realization. For this reason we summarize here, without proofs, the most interesting of them.

First of all, it is convenient to introduce the

(A.1) Definition - A sequence $\{w_i(t_1,\ldots,t_i)\}_2^\infty$ of kernels is said to be factorizable if there exist three analytic functions F(t), G(t), H(t), respectively, m×m, m×1, 1×m, such that

$$w_i(t_1,\ldots,t_i) = H(t_i)F(t_{i-1}-t_i)\ldots F(t_2-t_3)G(t_1-t_2) \tag{A.2}$$

The triplet {F(t), G(t), H(t)} is called the factorization, and m its dimension. A factorization {F(t), G(t), H(t)} is said to be minimal when its dimension assumes the smallest value over all the factorization of the sequence.

An interesting characterization of minimality can be given in terms of suitable matrices constructed from the pair F(t), G(t) and, respectively, from the pair F(t), H(t) with formulae similar to (I.3.1) and (I.3.18). More precisely, define

$$\begin{aligned} &\overline{P}_1(t_1) = G(t_1) \\ &\overline{P}_i(t_1,\ldots,t_i) = F(t_i)\overline{P}_{i-1}(t_1,\ldots,t_{i-1}) \quad (i=2,3,\ldots) \\ &P_i[F,G] = [\overline{P}_1(t_1)\overline{P}_2(t_1,t_2)\ldots\overline{P}_i(t_1,\ldots,t_i)] \\ &\qquad (i=1,2,\ldots) \end{aligned} \tag{A.3}$$

and

$$\overline{Q}_i(t_1) = H(t_1)$$

$$\overline{Q}_i(t_1,\ldots,t_i) = \overline{Q}_{i-1}(t_1,\ldots,t_{i-1})F(t_i) \qquad (i=2,3,..)$$

(A.4)
$$Q_i[F,H] = \begin{pmatrix} \overline{Q}_1(t_1) \\ \overline{Q}_2(t_1,t_2) \\ \vdots \\ \overline{Q}_i(t_1,\ldots,t_i) \end{pmatrix} \qquad (i=1,2,\ldots)$$

Then it is possible to prove that

(A.5) Theorem - An m-dimensional factorization $\{F(t), G(t), H(t)\}$ of a factorizable sequence of kernels is minimal if and only if the rows of $P_m[F,G]$ and the columns of $Q_m[F,H]$ are linearly independent over the interval $\Delta_m = \{(t_1,\ldots,t_m): 0 \le t_k \le 1, \forall k\}$. All minimal factorizations are a single equivalence class, i.e. $\{F_1(t),G_1(t),H_1(t)\} \sim \{F_2(t),G_2(t),H_2(t)\}$ if and only if

(A.6)
$$F_1(t) = TF_2(t)T^{-1}$$
$$G_1(t) = TG_2(t)$$
$$H_1(t) = H_2(t)T^{-1}$$

where T is a constant m×m nonsingular matrix.

In order to derive a minimal realization from a given non minimal one, we may use the following

(A.7) Lemma - If only $\tilde{m}<m$ rows of $P_m[F,G]$ {or, respectively only $\tilde{m}<m$ columns of $Q_m[F,H]$} are linearly independent over Δ_m, then there exists a constant non-singular m×m matrix T such that

(A.8)
$$T F(t) T^{-1} = \begin{pmatrix} F_{11}(t) & F_{12}(t) \\ 0 & F_{22}(t) \end{pmatrix}, \quad TG(t) = \begin{pmatrix} G_1(t) \\ 0 \end{pmatrix}$$

$$\{\text{or, } T\,F(t)\,T^{-1} = \begin{pmatrix} F_{11}(t) & 0 \\ F_{21}(t) & F_{22}(t) \end{pmatrix}, \quad H(t)T^{-1} = \begin{bmatrix} H_1(t) & 0 \end{bmatrix}\}$$

where the matrix $F_{11}(t)$ is $\tilde{m}\times\tilde{m}$ and $G_1(t)$ is $\tilde{m}\times 1$ $\{H_1(t)$ is $1\times\tilde{m}\}$. Moreover, $P_m[F_{11},G_1]$ has its $\tilde{m}$ rows $\{Q_m[F_{11},H_1]$ has its $\tilde{m}$ columns$\}$ linearly independent over Δ_m.

Thanks to this lemma it is quite easy to reduce any triplet $\{F(t),G(t),H(t)\}$ to a form that shows the typical pattern of zeros usually displayed by the canonical forms in linear systems. On the basis of this characterization of non-minimal factorizations it is possible to prove the following

(4.9) Theorem - Let $\{w_i(t_1,\ldots,t_i)\}_1^\infty$ be a sequence of bilinearly realizable kernels, and let $\{F_o(t),\ G_o(t),\ H_o(t)\}$ be a minimal factorization of the subsequence $\{w_i(t_1,\ldots,t_i)\}_2^\infty$. Then the bilinear realization associated, be means of (III.3.8), (III.3.9), (III.3.11) to a minimal linear realization of

$$\begin{pmatrix} w_1(t) & H_o(t) \\ G_o(t) & F_o(t) \end{pmatrix} \tag{A.10}$$

is minimal.

From this it follows a procedure for computing a minimal realization, alternative to the one described in section 3 of Chapter III. The steps are the following

(a) find a minimal factorization $\{F_o(t),\ G_o(t),\ H_o(t)\}$ from the given factorization $\{F(t),G(t),H(t)\}$

(b) arrange $w_1(t)$, $F_o(t)$, $G_o(t)$, $H_o(t)$ as in (A.10)

(c) find a minimal linear realization A,R,S of (A.10)

(d) compute the minimal bilinear realization A,N,B,C by the formulae (III.3.8), (III.3.9), (III.3.11).

We note again that the whole procedure operates through standard linear algebraic methods (reduction of the factorization) and techniques of linear realization.

APPENDIX II - A reduction algorithm

This is an algorithm for reducing a bilinear realization $(A,N_1,\dots,N_p,B,C;x_e)$ to minimal form, based on Remark (II.3.9).

The algorithm consists in performing elementary row and column operations on the matrix

$$[B_{eq}(A)_1(N_1)_1\dots(N_p)_1\dots(A)_n(N_1)_n\dots(N_p)_n] \tag{A.11}$$

where $B_{eq} = (B\, x_e)$ and $(M)_i$ denotes the i-th column of a matrix M. It is assumed that $B_{eq} \neq 0$. Let T_o be an n×n nonsingular matrix that triangularizes B_{eq}, i.e. that reduces B_{eq} to the form

$$T_o B_{eq} = \begin{pmatrix} B_o \\ 0 \end{pmatrix} \tag{A.12}$$

where B_o is k×(p+1) and of rank k, and replace matrix (A.11) with the matrix

$$[T_o B_{eq}(T_o A T_o^{-1})_1(T_o N_1 T_o^{-1})_1\dots(T_o N_p T_o^{-1})_1\dots$$
$$\dots(T_o A T_o^{-1})_n(T_o N_1 T_o^{-1})_n\dots(T_o N_p T_o^{-1})_n] \tag{A.13}$$

From this matrix carry out the following cycle of steps:

(a) If the first k(p+1) columns of the (n-k)×(np+n) submatrix in the lower rigth hand corner of (A.13) are zero, the process terminates. If not, let h_1 denote the first (from the left) column of (A.13) whose lower (n-k)×1 subcolumn is non zero.

(b) Perform on matrix (A.12) elementary row operations to bring a nonzero element of this subcolumn in the position $(k+1,h_1)$ and to annihilate all the elements $(k+2,h_1),\dots,(n,h_1)$. Let T_1 be the n×n nonsingular matrix that performs these operations.

(c) Replace matrix (A.13) with the matrix

$$[\tilde T_1 B_{eq}(\tilde T_1 A \tilde T_1^{-1})_1(\tilde T_1 N_1 \tilde T_1^{-1})_1\dots(\tilde T_1 N_p \tilde T_1^{-1})_1\dots$$
$$\dots(\tilde T_1 A \tilde T_1^{-1})_n(\tilde T_1 N_1 \tilde T_1^{-1})_n\dots(\tilde T_1 N_p \tilde T_1^{-1})_n] \tag{A.14}$$

where $\tilde{T}_1 = T_1 T_o$. It is readily verified that this operation does not destroy the pattern of zeros generated previously (i.e. an $(n-k)\times(h_1-1)$ zero submatrix in the lower left-hand corner and an $(n-k-1)\times 1$ zero subcolumn in the lower edge of column h_1).

Proceed in the same way as in (a), (b) and (c) on matrix (A.14) replacing k with k+1, h_1 and T_1 with suitable h_2 and T_2 and $\tilde{T}_1$ with $\tilde{T}_2 = T_2\tilde{T}_1$, etc. When the procedure terminates, at some cycle $m \leq n-k+1$, the overall transformation $\tilde{T}_{m-1}$ is such that $\tilde{T}_{m-1}B$, $\tilde{T}_{m-1}x_e$, $\tilde{T}_{m-1}A\tilde{T}^{-1}_{m-1}$, $\tilde{T}_{m-1}N_1\tilde{T}^{-1}_{m-1}, \ldots, \tilde{T}_{m-1}N_p\tilde{T}^{-1}_{m-1}$ assume the form

$$\text{(A.15)} \qquad \begin{pmatrix} A_{11} & A_{12} \\ 0 & A_{22} \end{pmatrix}, \ldots, \begin{pmatrix} N^{(i)}_{11} & N^{(i)}_{12} \\ 0 & N^{(i)}_{22} \end{pmatrix}, \ldots, \begin{pmatrix} B_1 \\ 0 \end{pmatrix}, \begin{pmatrix} x_{e1} \\ 0 \end{pmatrix}$$

The upper left-hand blocks are $\nu\times\nu$ ($\nu\times p$ and $\nu\times 1$ for the last two) while $\nu = k+m-1$.

It is now possible to prove that the state space of the bilinear system corresponding to the set of matrices $A_{11}, (N_1)_{11}, \ldots, (N_p)_{11}, B_1$, is spanned by the states reachable from x_{e1}. To this end it is sufficient to construct, from the aforementioned set of matrices, the matrix P_ν defined by eqs. (II.2.6) and to verify that its ν rows are linearly independent (*).

In fact, the first k rows of $\overline{P}_1$ are linearly independent and the remaining ones are zero. To examine the linear dependence of the rows in $(\overline{P}_1\ \overline{P}_2)$ it is sufficient therefore to examine that of the last $\nu-k$ rows of $\overline{P}_2$. In fact, denoting with $\tilde{A}, \tilde{N}_1, \ldots, \tilde{N}_p$ the $(\nu-k)\times k$ submatrices in the lower left-hand corner of each matrix A_{11}, $(N_1)_{11}, \ldots, (N_p)_{11}$, we have that

$$\text{(A.16)} \qquad (\overline{P}_1\ \overline{P}_2) = \begin{pmatrix} B_o & \cdot & \cdot & \cdots & \cdot \\ & & & & \\ 0 & \tilde{A}B_o & \tilde{N}_1 B_o & \cdots & \tilde{N}_p B_o \end{pmatrix}$$

By construction the first rows, say k_1 rows ($k_1 \geq 1$), of the matrix $(\tilde{A}\ \tilde{N}_1\ \ldots\ \tilde{N}_p)$ are linearly independent, while the remaining

(*) By Corollary (I.4.21), we may assume $\overline{P}_1 = (B\ x_e)$.

ones are zero. Therefore also the first k_1 rows of $(\tilde{A}B_o\ \tilde{N}_1B_o..\tilde{N}_pB_o)$ are linearly independent while the remaining ones are zero. From this we can conclude that the first $k+k_1$ rows of $(\overline{P}_1\ \overline{P}_2)$ are linearly independent and that the last $\nu-k-k_1$ are zero. By iterating this procedure we can prove that $(\overline{P}_1\overline{P}_2 \dots \overline{P}_\nu)$ has rank ν.

We observe that the present algorithm generalizes, to the case of bilinear systems, the one proposed by H.H. Rosembrock for reducing linear systems [11].

REFERENCES

[1] R.R.MOHLER, Bilinear control processes with applications to engineering, ecology and medicine, Academic Press, New York (1973).

[2] C.BRUNI, G.DI PILLO, G.KOCH, Bilinear systems: an appealing class of "nearly linear" systems in theory and applications. IEEE Trans. Automatic Control, AC 19 (1974), to appear.

[3] P.d'ALESSANDRO, A.ISIDORI, R.RUBERTI, Realization and structure theory of bilinear dynamical systems, SIAM J. Control, 12 (1974), to appear.

[4] P.d'ALESSANDRO, A.ISIDORI, R.RUBERTI, Lectures an Bilinear System Theory, Notes for a Course held at C.I.S.M., Udine (Italy), Springer Verlag (Wien), 1972.

[5] R.W.BROCKETT, On the algebraic structure of bilinear systems, Theory and Applications of Variable Structure Systems (R.R. Mohler and A. Ruberti, eds.), Academic Press, New York (1972), pp. 153-168.

[6] A.ISIDORI, Direct construction of minimal bilinear realizations from nonlinear input/output maps, IEEE Trans. Automatic Control, AC 18 (1973), to appear.

[7] A.ISIDORI, The computation of reduced forms for bilinear systems, Ricerche di Automatica, 3 (1972), pp. 296-299.

[8] C.BRUNI, G.DI PILLO, G.KOCH, On the mathematical models of bilinear systems, Ricerche di Automatica, 2 (1971), pp. 11-26.

[9] R.E.KALMAN, P.L.FALB, M.A.ARBIB, Topics in mathematical system Theory, Mc Graw Hill (New York), 1969, pp. 288-308.

[10] A.J.TETHER, Construction of linear state-variable models from finite input/output data, IEEE Trans. Automatic Control, AC 15 (1970), pp. 427-436.

[11] H.H.ROSENBROCK, Computation of minimal representation of rational transfer function matrix, Proc. IEE, 115 (1968), pp. 325-7.

AN INTRODUCTION TO STOCHASTIC DIFFERENTIAL EQUATIONS ON MANIFOLDS

J.M.C. Clark

Department of Computing and Control

Imperial College, London

These notes introduce the mathematical apparatus that is relevant for an understanding and precise description of the idea of a dynamical system driven by white noise. The main tools are the stochastic integral and stochastic differential equations of Ito; however the representations of Fisk and Stratonovich are also included, not only because they have a nice physical interpretation, but because they seem to provide a natural formalism for processes that lie on manifolds. The treatment is sketchy in places but we have tried to give at least the gist of the main arguments. Basic references are Ito [1,2,3], McKean [4] and Wong [5].

1. BROWNIAN MOTION

All the stochastic processes considered in these notes will be functionals of the process known as Brownian motion or, as it is sometimes called, the Wiener process. This is a process $w(\cdot) = \{w(t), t \geq 0\}$ satisfying the following properties

(1) $w(0) = 0$ and increments $w(t_2)-w(t_1)$, $w(t_3)-w(t_2)$, $w(t_4)-w(t_3),\ldots$, for $t_1 < t_2 < t_3 < t_4\ldots$ are independent.

(2) $w(t)-w(s)$ is normal with zero mean and variance $|t-s|$: along with the previous property this implies $w(\cdot)$ is Gaussian with zero mean and covariance function $E[w(t)w(s)] = \min(t,s)$.

(3) The sample paths $w(\cdot)$ are continuous a.s. (almost surely).

One of the simplest constructions of Brownian motion is based on the following idea [6]: the distributions of a Gaussian process of zero mean is determined by its covariance function and the formal derivative $\dot{w}(\cdot)$ of $w(\cdot)$ has $\delta(t-s)$ as its covariance function. If $\dot{w}(\cdot)$ can be expressed as an expansion of orthonormal functions $\psi_n(t)$ of $L^2[0,1]$ with random coefficients;

$$\dot{w}(t) = \sum_{n=0}^{\infty} a_n \psi_n(t),$$

then necessarily

$$\begin{aligned} E[a_m a_n] &= \int_0^1 \int_0^1 E[\dot{w}(t)\dot{w}(s)\psi_n(t)\psi_m(s)dt\, ds \\ &= \int_0^1 \int_0^1 \delta(t-s)\psi_n(t)\psi_m(s)dt\, ds \\ &= \delta_{mn} \end{aligned}$$

w(t) would then be the indefinite integral of the expansion. The precise version is as follows.

<u>Proposition (1)</u> If $\psi_n(t)$ is a complete orthonormal set of $L^2[0,1]$ and $a_0, a_1, \ldots$ is a sequence of independent normal random variables with zero mean and unit variance, then the series

$$w(t) = \sum_0^{\infty} a_n \int_0^t \psi_n(s)ds$$

is uniformly convergent a.s. for t in [0,1] and the limit is a Brownian motion.

<u>Outline of Proof</u> First observe that

$$\int_0^t \psi_n(s)ds = \langle I_t, \psi_n \rangle$$

where $\langle .,. \rangle$ is the inner product on $L^2[0,1]$ and

$$\begin{aligned} I_t(s) &= 1 \quad 0 \le s \le t \\ &= 0 \quad t < s \le 1. \end{aligned}$$

If $w^N(t) = \sum_0^N a_n \int_0^t \psi_n(s)ds$,

$$\begin{aligned} E[(w^N(t)-w^M(t))^2] &= \sum_{N+1}^{M} \langle I_t, \psi_n \rangle^2 \\ &\to 0 \quad M,N \to \infty. \end{aligned}$$

So $w^n(t)$ is a quadratic mean Cauchy sequence and has a limit w(t). But Parseval's relation implies

$$\begin{aligned} E[w(t)w(s)] &= \sum_0^{\infty} \langle I_t, \psi_n \rangle \langle I_s, \psi_n \rangle \\ &= \langle I_t, I_s \rangle \\ &= \min(s,t) \end{aligned}$$

So at least the limit process $w(\cdot)$ has the right covariance function. However inspection of the limits of the joint characteristic functions of $w^N(\cdot)$ shows, by an extension of this argument, (see [6]), the limit process has joint normal distributions, from which (1) and (2) follow. The continuity of the sample paths would follow from the a.s. uniform convergence of $w^N(.)$. The proof of this, carried out by Nisio for the general case, is rather more difficult; we give the proof of Levy and Cielsielski for the special case where the $\{Q_n\}$ are Haar functions. These are, for $2^n \leq k < 2^{n+1}$,

$$\begin{aligned} H_k(t) &= 2^{\frac{1}{2}n}, \quad 0 \leq k-2^n \leq 2^n t < k-2^n+\tfrac{1}{2} \\ &= -2^{\frac{1}{2}n} \quad k-2^n+\tfrac{1}{2} \leq 2^n t < k-2^n+1 \\ &= 0 \quad \text{otherwise.} \end{aligned}$$

The integrals $\Delta_k(t) \equiv \int_0^t H_k(s)ds$ are little tents sitting on intervals of length $1/2^n$, and of height $2^{-\frac{1}{2}n}$. So

$$\sum_{2^n \leq k < 2^{n+1}} |a_k|\ \Delta_k(t) \leq 2^{-\frac{1}{2}n} \max\{|a_k| : 2^n \leq k < 2^{n+1}\}$$

Denote the right-hand side by R_n. If we can show that $\sum_0^\infty R_n < \infty$ a.s. the series $\sum_0^\infty a_k \Delta_k(t)$ would converge <u>absolutely</u> uniformly a.s. Now by integration by parts

$$\begin{aligned} P\{|a_k| > b\} &= \frac{2}{\sqrt{\pi}} \int_b^\infty e^{-\frac{1}{2}x^2} dx \\ &< b^{e^{-\frac{1}{2}b^2}} \end{aligned}$$

and if $b = 2(\log k)^{\frac{1}{2}}$ for $k > 3$

$$P\{|a_k| > 2(\log k)^{\frac{1}{2}}\} < k^{-2}$$

and the series $\sum_1^\infty P\{|a_k| > 2(\log k)^{\frac{1}{2}}\}$ converges.

It follows from the Borel-Cantelli lemma that

$$P\{a_k < 2(\log k)^{\frac{1}{2}} \text{ for all } k \text{ sufficiently large}\} = 1$$

and since $R_n < 2^{-\frac{1}{2}n} \max\{|a_k|;\ k \leq 2^{n+1}\}$, that

$$P\{R_n < 2.2^{-\frac{1}{2}n}((n+1)\log 2)^{\frac{1}{2}} \text{ for } n \text{ sufficiently large}\} = 1.$$

So $\sum_0^\infty R_n$ converges and $w^N(\cdot)$ converges uniformly a.s. to $w(.)$.

Brownian motion on $[0,\infty)$ is obtained by piecing together independent Brownian motions on $[0,1], [1,2], \ldots,$. A vector Brownian motion is just a vector of independent Brownian motions.

Three further properties of Brownian motion are important in what follows. The first is that the quadratic variation of $w(\cdot)$ on $[0,t]$ is t. That is, if

$$Q_t^n \equiv \sum_{0 \leq k < 2^n t} [w(\tfrac{k+1}{2^n}) - w(\tfrac{k}{2^n})]^2$$

Then $Q_t^n \to t$ a.s.. For Q_t^n is a sum of independent random variables, each of which is independent of the others and has mean $1/2^n$ and variance $3/2^{2n}$. So if $t_n = \max\{k/2^n < t\}$,

$$E(Q_t^n - t_n)^2 = 2^n t_n \left(\frac{3}{2^{2n}} - \frac{1}{2^{2n}}\right) = \frac{2t_n}{2^n}$$

By Chebychev's inequality, for any $\varepsilon > 0$

$$\sum_1^\infty P\{|Q_t^n - t_n| > \varepsilon\} \leq \sum_1^\infty \frac{1}{\varepsilon^2} E(Q_n - t_n)^2$$
$$< \infty$$

and so by the Borel-Cantelli lemma $|Q_t^n - t_n| < \varepsilon$ for all but a finite number of n almost surely. Since ε is arbitrary Q_t^n converges to $t = \lim_n t_n$ a.s..

The second property of interest is that $w(\cdot)$ has infinite length (i.e. total variation) on $[0,t]$. This can be deduced from the previous property and the inequality

$$Q_t^n \leq \max_{k<2^n t} \left|w\left(\frac{k+1}{2^n}\right) - w\left(\frac{k}{2^n}\right)\right| . \sum_{k<2^n} \left| w\left(\frac{k+1}{2^n}\right) - w\left(\frac{k}{2^n}\right)\right|$$

Since $Q_t^n \to t$ a.s. and by the continuity of the sample paths of $w(\cdot)$, $\max_k |w(k+1/2^n) - w(k/2^n)| \to 0$ a.s., the limit inferior of the sum on the right of this inequality, and therefore the total variation, is infinite.

Finally the third property is that Brownian motion starts afresh at stopping times. We follow the description of McKean [4,p.10]. Suppose $\underline{\underline{F}}_t$ is the smallest σ-field for which the family $\{w(s),\ 0 \leq s \leq t\}$ are measurable A non-negative random variable T is a stopping time if $\{T < t\}$ is $\underline{\underline{F}}_t$-measurable. Constant time is a stopping time and also first passage times such as $\min\{t : w(t) \geq c\}$. Now define the random σ-field $\underline{\underline{F}}_T$ composed of all events A such that $A \cap \{T < t\} \in \underline{\underline{F}}_t$ for all t. If t is constant $\underline{\underline{F}}_{t+} = \bigcap_{\delta>0} \underline{\underline{F}}_{t+\delta}$. Roughly $\underline{\underline{F}}_{T+}$ is all the events up to (and just beyond) T. Both T and w(T) are $\underline{\underline{F}}_{T+}$-measurable. Brownian motion w(t) starts afresh at T in the sense that $w^+(t) = w(T+t)-w(T)$ is a Brownian motion independent of $\underline{\underline{F}}_{T+}$. This result is due to Dynkin and Hunt.

2. ITO INTEGRALS

Suppose w_t is a Brownian motion (we write w_t or $w(t)$ interchangeably). Then one cannot define sample path Stieltjes integrals of the form $\int_0^t \phi_s\, dw_s$ because the sample paths have infinite total variation. However it is possible to show that for certain classes of integrands especially selected Riemann sums will converge to a random variable, and this is how the Ito integral is constructed.

We consider three classes of integrand functions. A process $g = \{g_t,\ 0 \leq t \leq b\}$, jointly measurable on $\underline{\underline{B}} \times \underline{\underline{F}}_b$, ($\underline{\underline{B}}$ the Borel field on $[0,b]$) will be said to belong to L_p^2 if:

(1) it is nonanticipating; that is, g_t is $\underline{\underline{F}}_t$-measurable for each t,

(2) $\int_0^b g_t^2\, dt < \infty$ a.s.

The other classes are L^2, the class of g in L_p^2 such that

(3) $\int_0^b E\, g_t^2\, dt < \infty$,

and S, the class of step g in L^2 such that

(4) for some partition $T_n : 0 = t_1 < t_2 < \ldots < t_{n+1} = b$

$$g_t = g_{t_i} \text{ for } t_i \leq t < t_{i+1} .$$

The Ito integral for g in S is simply

$$I_b(g) = \sum_{i=1}^{n} g_{t_i}(w_{t_{i+1}} - w_{t_i})$$

With refinement of the partition the integral is not altered and so $I_b(g)$ is a linear random functional over S. Moreover since g_t is nonanticipating

$$E[I_b(g)] = E[\textstyle\sum_i E[g_{t_i}(w_{t_{i+1}} - w_{t_i}) \mid \underline{\underline{F}}_{t_i}]] = 0$$

and

$$E[I_b^2(g)] = \textstyle\sum_{ij} E[g_{t_i} g_{t_j}(w_{t_{i+1}} - w_{t_i})(w_{t_{j+1}} - w_{t_j})]$$

$$= \textstyle\sum_i E\, g_{t_i}^2 (t_{i+1} - t_i)$$

$$= \int_0^b E\, g_t^2\, dt$$

So $I_b(g)$ is an isometry between the Hilbert spaces L^2 with norm $\int_0^b E\, g_t^2 dt$ and H the space of random variables with finite second moment, and the domain of definition of I_b can be extended to the closure in L^2 of S. It turns out this is L^2 itself. Suppose $g \in L^2$ and is continuous in t. If $g_t^n \equiv g_{k/n}$, $k/n \leq t < k+1/n, k<nb$. Then $\int_0^b E(g_t - g_t^n)^2 dt \to 0$ and S is dense in the set of such g. But for any $g \in L^2$, the sequence of continuous g

$$g_t^m = \int_0^t m e^{m(t-s)} g_s ds$$

converges to g_t and so S is also dense in L^2. We can easily establish the following properties for $g \in L^2$

(5) $E\, I_b^2(g) = \int_0^b E\, g_t^2\, dt$.

(6) The indefinite integral $I_t(g) \equiv \int_0^b J(s \leq t) g_s dw_s$ where J(A) is the indicator function of A, is a <u>martingale</u>; that is, $I_t(g)$ is $\underline{\underline{F}}_t$-measurable and

$$E[I_t(g) \mid \underline{\underline{F}}_s] = I_s(g) \quad \text{a.s.} \quad s \leq t$$

For the extension of I to L_p^2 we introduce a metric on L_p^2 that is equivalent to convergence of the $L^2[0,b]$ norm in probability:

$$\rho_1(g) \equiv \inf\{\varepsilon : P(\textstyle\int_0^b g_t^2\, dt > \varepsilon) < \varepsilon\} , \; g \in L_p^2 ,$$

and a metric on H_o, the space of finite random variables, that is equivalent to convergence in probability:

$\rho_2(X) \equiv \inf\{\varepsilon : P(|X| > \varepsilon)<\varepsilon\}, X \in H_o$.

Then L^2 is dense in (L^2_p,ρ_1) and H is dense in (H_o,ρ_2). The second is obvious; to see the first, let J(A) denote the indicator function of the event A. Suppose $g \in L^2$ and $\rho_1(g) = \varepsilon$. Then by Chebyshev's inequality and property (5) of $I_b(g)$:

$$P(I_b(g) > \delta) \leq P(I_b(g) > \delta, \int_0^b g_t^2 \, dt < \varepsilon)+P(\int_0^b g_t^2 \, dt > \varepsilon)$$

$$\leq \frac{1}{\delta^2} E[I_b(g)J(\int_0^b g_t^2 \, dt < \varepsilon)]^2 + \varepsilon \qquad \dagger$$

$$\leq \frac{\varepsilon^2}{\delta^2} +\varepsilon$$

If $\delta = \varepsilon^{\frac{1}{2}} + \varepsilon$, the probability of the left $< \delta$, and so

$$\rho_2(I(g)) \leq \rho_1^{\frac{1}{2}}(g) + \rho_1(g)$$

and the integral can be continuously extended over the ρ_1-closure of L^2. This, however, is L^2_p, because for any $g \in L^2_p$, the sequence g^n:

$$g_t^n = g_t \; J(\int_0^t g_s^2 < n)$$

belongs to L^2 and obviously converges in ρ_1 to g.

Two other properties will be useful $(g \in L^2_p)$:

(7) The indefinite integral $I_t(g)$ can be defined to be continuous in t.

(8) For a stopping time $T \leq b$,

$$\int_0^T g_t dw_t = \int_0^b g_t \; J(t \leq T) dw_t \; .$$

Both of these results which are clear enough for step g and T taking discrete values, can be proved by a uniform convergence argument (see e.g. [6]p.11), but besides the Borel-Cantelli lemma this requires a martingale inequality of Doob's, which would take us too far afield to include. Note finally that b can be taken to be ∞ .

As an example of a stochastic integral consider $\int_0^t w_s dw_s$. By (5) $E[\int_0^t w_s dw_s] = 0 \neq E[\frac{1}{2} w_t^2]$ as one might expect. So the ordinary rules of calculus do not hold. In fact by direct computation we see that for $t = (m+1)/2^n$,

$$\tfrac{1}{2} w_t^2 = \sum_{k=0}^m w(k/2^n) D_k + \tfrac{1}{2}\sum_k D_k^2$$

where $D_k = w(k+1/2^n) - w(k/2^n)$; the first term clearly converges as $n \to \infty$ to $\int_0^t w_s dw_s$ and the second term to the quadratic variation $\frac{1}{2}t$. So

(9) $\int_0^t w_s dw_s = \frac{1}{2} w_t^2 - \frac{1}{2} t$.

<u>Stochastic differentials and Ito's rule</u> If f and g are non-anticipating processes that are respectively integrable and

† It can be shown that

$$\int_o^b g_t dw_t . J(\int_o^b g_t^2 dt<\varepsilon) = \int_o^b g_t J(\int_o^t g_s^2 ds<\varepsilon) \, dw_t . J(\int_o^b g_t^2 dt<\varepsilon)$$

square integrable almost surely in t, and x_o is an $\underline{\underline{F}}_o$-measurable variable, we can define the <u>integral</u> process

(10a) $$x_t = x_o + \int_0^t f_s ds + \int_0^t g_s dw_s$$

This expression is often written as a <u>stochastic differential</u>

(10b) $$dx_t = f_t dt + g_t dw_t$$

We can define integrals with respect to integral processes in an obvious fashion: if h is a nonanticipating process such that $h_t f_t$ is integrable a.s. and $h_t g_t$ is square-integrable a.s. we can define

$$\int_0^t h_s dx_s \equiv \int_0^t h_s f_s ds + \int_0^t h_s g_s dw_s$$

In particular since an integral process y is itself a nonanticipating process that is continuous, we can define the process $\int_0^t y_s\, dx_s$ which is also an integral process. Ito's rule shows that C^2 functionals of integral processes are integral processes.

<u>Ito's rule</u> Suppose $x_t = x_o + \int_0^t f_s ds + \int_0^t g_s dw_t$ is a vector of integral processes. Suppose h(t,x) is a function on $[0,\infty) \times R^n$ with continuous partial derivates $\partial h/\partial t$, $\partial h/\partial x^i$, $\partial^2 h/\partial x^i \partial x^j$. Then

(11) $$dh(t,x_t) = \frac{\partial h}{\partial t} dt + \sum_i \frac{\partial h}{\partial x^i} dx_t^i + \tfrac{1}{2} \sum_{i,j} \frac{\partial^2 h}{\partial x^i \partial x^j} (dx^i, dx^j)_t$$

where the terms are evaluated at (t,x_t), and $(dx^i,dx^j)_t$ denotes $g_t^i g_t^j dt$.

The proof is given in [1,p.187], but see also the abbreviated proof in [6]. The result also holds if $P(x_t \in B \text{ all } t) = 1$ where B is some open set and the derivatives of h with respect to x exist only on that set. If x depends on a vector of Brownian motions and g is a matrix then $(dx^i,dx^j)_t$ is to be interpreted as $\sum_k g_t^{ik} g_t^{jk} dt$. The formal rule to obtain this term is: expand the product of differentials, replace the terms $dw_t^k\, dt$, $dw_t^k dw_t^\ell$, $k \neq \ell$, by zero and the terms $(dw_t^k)^2$ by dt.

The simplest application of this is the product rule. If x,y are two integral processes

(12) $$d(x_t y_t) = x_t dy_t + y_t dx_t + (dx,dy)_t$$

in particular we get the differential form of (9): $d(w_t^2) = 2w_t dw_t + dt$.

It is a simple exercise in Ito's rule to prove that the stochastic differential equation

$$dx_t = x_t dw_t \qquad x_o = 1$$

has one and only one solution. First verify that $x_t = \exp(w_t - \tfrac{1}{2}t)$ is a solution; suppose y_t is another. Since $x_t > 0$ a.s. for all t, we can expand by Ito's rule

$$d\left(\frac{y_t}{x_t}\right) = \frac{dy_t}{x_t} - \frac{y_t dx_t}{x_t^2} - \frac{(dx,dy)_t}{x_t^2} + y_t \frac{(dx,dx)_t}{x_t^3}$$

Setting $dy_t = y_t dw_t$, $dx_t = x_t dw_t$, we see that $d(y_t/x_t) \equiv 0$ for all t.

3. STOCHASTIC DIFFERENTIAL EQUATIONS

We consider the processes that are solutions of the stochastic differential equation

(1) $$dx_t = f(x_t)dt + g(x_t)dw_t$$

where w_t may be taken to be a vector Brownian motion and f and g vector and matrix functions. A solution x_t of this equation is a diffusion process; that is, a continuous process with the strong Markov property that conditioned on the value at a stopping time future values are independent of the past. We clarify this later.

Proposition (1) Suppose x_o, $E\, x_o^2 < \infty$; is independent of $\{w_t, t \geq 0\}$ and for some constant K and continuous function K(x,y),

(2) Local Lipschitz condition
$$||f(x)-f(y)|| + ||g(x)-g(y)|| < K(x,y)||x-y|| \text{ all } x,y,$$

(3) Linear growth
$$||f(x)||^2 + ||g(x)||^2 \leq K^2(1 + ||x||^2),$$
where $||\cdot||$ are appropriate Euclidean norms. Then (1) has a unique solution x_t with the properties

(4) x_t is $\underline{\underline{F}}_t$-measurable where $\underline{\underline{F}}_t$ is the σ-field generated by $\{x_o, w_s;\ 0 \leq s \leq t\}$

(5) $$\int_0^t E\, x_s^2\, ds < \infty$$

(6) x_t is a diffusion process.

Outline of proof. We consider the scalar case and assume K(x,y) is bounded by K. The solution is constructed by successive approximation. Let

$$x_t^o = x_o\ , \quad t \geq 0$$
$$x_t^1 = x_o + \int_0^t f(x_s^o)ds + \int_0^t g(x_s^o)dw_s$$
$$\ldots = \ldots$$
$$x_t^{n+1} = x_o + \int_0^t f(x_s^n)ds + \int_0^t g(x_s^n)dw_s$$

It is easy to verify using the linear growth condition that each process x_t^n is continuous in t and satisfies (4) and (5). Then let

$$D_t^n = E(x_t^{n+1} - x_t^n)^2 \quad n = 0,1,2,\ldots$$

Since

$$x_t^{n+1} - x_t^n = \int_0^t f(x_s^n) - f(x_s^{n-1})\, ds$$
$$+ \int_0^t g(x_s^n) - g(x_s^{n-1})\, dw_s$$

it follows from the Lipschitz condition, the inequalities $(A+B)^2 \leq 2A^2 + 2B^2$, $(\int_0^t hs\, ds)^2 \leq t \int_0^t h_s{}^2\, ds$ and the properties of the stochastic integral that for $0 \leq t \leq b$

$$D_t^n \leq c_o \int_0^t D_s^{n-1}\ , \qquad c_o \equiv 2K^2(1+b).$$

Since $D_t^o \leq c_1 t$ where $c_1 = 2K^2(1+b)E\ x_o^2$, repeated integration gives

$$D_t^n \leq c_1 \frac{(c_o t)^n}{n!}$$

from which we can deduce that x_t^n is a Cauchy sequence and converges in quadratic mean uniformly in t to a $\underline{\underline{B}} \times \underline{\underline{F}}_b$-measurable process x_t^* satisfying (4) and (5). Gronwall's lemma shows that the solution is unique up to null sets. Substitution of x* into the right side of (1) gives a second version x_t of the limit which is continuous in t and which therefore solves the equation.

It remains to show x_t is a diffusion process. Suppose T is a stopping time of x_t; i.e. $(T < t) \in \underline{\underline{F}}_t$. Let x_t^+ be the solution of (1) with $x_o = x_T$ and w_t replaced by $w_t^+ = w_{T+t} - w_T$. The crucial point is that x_{T+t} is also a solution and so $x_{T+t} = x_t^+$. Since x_t^+ is a measurable function of $\{x_T, w_t^+\ t \geq 0\}$ and w^+ is independent of $\underline{\underline{F}}_T$ it follows that for any bounded function $h(x^+) = h(x_t^+, x_u^+, \ldots)$

$$E(h(x^+) | \underline{\underline{F}}_{T+}) = E(h(x^+) | x_T) \quad \text{a.s.}$$

which is the strong Markov property.

If K(x,y) increases with x and y we first construct a solution x^n to (1) with f and g multiplied by a smooth function that goes to zero smoothly outside a ball B_n of radius n, and then show two such solutions x^n, x^m, $n > m$ are the same until they first leave B_m. The solution to the original (1) is then the limit of x^n. Without the linear growth condition (3) a solution can still be defined, but possibly only up to a finite random time at which the solution "explodes" to infinity.

<u>Generators</u> Suppose x_t is a solution of a scalar form of (1) with $x_o = x$. If h is some bounded C^2 function on R^1 we can expand $h(x_t)$ by Ito's rule:

$$h(x_t) = h(x_o) + \int_0^t (fh' + \tfrac{1}{2}g^2 h'')ds + \int_0^t gh'dw_s$$

where h',h" are the derivatives of h and the argument of the integrands is x_s. By the martingale property of the Ito integral

$$E_x\ h(x_t) = h(x) + \int_0^t E_x(Dh(x_s))ds$$

where D is the second-order partial differential operator

$$(6) \quad D = f\frac{\partial}{\partial x} + \tfrac{1}{2}\ g^2 \frac{\partial^2}{\partial x^2}$$

and so
$$\lim_{t\to 0} \frac{1}{t}(E_x\ h(x_t) - h(x)) = Dh(x)$$

If T is a stopping time we also have

$$(7) \quad E_x(\int_0^T Dh(x_s)ds) = E_x h(x_T) - h(x)$$

which is Dynkin's formula [1, p.67]. D is called the <u>generator</u> of the process. The transition probability $P_t(x,A) \equiv P\{x_t \text{ is in } A\}$ can be shown to satisfy the <u>backward</u> equation

(8) $$\frac{\partial P}{\partial t} = DP$$

and the probability density $p_t(x,y) = dP\{x_t < y\}/dy$, if it exists, satisfies its dual form, the <u>forward</u> or <u>Fokker-Planck</u> equation

(9) $$\frac{\partial p}{\partial t} = -\frac{\partial}{\partial x}(fp) + \tfrac{1}{2}\frac{\partial^2}{\partial x^2}(g^2 p)$$

4. THE REPRESENTATIONS OF FISK AND STRATONOVICH

It is possible to define a stochastic integral in such a way that it obeys the normal rules of calculus. Fisk [7] and Stratonovich [8] have constructed this integral in different contexts. Rather than justify their construction we shall take the easier path of defining it via its relation to the Ito integral.

Suppose x_t, y_t are two integral processes in the sense of (10a) of Section 2. Then the product rule (12) of that section states that

$$x_t y_t - x_o y_o = \int_0^t x_s dy_s + \int_0^t y_s dx_s + \int_0^t (dx,dy)_s$$

If we share out the last term equally between the first two integrals we can define a new functional of x,y:

(1) $$\int_0^t x_s \bar{d}y_s \equiv \int_0^t x_s dy_s + \tfrac{1}{2}\int_0^t (dx,dy)_s$$

Then

$$x_t y_t - x_o y_o = \int_0^t x_s \bar{d}y_s + \int_0^t y_s \bar{d}x_s$$

which is like the product rule of ordinary calculus. We can extend this definition to C^1 functions of integral processes. Suppose h is C^1 and h' its first derivative. Define

(2) $$\int_0^t h(x_s)\bar{d}y_s \equiv \int_0^t h(x_s)dy_s + \tfrac{1}{2}\int_0^t h'(x_s)(dx,dy)_s$$

which is consistent with (1) if $h(x_t)$ is an integral process. Now suppose k is a C^2 function and expand $k(x_t)$ by Ito's rule. Identifying y_t with x_t in the last formula we have

$$k(x_t) = k(x_o) + \int_0^t k'(x_s)dx_s + \tfrac{1}{2}\int_0^t k''(x_s)(dx,dx)_s$$

$$= k(x_o) + \int_0^t k'(x_s)\bar{d}x_s$$

which is the ordinary expansion rule. The functional defined in (2) we shall call the Fisk-Stratonovich (F-S) integral. The original definition of Fisk and Stratonovich was as the limit in probability of symmetric Riemann sums

$$\int_0^t h(x_t)\bar{d}w_t = \lim_{d\to 0} \sum_{t_i<t} h(\tfrac{1}{2}(x_{t_{i+1}} + x_{t_i}))(w_{t_{i+1}} - w_{t_i})$$

where $d = \max_i |t_{i+1} - t_i|$. The two definitions coincide where they overlap. For vector processes $dx_t = f_t dt + g_t dw_t$ depending on a vector Brownian motion w_t the defining formula becomes in expanded form

(3) $$\int_0^t h(x_t)\bar{d}x_s^i = \int_0^t h(x_t)dx_s^i + \tfrac{1}{2}\sum_{jk}\int_0^t \frac{\partial h}{\partial x_j} g_s^{ik} g_s^{jk} ds$$

and again the ordinary rules of calculus apply in expansions. These expressions can be written in differential form but this requires caution. Though $\bar{d}x_t = 1.\bar{d}x_t = dx_t$ we <u>cannot</u> "multiply" both sides by h to get $h_t\bar{d}x_t = h_t dx_t$, as is clear from the definition.

<u>F-S differential equations</u> The processes described by Ito stochastic differential equations can equally well be described by F-S differential equations. Suppose a scalar process x_t is a solution of

$$dx_t = f(x_t)dt + g(x_t)dw_t$$

where f,g are C^1 and obey the linear growth and Lipschitz conditions of Section 3. Then this can be rewritten as

(4) $dx_t = a(x_t)dt + g(x_t)\bar{d}w_t$

where

(5) $a = f - \frac{1}{2} g g'$

This follows immediately from (2). Note that if a and g obey a <u>uniform</u> Lipschitz condition (K(x,y) = K), which implies linear growth, then f does also.

If x_t is a vector process, (4) still holds where now

(6) $a^i \equiv f^i - \frac{1}{2} \sum_{jk} g^{jk} \frac{\partial g^{ik}}{\partial x_j}$

The differential generator D associated with the process ((6) Section 3) now takes a quadratic form

(7) $D = \sum_i a^i \frac{\partial}{\partial x_i} + \frac{1}{2} \sum_j (\sum_i g^{ij} \frac{\partial}{\partial x_i})^2$

A consequence of the fact that F-S integrals obey the rules of ordinary calculus is that under a transformation of coordinates the coefficients a, g in (4) transform as though it was an ordinary differential equation.

<u>Systems driven by white noise</u> One of Stratonovich's original aims in formulating his integral was to obtain a natural idealisation of ordinary differential equations of the form

$$\dot{x}_t = a(x_t) + g(x_t)z_t$$

where z_t is a random disturbance that behaves like "white noise"; that is, its integral $\int_0^t z_s ds$ behaves like Brownian motion. A number of results have been obtained about the limiting behaviour of x_t as the noise becomes "whiter". Wong and Zakai [9] have obtained one of the most striking results. The following is a version of this

<u>Proposition (1)</u> Suppose z_t^n, n = 1,2,... are integrable processes w_t is a scalar Brownian motion and a and g are C^1 n-vectors on R^n with $g(0) \neq 0$. Let $x^n(t)$ and $x(t)$ be the respective solutions of the ordinary differential equation

(8) $\dot{x}^n = a(x^n) + g(x^n)z_t^n \quad x^n(0) = 0$

and the F-S equation

(9) $dx(t) = a(x)dt + g(x)\bar{d}w_t \quad x(0) = 0$

Let $R, (R_n)$ be the first time $||x(t)|| \geq r$ $(||x^n(t)|| \geq r)$. If, as

$n \to \infty$

$$w_t^n \equiv \int_0^t z_s^n \, ds \to w_t \quad \text{uniformly for } t \text{ in } [0,t_1] \text{ a.s.}$$

then for some $r > 0$

$$x^n(t \wedge R_n) \to x(t \wedge R) \text{ uniformly for } t \text{ in } [0,t_1] \text{ a.s.}$$

Proof For sufficiently small r we can choose a new coordinate system y on the ball B_r of radius r about the origin in R^n so that g is nonzero on B_r and transforms into the vector $I=(1,0,..,0)$. If $a \to b$, $x^n(t) \to y^n(t)$ $x(t) \to y(t)$, then the equations in integral form become

$$y^n(t) = \int_0^t b(y^n(s))ds + I\, w_t^n$$

$$y(t) = \int_0^t b(y(s))ds + I\, w_t$$

For sufficiently small r, b is uniformly Lipschitz, with constant K and by a standard argument for such equations

$$\sup||y^n(t\wedge R_n) - y(t\wedge R)|| \leq \sup||w^n(t\wedge R_n) - w(t\wedge R)||e^{kt_1}$$

where the suprema are over $[0,t_1]$ and $t\wedge R \equiv \min(t,R)$. The result then follows from the continuity of the inverse transformation.

The preceding argument cannot be used to prove convergence for processes with a vector of noise disturbances. The difficulty is that in general there is no transformation that takes the coefficient vectors simultaneously into constant vectors. To see what can happen consider the following example.

Let v^1, v^2 be independent Brownian motions, let w^{n1}, w^{n2} be polygonal processes that are linear between the values

$$w_{r/n}^{n1} = v_{r/n}^1 + v_{r/n}^2 \qquad r = 0,1,\ldots$$

$$w_{r/n}^{n2} = v_{r/n}^1 - v_{r+1/n}^2$$

and let z^{n1}, z^{n2} be their derivatives. Then w_t^{n1} and w_t^{n2} converge uniformly to the independent Brownian motions (variance 2t at t)

$$w_t^1 = v_t^1 + v_t^2 , \quad w_t^2 = v_t^1 - v_t^2$$

If $x_t \equiv \int_0^t w_s^1 \, dw_s^2$, $x_t^n \equiv \int_0^t w_s^{n1} z_s^{n2} ds$, then $E\, x_t = 0$.

But it is easy to show that $E\, x_t^n = \frac{1}{2} t$ for $t = r/n$ and so x_t^n does not converge to x_t. In fact it converges to $x_t + \frac{1}{2} t$. This lack of convergence stems from the fact that the time interval on which the crosscorrelation function of z^{n1}, z^{n2} is nonzero is of the same size as the interval on which the autocorrelations are nonzero. In the next result (McShane [10] Theorem 9.1) the hypotheses imply no correlation between the noises. We omit the proof.

Proposition (2) Suppose in vector versions of (8) and (9) a and g are vector and matrix functions C^1 on R^n with bounded first derivatives and w is a vector Brownian motion. Let

$$z_t^n = \frac{w(t_{i+1}) - w(t_i)}{t_{i+1} - t_i} , \quad t_i \leq t < t_{i+1}$$

for a partition $0 = t_o < t_1 < \dots < t_n = T$. Let $x_o^n = x_o$. Then as $\max|t_{i+1}-t_i| \to 0$, $x_t^n \to x_t$ in probability uniformly for $0 \leq t \leq T$.

Stratonovich [11] and Khas'minskii [12] have other convergence results in the multidimensional case. A reasonable conjecture would be that $x_t^n \to x_t$ if the component sequences $(z_t^{ni}\ n = 0,1,..)$ are independent of each other and $\int_0^t z_s^{ni} ds \to w_t^i$ uniformly in t. For discussions of the approximation question see [5,10,11].

5. THE EXISTENCE OF TRANSITION PROBABILITY DENSITIES

The results in the previous section suggest that any structural property of a controlled system

$$\dot{x} = a(x) + g(x)u ,$$

such as some form of controllability, would have its reflection in the behaviour of the diffusion process

$$dx_t = a(x_t)dt + g(x_t)\bar{d}w_t .$$

This line of reasoning has been pursued by Elliott [13] who among other results has shown that an algebraic condition for controllability of the first equation is sufficient for the diffusion process to possess densities. That it is also necessary is a consequence of the following result.

Suppose x_t is a solution of the F-S equation

$$\text{(1)}\quad dx_t = a(x_t)dt + \sum_r g_r(x_t)\bar{d}w_t^r ,\ x_o = \text{a constant}$$

where r ranges over $1,\dots,m$, the Brownian motions $w^1,\dots,w^m$ are independent, and where $a, g_1,\dots g_m$ are now C^∞ n-vector functions on R^n with bounded first derivatives. Consider the vector fields

$$A = \sum_i a^i \frac{\partial}{\partial x_i} ,\quad G_r = \sum_i g_r^i \frac{\partial}{\partial x_i} \quad r = 1,\dots,m.$$

Regarding these and $\partial/\partial t$ as vector fields on R^{n+1} with element $x^* = (x,t)$ we can introduce the Lie algebra of vector fields

$$H = LA(A + \frac{\partial}{\partial t} , G_1,\dots G_m)$$

generated by $A + \partial/\partial t$, $G_1,\dots G_m$. Assume that H is nonsingular in the sense that the dimension of the tangent spaces of H is constant for all x^*. This is the <u>rank</u> of H. The global version of Frobenius' complete integrability theorem implies the existence of a maximal connected integral manifold I of H containing $x_o^* \equiv (x_o,0)$, (see e.g. [14]). Let I_t denote the slice of I at constant t. If rank(H) = r then I is of dimension r and since {t = constant} is not an integral manifold of H, I_t is of dimension r-1.

<u>Proposition (1)</u> If rank(H) = r then I_t is an (r-1)-dimensional manifold and the solution x_t of (1) satisfies

$$P\{x_t \in I_t \text{ for all finite } t\} = 1$$

<u>Proof</u> Let $x_t^* = (x_t,t)$. Then

$$\text{(2)}\quad dx_t^* = a^*dt + \sum_r g_r^* \bar{d}w_t^r \quad x_o^* = (x_o,0)$$

where $a^* = (a,1)$, $g_r^* = (g_r,0)$, $r = 1,\dots m$. The associated vector fields are $A + \partial/\partial t$, $G_1,\dots,G_m$. The process x_t has continuous sample paths, I is a smooth manifold and for any $b > 0$ the random time

$$T \equiv \text{first time} \le b \text{ that } x_t^* \text{ leaves } I$$
$$\equiv b \quad \text{otherwise,}$$

is a stopping time. Now let $y = y(x^*)$ be a local coordinate system in R^{n+1} with domain U such that the sets on which $y_{r+1},\ldots,y_{n+1}$ are constant are the integral manifolds of H in U. The local form of Frobenius' theorem shows that such a coordinate system exists about any point. Now let z_o be x_T^* if this lies in U and an arbitrary fixed point of U otherwise. Then z_o is $\underline{\underline{F}}_{T+}$-measurable. Let z_t be the solution of (2) with $\bar{d}w_t$ replaced by $\bar{d}w_{T+t}$, starting at z_o. If S is the first time $\leqq$ b-T that z_t leaves U, then

$$x_{T+t} = z_t \quad \text{for } 0 \le t < S, \quad \text{if } T < b$$

by the uniqueness of solution of (2), but for any i : $r+1 \le i < n+1$,

$$dy_i(z_t) = Ay_i dt + \textstyle\sum_j G_j y_i \; \bar{d}w^j_{T+t}$$
$$= 0$$

since y_i is an integral function of H on U. So

$$y_i(x_{T+t}) = y_i(x_T) \qquad \text{for } 0 \le t \leqq S \quad \text{if } T < b.$$

Therefore $P\{x_T \text{ is in } U\} = 0$. Since I can be covered by a countable family of such U,

$$P\{\text{the first time } x_t^* \text{ leaves } I \ge b\} = 1.$$

However b can be chosen arbitrarily large, and the result follows.

If $r \le n+1$, the manifold I_t is of dimension strictly less than n and therefore has, as a set in R^n, Lebesgue volume measure zero. Since $P\{x_t \in I_t\} = 1$, the transition probability measure $P_t(x,B)$ is singular with respect to Lebesgue measure. Consequently the condition rank(H) = n+1 is necessary for a density to exist. The following result of Elliott ([13] Theorem 3.5) shows it is also sufficient. We omit the proof.

<u>Proposition (2)</u> Suppose rank(H) = n+1, then x_t with $x_o = x$ possesses a transition probability density $P_t(x,y)$ that is C^∞ on $(0,\infty) \times R^m \times R^n$. Moreover it is a solution of the Fokker-Planck equation

$$\frac{\partial}{\partial t} P_t(x,y) = - \sum_i \frac{\partial}{\partial y_i}(f^i(y)P_t) + \tfrac{1}{2} \sum_{ij} \frac{\partial^2}{\partial y_i \partial y_j}(b^{ij}(y)P_t) \tag{3}$$

where

$$f^i = a^i + \tfrac{1}{2} \sum_{jr} g^j_r \frac{\partial}{\partial x_j} g^i_r$$
$$b^{ij} = \sum_r g^i_r g^j_r \, . \tag{4}$$

The differential operator on the right-hand side is the dual of the generator of the process. See (6) Section 3.

The conclusions of the last two propositions can be proved directly for the simple Gaussian process (Ito and F-S forms are the same)

$$dx_t = Fx + g\, dw_t$$

where F is a constant matrix and g a constant vector. The rank condition rank $LA((Fx)^T\partial/\partial x + \partial/\partial t, b^T\partial/\partial x) = n+1$, where $\partial/\partial x$ is the gradient operator, becomes the "complete controllability" criterion

(5) $\quad \mathrm{rank}(g, Fg, \ldots F^{n-1}g) = n,$

which is known to be equivalent to the positive definiteness of the symmetric matrix

$$\int_0^t e^{Fs}\, bb^T\, e^{F^T s}\, ds,$$

but this in turn is just the covariance of x_t as we see from the explicit expression

$$x_t = e^{Ft}x_o + \int_0^t e^{F(t-s)} b\, dw_s$$

Since x_t is Gaussian, it is easy to verify that the Fokker-Planck equation is satisfied. If in (5) the rank is $r < n$, then x_t sits on a moving r-dimensional hyperplane I_t passing through $e^{Ft}x_o$.

6. STOCHASTIC DIFFERENTIAL EQUATIONS ON A MANIFOLD

Since it transforms like an ordinary differential equation, the F-S stochastic differential equation (1) Section 5 can be thought of as a specification in local coordinates of the integral curve of the "randomised" vector field

$$dz_t = A\, dt + \sum_r G_r\, \bar{d}w_t^r$$

on the manifold R^n with element z. This suggests that stochastic processes on more general manifolds could also be generated by randomised vector fields. One way of making the extension that is sometimes possible is to imbed the manifold in a larger dimensional Euclidean space R^k, extend the vector fields on the manifold over R^k, and then identify the process on the manifold as a solution of the corresponding F-S equation in R^k. As Proposition (1) Section 5 shows, solutions that start on the manifold remain on it. A difficulty with this approach is that for the process to be conservative in the sense that it does not explode in finite time some such condition as linear growth is required on the vector fields in R^k and this will depend on the imbedding as well as the vector fields on the manifold. However it obviously works for matrix Lie groups imbedded in $R^n \times R^n$ the vector fields then satisfying trivially Lipschitz and linear growth conditions.

For more general manifolds we give a construction of the solution of a stochastic differential equation that is a corollary of a theorem of Ito [2].

Suppose M is a C^∞ n-dimensional manifold and $A, G_1, \ldots, G_m$ are C^∞-vector fields on it. Suppose z(t) is a stochastic process taking values on M. For any local coordinate system (x,U) mapping onto the open unit ball B_1 in R^n, let x_t be x(z(t)) if $z(t) \in U$ and zero otherwise. Let $w^1, \ldots w^n$ be independent Brownian motions, z_o an independent variable on M and $\underline{\underline{F}}_t$ the σ-field generated by $\{z_o, w_s^1, \ldots, w_s^m,\ 0 \le s \le t\}$. We shall say z(t) is a solution of the differential equation

(1) $dz_t = Adt + \sum_r G_r \bar{d}w_t^r \qquad z(0) = z_o$

if z(t) is continuous, $\underline{\underline{F}}_t$-measurable for each t, and if for any coordinate system (x,U) and stopping time S, there is a stopping T, > S if $z(S) \in U$, = S if $z(S) \notin U$, such that the coordinate representation x_t of z(t) is a solution of the F-S equation, for $S \leq t \leq T$

(2) $dx_t = a(x_t)dt + \sum_r g_r(x_t)\bar{d}w_t^r$

where $x_S = x(z(S))$ if $z(S) \in U$, = 0 otherwise, and where a, g_r are the local representations of A, G_r (i.e. $A = \sum a^i\partial/\partial x_i$ etc.,). This definition is consistent with the usual definition on R^n.

The Lipschitz and linear growth conditions of R^n are replaced by a boundedness condition. A vector field A is said to be <u>bounded</u> on M if for any $z \in M$ there is a coordinate system (x,U) taking a neighbourhood U of z onto B_1 and taking z into the origin such that the local representation a of A satisfies

$$||a|| < K \quad \text{for } ||x|| < 1,$$

where K is independent of z. Note that in R^n this is no more restrictive than linear growth: about any point z^o in R^n take as U the ball of radius $||z^o||$ with centre z^o and set $x = ||z^o||^{-1}(z-z^o)$. Then a z-vector field satisfying a linear growth condition transforms into a vector field bounded on B_1.

<u>Proposition (1)</u> Suppose A, $G_1,\ldots,G_m$ are bounded on M, then (1) has a unique solution and this is a diffusion process.

Ito stated his Theorem 3.1 [2] in terms of Ito stochastic differential equations. We have stated it in terms of a F-S equation because of its simpler vector-field interpretation. The following proof is basically that of Ito's in [2] though modified by a stopping time argument of McKean [4] p.91.

<u>Proof</u> Suppose (x,U) is a coordinate system mapping onto B_1. Let V,W denote the smaller domains in U that are the inverse images of $B_{\frac{1}{4}}$, $B_{\frac{3}{4}}$. Now choose a countable covering $\{U_i\}$ of M so that the the corresponding family $\{V_i\}$ also cover M. Set $D_i = V_i - \bigcup_{j<i} V_j$, so that the D_i are disjoint and cover M.

Now to construct z(t) let x_t^i be the solution of (2) starting at $x_o^i = x^i(z_o)$ if z_o is in D^i and zero if it is not, and stopping at the first time T_1^i that x_t^i leaves $B_{\frac{3}{4}}$. If it does not, set $T_1^i = \infty$. On $\{z_o \in D_i\}$ i = 1,2,..., define T_1 to be T_1^i and z(t) to be the inverse image under x^i of x_t^i. Then z(t) is defined up to T_1. Now repeat the procedure, starting at $z(T_1)$ at time T_1 if $T_1 < \infty$, and, if $z(T_1) \in D^i$, at $x_{T_1}^i = x^i(z(T_1))$. In this way we can construct z(t) piece by piece up to each of an increasing family of stopping times $\{T_n\}$. If $T_{n+1} = T_n$ then necessarily $T_n = T_{n+1} = \ldots = \infty$. We want to show that even if T_n is strictly increasing $\sup_n T_n = \infty$. We use the following lemma of McKean [4] p.93, which we give in a form suitable for our purpose:

<u>Lemma</u> If x_t in B_1 is a solution of (2) and $x_o = 0$, then for sufficiently small t

$$P\{\max_{0 \le s \le t} ||x_t|| > R\} \le \exp(-\frac{R^2}{ct})$$

where $c = 2n^2K^2$.

Now if T_n and T_{n+1} are finite and $z(T_n) \in D_i$, then x_t^i crosses a gap of width $\frac{1}{4}$. So for T_n finite,

$$P\{T_{n+1} - T_n \le \frac{1}{n} | z(T_n) \in D_i\}$$

$$\le P\{\max_{0 \le s \le \frac{1}{n}} ||x(T_n+s) - x(T_n)|| \ge \tfrac{1}{4} | z(T_n) \in D_i\}$$

$$\le \exp(-\frac{n}{16c})$$

So

$$P\{T_{n+1} - T_n \le \frac{1}{n}\} < \exp(-\frac{n}{16c}) ,$$

But the sum over n of these probabilities is convergent and so by the Borel-Cantelli lemma, if all T_n are finite then $T_{n+1} - T_n > \frac{1}{n}$ for all n sufficiently large almost surely. $\sum 1/n$ is however divergent and so $\sup T_n = \infty$ a.s. So the construction defines z(t) for all finite time.

It is clear that z(t) is continuous and $\underline{\underline{F}}_t$-measurable for each t. To prove it solves (1) is more involved. Suppose (x,U) is some coordinate system and S a given stopping time. Let $x_t = x(z(t))$ if $z(t) \in U$ and $= 0$ if not. Let y_t^i be the (x^i, U_i) representation of z(t) if for some n, t lies in the half-open interval $T_n \le t < T_{n+1}$ and $z(T_n) \in D_i$, and zero if not. Let T be the first time $||x_t|| \ge \frac{1}{2}(1 + ||x_S||)$, infinity if this never happens. If $z(S) \in U$, $T_n \le S \le T_{n+1}$ and $z(T_n) \in D^i$, z(S) lies in both U and U^i and as x_t is a coordinate transformation of y_t^i which satisfies (2), x_t also satisfies (2) in the x coordinate for $S \le t < \min(T, T_n)$. Repeat for all i and we have the result. Uniqueness follows by the argument used in Proposition (1) Section 2: consider a second solution q(t) and the time $T = \inf\{t: q(t) \ne z(t)\}$, and show that $q(t) = z(t)$ for $t > T$; this implies $T = \infty$ a.s.

Finally the strong Markov property follows as in the case in R^n from the fact that for any stopping time T z(T+t) can be considered as function of z(T) and the independent processes $w_{T+t}^r - w_T^r$.

By taking the expectations of bounded functions defined on local coordinate domains we can establish that the differential generator of z(t) has the local representation

$$D = \sum_i a^i \frac{\partial}{\partial x_i} + \tfrac{1}{2} \sum_r (\sum_i g_r^i \frac{\partial}{\partial x_i})^2$$

$$= \sum_i f^i \frac{\partial}{\partial x_i} + \tfrac{1}{2} \sum_{ij} (B^{ij} \frac{\partial^2}{\partial x_i \partial x_j}$$

where f^i is given by (6) Section 4 and $B^{ij} = \sum_r g_r^i g_r^j$.

The question arises whether it is possible to define a

stochastic differential equation on a manifold so that its solution is governed by a given generator. In R^n this is possible using no more than n Brownian motions: if f and B are given and g is the symmetric matrix square root of B, then the solutions of the Ito equation $dx_t = f\,dt + g\,dw_t$ are governed by D. The generalisation of this procedure for n-dimensional manifolds would be to identify the columns g_r of g with local representations of vector fields G_r; but this does not always work as it is not generally possible to span the Lie algebra of vector fields on a manifold with just n vector fields: for instance this is not possible on the surface of a sphere. If the restriction on the number of Brownian motions is removed, then stochastic differential equations to produce a given generator can be defined on compact manifolds; whether it is possible in general is not known.

Ito [3] and McKean [4] have shown however how to construct a diffusion process on a manifold that is governed by a given generator. The resulting process is not the solution of just one stochastic differential equation but is obtained by random switching between local solutions. See [3] and [4] p.90.

REFERENCES

[1] K. Ito, Lectures on Stochastic Processes (Notes by K.M. Rao) Tata Institute of Fundamental Research, Bombay (1961).

[2] K. Ito, On stochastic differential equations on a differentiable manifold 1, Nagoya Math. J. 1 35-47 (1950).

[3] K. Ito, On stochastic differential equations on a differentiable manifold 2, Mem. Coll. Sci. Univ. Kyoto, A. 28 82-85 (1953).

[4] H.P. McKean Jr., Stochastic Integrals, Academic Press, New York (1969).

[5] E. Wong, Stochastic Processes in Information and Dynamical Systems, McGraw-Hill, New York (1971).

[6] J. Lamperti, Probability, W.A. Benjamin, New York (1966).

[7] D.L. Fisk, Quasi-martingales and stochastic integrals, Michigan State Univ. Dept. of Stat.Tech.Report No.1.(1963).

[8] R.L. Stratonovich, A new representation for stochastic integrals and equations, J. SIAM Control 4 362-371 (1966). Originally published in Russian in Vestnik Moskov, Univ. Ser.1. Mat.Meh.1. 3-12 (1964).

[9] E. Wong and M. Zakai, On the relationship between ordinary and stochastic differential equations, Int.J.Engng.Sci. 3 213-229 (1965).

[10] E.J. McShane, Stochastic differential equations and models of random processes, Proc. Sixth Berkeley Symposium on Math.

Stat. and Prob. 3 263-294 (1972).

[11] R.L. Stratonovich, Topics in the Theory of Random Noise vol.I. Gordon Breach, New York (1963).

[12] R.Z. Khas'minskii, A limit theorem for the solutions of differential equations with random right-hand sides, Th. Prob. Applic. XI 390-406 (1966).

[13] D.L. Elliott, Controllable Nonlinear Systems Driven by White Noise, Ph.D. Dissertation, Univ. California, Los Angeles (1969). See also Elliott's paper in this volume.

[14] R. Hermann, Differential Geometry and the Calculus of Variations, Academic Press, New York (1968).

GENERAL THEORY OF GLOBAL DIFFERENTIAL DYNAMICS

Lawrence Markus

University of Warwick, Mathematics Institute

1. Why global differentiable dynamics?

In the theory of local dynamical systems we study differential equations in an open subset of the real number space R^n; whereas in global dynamics the space is a general differentiable manifold M^n. We specify a global dynamical system as a tangent vector field v on M^n; and in any local chart $(x^1, \ldots, x^n)$ on M^n we denote the dynamical system v by its components $v^i(x^1, \ldots, x^n)$, say

$$v) \quad \frac{dx^i}{dt} = v^i(x^1, \ldots, x^n) \qquad i = 1,2, \ldots, n$$

or

$$v) \quad \dot{x} = v(x).$$

There are two basic motivations for studying differential systems on general differentiable manifolds.

i) Mathematical motivation. It is of interest to study differential systems within the most general context for which the concepts of differentiation and mathematical analysis are meaningful. Thus we take the ambiant space M^n to be of arbitrary finite dimension n, and locally differentiably equivalent to R^n. That is, M^n is a differentiable n-manifold (separable, metrizable C^∞-manifold without boundary), for instance R^n, or the n-sphere S^n, or the n-torus T^n.

ii) Physical motivation. Physical dynamical systems are often described by first-order vector differential equations involving the displacements, velocities, angles, and other generalized coordinates. If the generalized coordinates are unrestricted real variables, then the space in which the system evolves in time is some real vector space R^n. However, frequently the generalized coordinates are restricted by constraint or energetic equalities, or account for some angular periodicities of the physical configuration, and in these cases the space of the system is some differentiable manifold M^m.

For instance, an ordinary planar pendulum has a configuration space of a circle S^1, and a velocity-phase space of a product cylinder $S^1 \times R^1$. A spherical pendulum has a configuration space of a sphere S^2, and a velocity-phase space that is the tangent bundle TS^2. The configuration space of a rigid rotor is the rotation matrix group SO(3), which is diffeomorphic to the real projective space P^3, and the velocity-phase space is the 6-manifold TP^3 (which incidentally is the product $P^3 \times R^3$).

Besides the greater generality and applicability, the main advantages of the global viewpoint for dynamics are:

i) The notation and methodology of global differential geometry (involving manifolds, vector fields, trajectory curves, etc.) are highly suited to the requirements of the problems of dynamics. Old problems can be carefully phrased and solved, and new problems and concepts are suggested. For example, a careful discussion of the spherical pendulum requires knowledge that the tangent bundle TS^2 is not the product $S^2 \times R^2$. New concepts of structural stability and genericity arise naturally when various dynamical systems are compared globally.

ii) The global viewpoint emphasizes the unified family of all the trajectories as a portrait of a given dynamical system, rather than singling out special trajectories by their initial data. This is particularly important in physical systems where we wish to classify and compare the diverse modes of asymptotic behavior of the trajectories of the system.

2. Fundamental methods and concepts of global differentiable dynamics.

Let M^n be a differentiable manifold and let v be a tangent C^r-vector field on M^n. Then in overlapping charts $(x^1, \ldots, x^n)$ and $(\bar{x}^1, \ldots, \bar{x}^n)$ on an open set of M^n, the components of v are

$$\text{v)} \quad \dot{x}^i = v^i(x) \quad \text{and} \quad \dot{\bar{x}}^i = \bar{v}^i(\bar{x})$$

where the contravariant vector transformation law holds for the C^r-functions $(r = 1,2, \ldots, \infty)$ v^i and $\bar{v}^i$,

$$\bar{v}^i = \frac{\partial \bar{x}^i}{\partial x^j} v^j \quad (\text{sum on } j) \quad i = 1, \ldots, n.$$

A solution or trajectory of this vector field or differential system v is a C^1-curve

$$I \to M^n : t \to P_t \quad (I \text{ open interval in } R)$$

whose tangent vector at each point coincides with the vector of the field v.

The usual local existence, uniqueness, regularity results are valid. That is, for each initial point $P_o \in M^n$ there exists a unique trajectory P_t of v through P_o at $t = o$ (and defined on some maximal time duration I). If M^n is compact (and we shall assume this henceforth for simplicity of exposition), the maximal interval I is all R. In this case the solutions of v define a C^r-flow or action of R on M^n. That is there is a C^r-map

$$\Phi : R \times M^n \to M^n : (t, P_o) \to P_t,$$

and for fixed $t \in R$

$$\Phi_t : P_o \to P_t$$

is a C^r-diffeomorphism of M^n onto itself, and the group property holds for all times $t_1, t_2 \in R$

$$\Phi_{t_1} \circ \Phi_{t_2} = \Phi_{t_1 + t_2}, \quad \Phi_o = \text{Identity}.$$

Every trajectory of v on M^n is either

i) a point P_o

ii) a C^r-diffeomorphic image of a circle S^1

iii) a bijective regular C^r-differentiable image of a line R^1.

The case i) is a critical point where v vanishes at P_o, case ii) corresponds to a periodic solution or closed orbit, but case iii) can lead to curves that are not topological lines, say for an irrational flow on the torus T^n.

An invariant set Σ of v on M^n is a subset that is the union of whole trajectories of v. That is, a subset $\Sigma \subset M^n$ is invariant for the flow of v in case each trajectory initiating in Σ remains forever in Σ. Of course, a critical point or a closed orbit is necessarily an invariant set. Also, for each initial point P_o in the compact manifold M^n, the past (negative) and future (positive) limit set of the trajectory P_t

$$\alpha(P_o) = \bigcap_{\tau<o} \overline{\bigcup_{t<\tau} P_t} \quad \text{and} \quad \omega(P_o) = \bigcap_{\tau>o} \overline{\bigcup_{t>\tau} P_t}$$

are each compact connected invariant sets. Clearly $\alpha(P_o)$ and $\omega(P_o)$ depend only on the trajectory P_t and not on the initial point P_o. If $P_o \in \omega(P_o)$ then P_t is called future recurrent or Poisson stable, and P_o is recurrent if it is both past and future recurrent. A rather weaker property is regional recurrence of P_o - namely each neighborhood U of P_o in M^n has a trajectory U_t that meets U for some arbitrarily large past and future times. The set Ω of all regionally recurrent points, usually called the nonwandering set, is a compact invariant set containing all critical points, periodic orbits, and recurrent trajectories of the dynamical system v.

An invariant set Σ for the flow v in M^n is called future stable in case: for each neighborhood W of Σ in M^n there exists a subneighborhood $W_1 \subset W$ such that $P_o \in W_1$ implies that $P_t \in W$ for all future times $t > o$. If in addition $\omega(P_o) \subset \Sigma$ then the set Σ is future asymptotically stable (and analogous statements hold for past times).

We illustrate these concepts of recurrence and stability by some examples of invariant sets for flows in vector spaces and cylinders.

<u>Example 1</u>. $\dot{x} = Ax$ for $x \in R^n$ and A real constant matrix. If the matrix A is nonsingular, with complex eigenvalues $\lambda_1, \lambda_2, \ldots, \lambda_n$ not zero, then the origin $x = o$ is the unique critical point.

If no eigenvalue of A is pure imaginary, that is $Re\lambda_j \neq o$, then the origin $x = o$ is called a hyperbolic critical point. Define the attractor (stability) set of all points $P \in R^n$ for which $\omega(P) = o$, and similarly the repellor (instability) set by

$\alpha(P) = o$. Then the attractor set is a linear subspace whose dimension equals the number of eigenvalues whose real parts are negative. Also $x = o$ is asymptotically stable just in case the attractor space is all R^n.

Now consider a C^1-differential system v on a differentiable manifold M^n. Let P_o be a critical point for v and take a local chart (x) centered at P_o to write the differential system

$$v) \quad \dot{x} = A x + \ldots$$

We define P_o to be a hyperbolic critical point of v in case no eigenvalue of A is pure imaginary. In this case we define the attractor and repellor sets and prove that these are each C^1-differentiable submanifolds of M^n, in fact they are each regular bijective images of vector spaces of the appropriate dimensions.

<u>Example 2</u>. $\dot{x} = Ax$ and $\dot{\theta} = 1$, where $x \in R^{n-1}$ and $\theta \in S^1$. Here the manifold $M^n = R^{n-1} \times S^1$ is a cylinder. There are no critical points but the circle $x = o$, $o \leq \theta < 1$ is a periodic orbit σ. The Poincaré map around this periodic orbit σ is given by $x_o \to e^A x_o$, and its eigenvalues $\mu_1, \ldots, \mu_{n-1}$ are the (non-trivial) characteristic multipliers.

If none of the characteristic multipliers has a modulus of unity, $|\mu_j| \neq 1$ for $1 \leq j < n-1$, then the periodic orbit is defined to be hyperbolic. Again we define the attractor set by $\omega(P) \subset \sigma$ and the repellor set by $\alpha(P) \subset \sigma$, and it is clear that these are each cylinders of dimensions specified by the moduli of the characteristic multipliers.

Now consider a C^1-differential system u on a differentiable manifold M^n. Let σ be a periodic orbit of v with Poincaré map of a transversal section yielding the characteristic multipliers $\mu_1, \ldots, \mu_{n-1}$. If all $|\mu_j| \neq 1$ then σ is a hyperbolic periodic orbit. Again the attractor and repellor sets are C^1-submanifolds which are regular bijective images of either cylinders or generalized nonorientable Mobius bands, depending on the moduli and arguments of the complex characteristic multipliers.

<u>Example 3</u>. Consider a hyperbolic periodic orbit σ of a dynamical system **v** in R^3 with attractor and repellor manifolds each a cylinder $R^1 \times S^1$. Assume that the attractor and repellor manifolds have a nontrivial intersection (not just the orbit σ). This situation is called a homoclinic cycle and leads to a very complicated type of Poincaré map of a plane section, as indicated in the figure (still more complicated homoclinic cycles arise when several critical points or periodic orbits are involved).

The importance of the homoclinic cycle was recognized by Poincaré and G.D. Birkhoff. More recently it was brought forward by Smale.

The arc ab of the repellor manifold must have an image a'b', under the Poincaré map (or some iterate) Ψ. In particular a band-neighborhood N of ab is mapped by Ψ onto an elongated <u>horseshoe</u> Ψ N that crosses the band N. It can be shown that there is a compact set C x C ⊂ N (where C is the linear Cantor set) which is invariant under Ψ. Moreover Ψ acts as a discrete dynamical system on C x C by a shift-translation.

In the full space R^3 there is a connected compact invariant set H generated by the trajectories through C x C. Namely, H is topologically (C x C) x I with the endpoints of the interval I identified by means of the action of Ψ on C x C. The set H is called a "shoe" by Zeeman in his recent studies of structural stability.

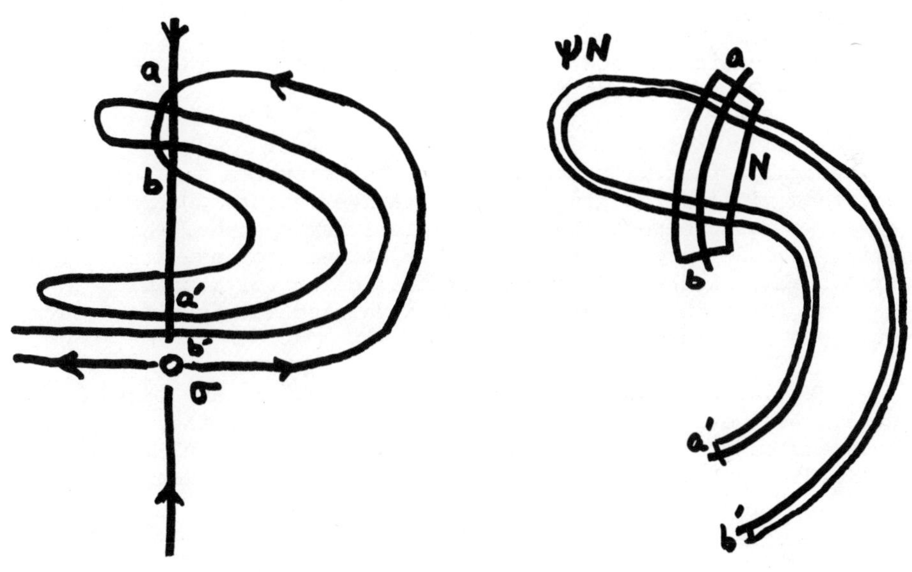

HOMOCLINIC CYCLE AND HORSESHOE MAP

3. Structural stability and generic dynamical systems.

Let M^n be a compact differentiable manifold and let V be the set of all C^1-vector fields on M^n. We consider an equivalence relation and two topologies on V.

Two vector fields v and u of V are topologically equivalent in case there exists a topological map of M^n onto M^n carrying the sensed (but not time-parametrized) solution curves of v onto those of u. Under such a topological equivalence the critical points of v and u correspond, as do the periodic orbits (but not the magnitude of the periods), as well as other types of invariant sets of v and u together with all their stability properties.

The C^o-topology on V demands that v is near u in case their vector components are uniformly near (in some finite set of covering charts for the compact manifold M^n). The C^1-topology on V demands that v is near u only in case their respective components, and the first partial derivatives of these components, are uniformly near on M^n. Both topologies make V into a separable metric space, but only in the C^1-topology is V complete. In this sense a subset of V which is C^1-residual (complement has Baire category 1) is called generic.

Definition. A dynamical system $v \in V$ on a compact differentiable manifold M^n is structurally stable in case: there exists a C^1-neighborhood N of v consisting of systems $u \in N$ each topologically equivalent to v.

Roughly speaking a structurally stable system can be perturbed without changing its qualitative features. For this reason we could expect that real engineering systems, which always include varying dissipative frictional forces, are structurally stable. On the other hand celestial mechanics where no friction is present in the idealized mathematics, would not yield structurally stable systems.

The most general structurally stable systems are very complicated; even though they are not generic in V they may contain shoes and be quite bewildering. However a very useful model for simple dissipative systems is provided by the Morse-Smale class.

Definition. A dynamical system $v \in V$ is of Morse-Smale type on the compact manifold M^n in case:

i) the nonwandering motions Ω consist of a finite set of critical points and periodic orbits, each hyperbolic

ii) attractor and repellor manifolds meet transversally wherever they intersect.

Morse-Smale systems have many pleasing properties; for instance a Morse-Smale system v contains no homoclinic cycles and hence no shoes. Also each future limit set is precisely one critical point or one periodic orbit. Moreover each attractor and repellor manifold is topologically embedded in M^n. In this sense Morse-Smale systems have quite simple structural properties.

Theorem. Each Morse-Smale system $v \in V$ on a compact differentiable manifold M^n is structurally stable.

In very recent research (still unpublished) Zeeman has defined a Smale system $v \in V$ by allowing Ω to contain hyperbolic shoes (saddleshoes?) as well as critical points and periodic orbits. He then obtains analogous theorems, but with the additional remarkable result: Smale systems are C^0-dense in V.

4. Relative genericity.

Let M^n be a compact differentiable manifold with C^1-vector fields V. We may wish to study a special subset V_1 of V which is significant with reference to some special subatlas of the differentiable charts of M^n. For example if we wish to study linear autonomous differential systems on M^n, this theory would be meaningful relative to an affine atlas of charts which overlap with affine rather than differentiable coordinate transformations. If we wish to study Hamiltonian systems $H \subset V$ on M^n, we specify them by the format

$$\dot{x} = \frac{\partial H}{\partial y}, \quad \dot{y} = -\frac{\partial H}{\partial x}$$

in terms of the charts (x,y) of the symplectic or canonical atlas on M^n.

The concepts of structural stability, perturbation, approximation, and genericity can be understood within a relative class $V_1 \subset V$. For Hamiltonian systems some results are known, and many others are conjectured.

REFERENCES

1. Markus, L. Lectures in Differentiable Dynamics, Regional Conference Series in Mathematics No. 3, (1971)

2. Markus, L. and Meyer, K. Generic Hamiltonians are neither Integrable nor Ergodic, Memoir A.M.S. (1974)

3. Robinson, R.C. Lectures on Hamiltonian Systems, I.M.P.A. (1971)

4. Smale, S. Differentiable Dynamical Systems, B.A.M.S. (1967)

5. Zeeman, E.C. Report of Warwick University, to appear.

TWO PROOFS OF CHOW'S THEOREM

P. Stefan

School of Mathematics and Computer Science,

U.C.N.W., Bangor LL57 2UW.

1. ACCESSIBLE SETS

Throughout this paper, M is a finite-dimensional C^q-manifold $(1 \leq q \leq \omega)$ and the word 'differentiable' refers to this fixed class C^q. (As usual, C^ω stands for 'real analytic' and we assume $\infty < \omega$.) If X is a differentiable vectorfield on M, $t \to \exp X^t.x$ denotes the integral curve of X passing through x at $t = 0$; for each fixed $t \in R$, $\exp X^t$ is a diffeomorphism of an open (possibly empty) subset of M onto another open subset of M.

Let S be a collection of differentiable vectorfields on M. We write $x = y \bmod S$ (resp. $x \to y \bmod S$) if

$$y = \exp X_1^{t_1} \circ \exp X_2^{t_2} \circ \ldots \circ \exp X_p^{t_p}.x$$

for some $X_i \in S$ and $t_i \in R$ (resp. $t_i \geq 0$). Thus $x = y \bmod S$ if and only if $x \to y \bmod \pm S$, where $\pm S = S \cup (-S) = \{\pm X : X \in S\}$. It is clear that $x = y \bmod S$ is an equivalence relation on M; its equivalence classes are termed here the <u>accessible sets</u> of S and

$$xS = \{y : x = y \bmod S\}$$

denotes the accessible set of S containing the point x.

Let S(x) denote the vector subspace of T_xM spanned by the vectors X(x), $X \in S$, and let $\bar{S}(x)$ denote the subspace of T_xM spanned by the vectors of the form $\phi^*(y).X(y)$, where $X \in S$,

$$\phi = \exp X_1^{t_1} \circ \exp X_2^{t_2} \circ \ldots \circ \exp X_p^{t_p}$$

for some $p \geq 1$, $X_i \in S$ and $t_i \in R$, $\phi(y) = x$ and $\phi^*(y) : T_yM \to T_xM$ is the differential of ϕ evaluated at y. It is clear that $S(x) \subset \bar{S}(x)$; the example $M = R^3$, x = the origin and $S = \{\partial/\partial\xi, \xi.\partial/\partial\eta, \xi\eta.\partial/\partial\zeta\}$

shows that the $\bar{S}(x)$ can be strictly greater than $S(x)$ and that we must allow $p > 1$ in general.

It has been proved recently by H. Sussmann [8], [9] and, independently, by the author [6], [7] that the accessible sets of S are immersed submanifolds of M. The details are given in the following theorem.

THEOREM 1. (i) If L is an accessible set of S, then there exists a unique differentiable structure σ on L such that

(a) (L,σ) is an immersed submanifold of M and

(b) $T_x(L,\sigma) = \bar{S}(x)$ for every $x \in L$.

(ii) The differentiable structure σ has the following additional properties

(c) If N is a differentiable manifold and $f : N \to M$ is a differentiable function such that $f(N) \subset L$ and $f(x) = y$ implies $f^*(x).T_xN \subset T_y(L,\sigma)$, then $f : N \to (L,\sigma)$ is differentiable.

(d) The partition of M into the accessible sets of S equipped with the said differentiable structures is a foliation with singularities.

(iii) If the connected components of M are separable, then also

(e) σ is the unique differentiable structure which makes L into a connected immersed submanifold of M.

(f) If P is a locally connected topological space, $f : P \to M$ is continuous and $f(P) \subset L$ then $f : P \to (L,\sigma)$ is continuous. If P is a manifold and f is differentiable, then $f : P \to (L,\sigma)$ is also differentiable.

REMARK. By (b), $T_x(L,\sigma) = S(x)$ for every $x \in L$ if and only if $(\exp X^t)*(x).S(x) \subset S(y)$ for every X, x and t such that $X \in S$, $x \in L$ and $\exp X^t.x = y$. Hence various sufficient conditions for integrability of 'distributions with singularities' (see [6], [7] and [8]).

Let now $\underset{\sim}{P}$ be an arbitrary partition of M into immersed submanifolds (possibly of varying dimension) and let $(M,\underset{\sim}{P})$ denote the disconnected sum of members of $\underset{\sim}{P}$: this is a differentiable manifold with the same underlying set as M. Let $V(\underset{\sim}{P})$ denote the collection of all the differentiable vectorfields on M whose integral curves leave the partition $\underset{\sim}{P}$ invariant. We record the following simple observations.

LEMMA 1. (a) $X \in V(\underset{\sim}{P})$ if and only if $X(x) \in T_x(M,\underset{\sim}{P})$ for every $x \in M$.

(b) If $q = \infty$ or ω, then $V(\underset{\sim}{P})$ is closed under formation of the Lie bracket.

PROOF. For (a), see [10], Lemma 2.4. The assertion (b) follows from the fact that $(M,\underset{\sim}{P})$ is an immersed submanifold of M and the differential of the inclusion mapping preserves Lie brackets [1].

REMARKS. (i) The assertion (a) is also valid for the time-dependent vectorfields: all we need is the existence and uniqueness theorem for ordinary differential equations. (ii) The assertion (b) is valid for the partitions of M into arbitrary <u>subsets</u>, see [6].

2. CHOW'S THEOREM

Throughout the rest of this paper we assume that $q = \infty$ or ω. S denotes a collection of differentiable vectorfields on M and [S] is the smallest set of vectorfields that contains S and is closed under formation of the Lie bracket.

In 1939, Wei-Liang Chow [2] published results that must be regarded as the first germ of Theorem 1 above. The main difference lies in Chow's assumption that the 'homogeneous envelope' of S is of <u>constant rank</u> (in a neighbourhood of the point in question: the assumptions and results of [2] are formulated only locally).

R. Hermann [4] pointed out the importance of [2] for Control Theory (see also [3], [5] and [10]). The following theorem, which can be deduced from the results of [2], seems particularly useful.

THEOREM 2 (Chow). If $\dim[S](x) = \dim M$, then xS is an open subset of M.

PROOF. Let L = xS and let $\underset{\sim}{P}$ be the partition of M into the accessible sets of S. Let $V(\underset{\sim}{P})$ be defined as above. By Theorem 1 and Lemma 1(b), $V(\underset{\sim}{P})$ is closed under formation of the Lie bracket. Since $S \subset V(\underset{\sim}{P})$, it follows that $[S] \subset V(\underset{\sim}{P})$. By Lemma 1(a), $[S](x) \subset T_x(L,\sigma)$ and so $\dim L \geq \dim[S](x) = \dim M$.

REMARK. It is not difficult to construct an example with $xS = L = M = R^3$ and $\dim[S](x) = 1$ or 2 for every $x \in L$. On the other hand, it follows from Theorem 1 and the classical Frobenius theorem that $\dim[S](x) = \dim L$ if $\dim[S](x)$ is constant along L. (This condition is automatically satisfied in the real analytic case; see [5] and [6]).

3. A DIRECT PROOF OF THEOREM 2

In this section we outline a proof of Theorem 2 which does not depend on Theorem 1 or the results of [2]. The details are given in [6].

Let f be a differentiable function $R^k \to R^n$ defined in the neighbourhood of the origin and assume that $f(t) = f(0)$ whenever at least one component of $t = (t_1, t_2, \ldots, t_k)$ is zero. Then

$$(1) \qquad f(t) = f(0) + t_1 t_2 \ldots t_k \bar{D}f(0) + \ldots$$

where $\bar{D} = D_1 D_2 \ldots D_k$ is the k-th mixed partial derivative ($D_i = \partial/\partial t_i$). If $\phi : R^n \to R^m$ is differentiable and defined in the neighbourhood of $x = f(0)$, then

$$(2) \qquad \bar{D}(\phi \circ f)(0) = \phi^*(x).\bar{D}f(0),$$

so that $\bar{D}f(0)$ transforms as the first derivative. In particular, $\bar{D}f(0)$ is a well-defined vector in $T_x M$ if R^n is replaced by the manifold M.

A differentiable function $a : R^k \times M \to M$ is a multiarrow of order k if (i) it is defined on a neighbourhood of $0 \times M$; (ii) for every $t \in R$, $a^t = a(t,-)$ is a diffeomorphism of open subsets of M; and (iii) $a(t,x) = x$ whenever (t,x) belongs to the domain of a and at least one component of t is zero. It follows from (1) and (2) that the equation

$$(3) \qquad \bar{D}a(x) = D_1 D_2 \ldots D_k a(0,x)$$

defines a differentiable vectorfield $\bar{D}a$ on M and that

$$(4) \qquad D_1 a(0, t_2, t_3, \ldots, t_k, x) = t_2 t_3 \ldots t_k \bar{D}a(x) + \ldots$$

The bracket $[a,b]$ of a k-arrow a and an ℓ-arrow b is the $k+\ell$-arrow given by

$$(5) \qquad [a,b](t,s,x) = (b^s)^{-1} \circ (a^t)^{-1} \circ (b^s) \circ (a^t).x \; .$$

LEMMA 2.

$$\bar{D}[a,b] = [\bar{D}a, \bar{D}b] \; .$$

The proof is by formal differentiation and consists chiefly of the repeated use of (2).

Let $\sim$ be an equivalence relation on M. We say that a multiarrow a respects $\sim$ if $a(t,x) \sim x$ for all (t,x) in the domain of a. Let $\tilde{A}$ denote the set of all the multiarrows on M which respect $\sim$.

LEMMA 3. Let $\tilde{V} = \bar{D}\tilde{A}$ be the collection of all vectorfields of the form $\bar{D}a$, $a \in \tilde{A}$. Then $\tilde{V}$ is closed under formation of the Lie bracket.

This follows at once from Lemma 2.

LEMMA 4. If $\dim \tilde{V}(x) = d$, then there exists an immersion $\psi : (R^d,0) \to (M,x)$ such that $\psi(t) \sim x$ for every t in the domain of ψ.

PROOF. Choose $a_i \in \tilde{A}$ so that the vectors $\bar{D}a_1(x)$, $\bar{D}a_2(x)$, $\ldots, \bar{D}a_d(x)$ form a basis of $\tilde{V}(x)$. It follows from (4) that there exist $\tau_i = (t_{2i}, t_{3i}, \ldots)$ such that the first derivatives $D_1a_1(0,\tau_1,x)$, $D_1a_2(0,\tau_2,x), \ldots$ are also linearly independent. Put $b_i^t = a_i(t,\tau_i,-)$ and let

$$\psi(t_1,t_2,\ldots,t_d) = b_1^{t_1} \circ b_2^{t_2} \circ \ldots \circ b_d^{t_d}.x \quad .$$

Then $D_i\psi(0) = D_1a_i(0,\tau_i,x)$, so that ψ has rank d at the origin and restricts to an immersion on its neighbourhood.

PROOF OF THEOREM 2. Let $\sim$ be the relation $x = y \bmod S$. It is clear that $S \subset \tilde{V}$ and so, by Lemma 3, $[S] \subset V$. Hence $\dim \tilde{V}(x) = \dim M$ and Lemma 4 implies the existence of an open set H such that $x \in H \subset xS$. If $y \in xS$, then $y = \phi(x)$, where

$$\phi = (\exp X_1^{t_1}) \circ (\exp X_2^{t_2}) \circ \ldots \circ (\exp X_p^{t_p})$$

for some $X_i \in S$ and $t_i \in R$. We may assume that H lies in the domain of ϕ. It remains to note that $y \in \phi(H) \subset xS$ and $\phi(H)$ is open in M.

REFERENCES

[1] CLAUDE CHEVALLEY, Theory of Lie Groups, Princeton University Press, Princeton, New Jersey, 1946.

[2] W.L. CHOW, Uber Systeme von linearen partiellen Differentialgleichungen erster Ordnung, Math. Ann., 117 (1939), pp. 98-105.

[3] G.W. HAYNES and H. HERMES, Nonlinear Controllability Via Lie Theory, SIAM J. Control, 8 (1970), pp. 450-460.

[4] R. HERMANN, On the accessibility problem in control theory, Internat. Sympos. Nonlinear Differential Equations and Nonlinear Mechanics, Academic Press, New York 1963, pp. 325-332.

[5] C. LOBRY, Contrôlabilité des systèmes non linéaires, SIAM J. Control, 8 (1970), pp. 573-605.

[6] P. STEFAN, Accessible sets, Orbits and Foliations with Singularities, to appear.

[7] P. STEFAN, On tangent spaces of accessible sets, duplicated notes, University College of North Wales, Bangor, August 1973.

[8] H.J. SUSSMANN, Orbits of families of vector fields and integrability of systems with singularities, Bull. Amer. Maths. Soc. 79 (1973), pp. 197-199.

[9] H.J. SUSSMANN, Orbits of families of vector fields and integrability of distributions, to appear in Trans. Amer. Math. Soc.

[10] H.J. SUSSMANN and V. JURDJEVIC, Controllability of Nonlinear Systems, J. Differential Equations 12 (1972), pp. 95-116.

On Necessary and Sufficient Conditions for Local Controllability Along a Reference Trajectory

Henry Hermes†

Department of Mathematics

University of Colorado

Introduction. Consider an n-dimensional control system modelled by the differential equations

$$\dot{x}(t) = f(x(t),u(t)) , \quad (\dot{x}(t) = dx/dt) \tag{1}$$

where f is smooth and an admissible control u is a piecewise continuous function taking values in a given set U having non-empty interior in $\mathbb{R}^m$. Denote by $\mathcal{A}(t,q)$ the set of all points attainable at time t by solutions of (1) corresponding to admissible controls and initiating from q at time 0. Let u^* be a given control which generates a reference trajectory φ with $\varphi(0) = p$. The system (1) is <u>locally controllable along</u> φ <u>at</u> p if for all $\varepsilon > 0$, $\varphi(\varepsilon)$ is an interior point of $\mathcal{A}(\varepsilon,p)$. Loosely speaking, this implies the ability to control the system to a full neighborhood of the reference trajectory over an arbitrarily small interval of time.

The essence of the problem of local controllability for the general system (1) may be reduced to the study of the simpler system

$$\dot{x}(t) = X(x(t)) + Y(x(t))u(t) . \tag{2}$$

Let M be an analytic n-manifold with tangent space at x denoted TM_x; X,Y be analytic tangent vector fields on M, and $\mathcal{S}$ the collection of vector fields defined by

---

† This research was supported by the National Science Foundation under grant GP27957

$$\mathcal{D} = \{X + \alpha Y: -1 \leq \alpha \leq 1\} .$$

For any collection of vector fields $\mathcal{C}$, we denote by $\mathcal{C}_x = \{X(x) \in TM_x: X \in \mathcal{C}\}$. A <u>solution of</u> $\mathcal{C}$ is a piecewise continuously differentiable function φ such that $\dot{\varphi}(t) \in \mathcal{C}_{\varphi(t)}$ almost everywhere on some interval. The set of points attainable at time $t \geq 0$, initiating from q at time 0, will be denoted by $\mathcal{A}(\mathcal{C},t,q)$, or just $\mathcal{A}(t,q)$ if the collection under consideration is obvious from the context.

We consider the set of all analytic vector fields, $V(M)$, on M as a real lie algebra with product the lie product, denoted $[X,Y]$. Following the notation of [1], for $\mathcal{C} \subset V(M)$, $\mathcal{J}(\mathcal{C})$ denotes the smallest subalgebra which contains $\mathcal{C}$.

Returning to local controllability, let the vector field $X \in \mathcal{D}$ (i.e., $\alpha = 0$) generate the reference trajectory denoted $T^X(\cdot)p$, where $T^X(0)p = p \in M$. A necessary condition that the system $\mathcal{D}$ be locally controllable along T^X at p is that for all $\varepsilon > 0$, $\mathcal{A}(\varepsilon,p)$ have nonempty interior. (This is not sufficient since $T^X(\varepsilon)p$ may be on the boundary of $\mathcal{A}(\varepsilon,p)$ for all $\varepsilon > 0$.) From [1, th. 3.2], a necessary and sufficient condition that $\mathcal{A}(\varepsilon,p)$ have nonempty interior is that $\dim \mathcal{J}_0(\mathcal{D}) = n$, where $\mathcal{J}_0(\mathcal{D})$ is an ideal in $\mathcal{J}(\mathcal{D})$; specifically in our case let $\mathcal{J}'(\mathcal{D})$ denote the derived algebra of $\mathcal{J}(\mathcal{D})$, then

$$\mathcal{J}_0(\mathcal{D}) = \mathcal{J}\{Y, W: W \in \mathcal{J}'(\mathcal{D})\} .$$

(The necessity of this condition, in the analytic case, follows from [2, prop. 1.3.6]. In the C^∞ case, it is not necessary and the proof of sufficiency given in [1] also relies heavily on analyticity.) Let $(\mathrm{ad}X,Y) = [X,Y]$, $(\mathrm{ad}^2X,Y) = [X,[X,Y]]$, etc. and define

$$\mathcal{S} = \{Y, (\mathrm{ad}X,Y), (\mathrm{ad}^2X,Y), \ldots\} . \tag{3}$$

A well-known sufficient condition for $\mathcal{D}$ to be locally controllable along T^X at p (see [3, section 19], [4], or [5]), which is not necessary, is that

$$\dim \operatorname{span} \mathcal{S}_p = n . \tag{4}$$

Summarizing, a necessary and sufficient condition that for all $\varepsilon > 0$, $\mathcal{A}(\varepsilon,p)$ have nonempty interior is that

$$\dim \mathcal{T}_0(\mathcal{D})_p = n \tag{5}$$

while (4) is a sufficient condition that $\overset{X}{T}(\varepsilon)p$ be interior to $\mathcal{O}(\varepsilon,p)$. Clearly (4) implies (5), as expected.

Now define

$$\mathcal{S}^0 = \{Y,[Y,X],(\mathrm{ad}X,[Y,X]),(\mathrm{ad}^2X,[Y,X]),\ldots\}$$

$$\mathcal{S}^1 = \{Y,(\mathrm{ad}^2Y,X),(\mathrm{ad}X,(\mathrm{ad}^2Y,X)),(\mathrm{ad}^2X,(\mathrm{ad}^2Y,X)),\ldots\}$$

$$\mathcal{S}^2 = \{Y,(\mathrm{ad}^3Y,X),(\mathrm{ad}X,(\mathrm{ad}^3Y,X)),\ldots\} \quad \text{etc.}$$

(Note that $\dim \operatorname{span} \mathcal{S}^0_p = \dim \operatorname{span} \mathcal{S}_p$; essentially the elements of $\mathcal{S}^0$ and $\mathcal{S}$ only differ by a minus sign). Our main result is

THEOREM 1. (Dimension $n = 2$). Assume $\mathcal{T}_0(\mathcal{D})_p = 2$ and $X(p)$, $Y(p)$ are linearly independent. Then there exists an integer $m \geq 0$ such that $\dim \operatorname{span} \mathcal{S}^m_p = 2$. A necessary and sufficient condition for the system (2) to be locally controllable along T^X at p is that the smallest integer $m \geq 0$, for which $\dim \operatorname{span} \mathcal{S}^m_p = 2$, is even. (We consider $m = 0$ as even.)

An indication of the proof is given in section 2; details, and extensions to higher dimensions, will appear elsewhere.

1. Examples. We next construct several examples for which local controllability cannot be determined by the sufficient condition (4), which is equivalent to testing complete controllability of the linearized (variational) equation along the reference trajectory. It was shown in [6] that (4) provides no information when the reference trajectory in (2) is a singular arc of a time optimal problem, thus we construct examples of this type. We briefly describe the Greens theorem approach which is useful in constructing and analyzing such examples. (See [3, section 22] for details.) Let $n = 2$, $x = (x_1,x_2)$, $X(x) = \begin{pmatrix} a_1(x) \\ a_2(x) \end{pmatrix}$, $Y(x) = \begin{pmatrix} b_1(x) \\ b_2(x) \end{pmatrix}$ so equation (2) becomes

$$\begin{aligned} \dot{x}_1 &= a_1(x) + b_1(x)u \\ \dot{x}_2 &= a_2(x) + b_2(x)u \end{aligned} \tag{6}$$

Multiplying the first equation by $-b_2(x)$, the second by $b_1(x)$ and adding gives $-b_2(x)\dot{x}_1 + b_1(x)\dot{x}_2 = \Delta(x)$ where

$$\Delta(x) = -a_1(x)b_2(x) + a_2(x)b_1(x) \quad . \tag{7}$$

Let Γ^1, Γ^2 be two arcs, which can be realized as the graphs of solutions of (6), joining an initial point p to a terminal point q and having only these points in common. If $\Delta(x) \neq 0$ for $x \in \Gamma^1 \cup \Gamma^2$ the controls which determine these solutions are unique and hence we can assign a unique time of transfer, $t(\Gamma^i)$, to each such arc. By Greens theorem, taking line integrals in a counterclockwise direction, we obtain

$$t(\Gamma^1) - t(\Gamma^2) = \int_{\Gamma^1 \cup \Gamma^2} -\frac{b_2(x)}{\Delta(x)}\, dx_1 + \frac{b_1(x)}{\Delta(x)}\, dx_2 = \iint_{\mathfrak{R}} w(x)\, dx$$

where $\mathfrak{R}$ is the closed region bounded by $\Gamma^1 \cup \Gamma^2$, while

$$w(x) = \frac{\partial}{\partial x_1}(b_1/\Delta) + \frac{\partial}{\partial x_2}(b_2/\Delta)$$

and we assume $\Delta \neq 0$ in $\mathfrak{R}$. The sign of w in $\mathfrak{R}$ can be used to compare the transfer times $t(\Gamma^1)$ and $t(\Gamma^2)$; it is known, [7], that an admissible arc Γ is singular if and only if $w(x) = 0$ for $x \in \Gamma$. Furthermore $w \not\equiv 0$ implies $\mathfrak{A}(\varepsilon, p)$ has interior for $\varepsilon > 0$, in the analytic case, as can easily be verified by the linear test (4), i.e., $Y(p), [X,Y](p)$ are linearly independent if $w(p) \neq 0$.

In the following three examples, the reference trajectory will be taken as T^X, i.e., $u \equiv 0$.

<u>Example 1.1</u>. Here we shall have $\dim \mathcal{T}_0(\mathcal{D})_p = 2$, $\dim \operatorname{span} \mathscr{S}_p^0 = 1$, and the system will be locally controllable along T^X at p. Let $X(x) = \begin{pmatrix} 8 \\ x_2 \end{pmatrix}$, $Y(x) = \begin{pmatrix} x_2^2 \\ 1 \end{pmatrix}$ and $p = \begin{pmatrix} 0 \\ 0 \end{pmatrix}$. Then $\Delta(x) = -8 + x_2^3$; $\Delta^2(x)w(x) = -3x_2^2$ hence $T^X(t)p = \begin{pmatrix} 8t \\ 0 \end{pmatrix}$ is a singular arc. Computation shows $(\operatorname{ad}^k X, Y)(x) = \begin{pmatrix} (-2)^k x_2^2 \\ 1 \end{pmatrix}$ hence $\dim \operatorname{span} \mathscr{S}_p^0 = 1$. Also, $(\operatorname{ad}^2 Y, X)(x) = \begin{pmatrix} -6x_2 \\ 0 \end{pmatrix}$; $(\operatorname{ad}^k X, (\operatorname{ad}^2 Y, X))(x) = \begin{pmatrix} (-1)^{k+1} 6x_2 \\ 0 \end{pmatrix}$ hence $\dim \operatorname{span} \mathscr{S}_p^1 = 1$.

However $(\mathrm{ad}^3 Y, X)(x) = \begin{pmatrix} 6 \\ 0 \end{pmatrix}$ which is linearly independent of $Y(p)$ hence $\dim \operatorname{span} \mathscr{S}^2_p = 2$ and theorem 1 yields local controllability along T^X at p.

For later use in understanding the proof of theorem 1, it is instructive to compute

$$T^Y(t) \circ T^X(\tau)p = \begin{cases} t^3/3 + 8\tau \\ t \end{cases}, \quad T^{[X,Y]}(t) \circ T^X(\tau)p = \begin{cases} -2t^3/3 + 8\tau \\ t \end{cases}$$

the first having graph $x_1 = x_2^3/3 + 8\tau$, the second $x_1 = -2x_2^3/3 + 8\tau$.

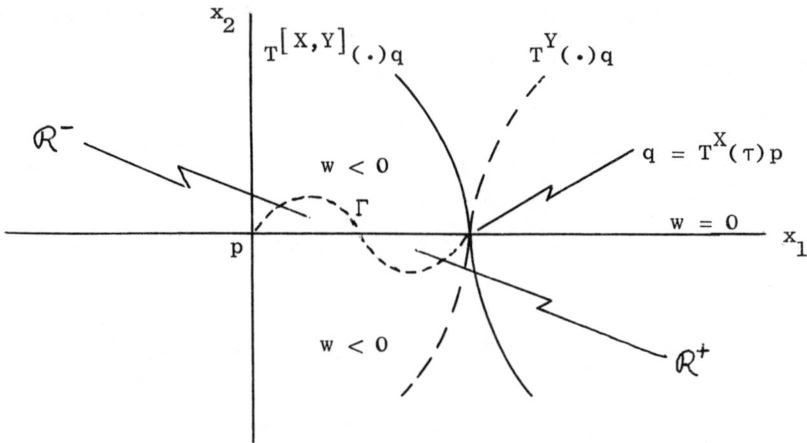

FIGURE 1.

By the Greens theorem method, one may easily check that $T^X(\cdot)p$ is not time optimal, say from p to $p^1 = \begin{pmatrix} 4 \\ 0 \end{pmatrix}$. Also at any point $q = T^X(\varepsilon)p$, $\varepsilon > 0$, one can join p to q by an admissible trajectory other than $T^X(\cdot)p$ <u>which also makes the transfer in exactly time</u> ε. (This strongly uses $\Delta(p) \neq 0$.) Indeed consider the "dotted" curve Γ in figure 1; if $\iint_{\mathcal{R}^-} w(x)\,dx = \iint_{\mathcal{R}^+} w(x)\,dx$ the transfer time along Γ, from p to q, is exactly ε. <u>Note that this is possible if and only if</u> w <u>has the "same sign on either side" of</u> $T^X(\cdot)p$! But there are points $x \in \Gamma$ at which $w(x) \neq 0$ and at these (see [6] or section 2 of this paper) $Y(x)$

and $(\mathrm{ad}X,Y)(x)$ are linearly independent. Thus (4) implies local controllability along Γ at p and hence $q = T^X(\varepsilon)p$ is an interior point of $\mathcal{A}(\varepsilon,p)$ for all $\varepsilon > 0$.

Note that in this example, $T^Y(\cdot)q$ and $T^{[X,Y]}(\cdot)q$ "cross" at $q = T^X(\tau)p$.

Example 1.2. Here we shall have $\dim \mathcal{T}_0(\mathcal{D})_p = 2$, $\dim \operatorname{span} \mathcal{S}_p^0 = 1$ and the system will not be locally controllable along T^X at p.

Let $X(x) = \begin{pmatrix} 4 \\ x_2 \end{pmatrix}$, $Y(x) = \begin{pmatrix} x_2 \\ 1 \end{pmatrix}$ and $p = \begin{pmatrix} 0 \\ 0 \end{pmatrix}$. Then $\Delta(x) = -4 + x_2^2$, $\Delta^2(x)w(x) = -2x_2$ hence $T^X(\cdot)p$ is again a singular arc. Computing shows $(\mathrm{ad}^k X,Y)(x) = \begin{pmatrix} (-1)^k x_2 \\ 1 \end{pmatrix}$ hence $\dim \operatorname{span} \mathcal{S}_p^0 = 1$. However, here $(\mathrm{ad}^2 Y,X)(p) = \begin{pmatrix} -2 \\ 0 \end{pmatrix}$ and is linearly independent of Y hence $\dim \operatorname{span} \mathcal{S}_p^1 = 2$ thus by theorem 1, we do not have local controllability along T^X at p. Indeed, here $T^X(\cdot)p$ is time optimal to any point of the form $q = \begin{pmatrix} c \\ 0 \end{pmatrix}$, $c > 0$ and this would necessitate $T^X(\varepsilon)p$ being on the boundary of $\mathcal{A}(\varepsilon,p)$. Again, we compute

$$T^Y(t) \circ T^X(\tau)p = \begin{cases} t^2/2 + 4\tau \\ t \end{cases}, \quad T^{[X,Y]}(t) \circ T^Y(\tau)p = \begin{cases} -t^2/2 + \tau \\ t \end{cases}$$

which have graphs, respectively, $x_1 = x_2^2/2 + 4\tau$ and $x_1 = -x_2^2/2 + 4\tau$.

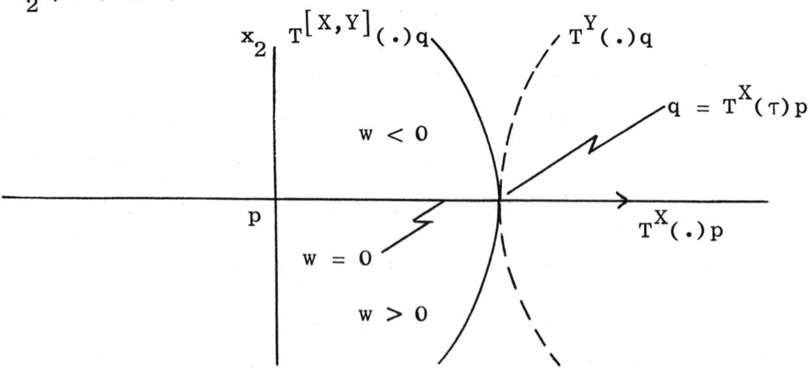

FIGURE 2.

Here we should note that the curves $T^Y(\cdot)q$ and $T^{[X,Y]}(\cdot)q$ do not cross at q. Also w changes sign as we cross $T^X(\cdot)p$ which forces $T^X(\cdot)p$ as the unique trajectory joining p and $T^X(\varepsilon)p$ in time ε.

Example 1.3. In this example we shall again have $\dim \mathcal{T}_0(\mathcal{D})_p = 2$, $\dim \operatorname{span} \mathcal{S}^0_p = 1$ and have local controllability along T^X at p. However here we shall have $[X,Y](T^X(\tau)p) \equiv 0$; in fact $(\mathrm{ad}^k X,Y)(T^X(\tau)p) \equiv 0$ for all $k = 1,2,\ldots$.

Let $X(x) = \begin{pmatrix} 0 \\ 4 + x_1^3 x_2 \end{pmatrix}$, $Y(x) = \begin{pmatrix} 1 \\ 0 \end{pmatrix}$ and $p = \begin{pmatrix} 0 \\ 0 \end{pmatrix}$. Then

$\Delta(x) = 4 + x_1^3 x_2$, $\Delta^2(x)w(x) = -3x_1^2 x_2$ hence $T^X(t)p = \begin{cases} 0 \\ 4t \end{cases}$ is singular but w has the same sign on both sides of this arc. Computing, $(\mathrm{ad}X,Y)(x) = \begin{pmatrix} 0 \\ 3x_1^2 x_2 \end{pmatrix}$ and in general $(\mathrm{ad}^k X,Y)(T^X(\tau)p)$ $= 0$, thus $\dim \operatorname{span} \mathcal{S}^0_p = 1$. Also, $(\mathrm{ad}^2 Y,X)(x) = \begin{pmatrix} 0 \\ 6x_1 x_2 \end{pmatrix}$ and one can easily verify that $(\mathrm{ad}^k X,(\mathrm{ad}^2 Y,X))(p) = 0$ for $k = 0,1,\ldots$, thus $\dim \operatorname{span} \mathcal{S}^1_p = 1$. Continuing, $(\mathrm{ad}^3 Y,X)(x) = \begin{pmatrix} 0 \\ -6x_2 \end{pmatrix}$ and $[X,(\mathrm{ad}^3 Y,X)](p) = \begin{pmatrix} 0 \\ 24 \end{pmatrix}$ thus $\dim \operatorname{span} \mathcal{S}^2_p = 2$ and Theorem 1 shows local controllability along T^X at p.

2. The Main Ideas in the Proof of Theorem 1.

The assumption that $X(p), Y(p)$ are linearly independent implies $\Delta(p) \neq 0$ hence $\Delta(x) \neq 0$ in some neighborhood of p. Define $Y^\perp(x) = \begin{pmatrix} b_2(x) \\ -b_1(x) \end{pmatrix}$ and $Z(x) = (1/\Delta(x))Y^\perp(x)$ so the inner products

$$(Z(x),Y(x)) \equiv 0\,, \quad (Z(x),X(x)) \equiv -1 \tag{8}$$

hold in some nbd. of p. Next define

$$h(t,\tau) = (Z(T^Y(t) \circ T^X(\tau)p)\,,\; [X,Y](T^Y(t) \circ T^X(\tau)p))\,,$$

noting that the map $(t,\tau) \to T^Y(t) \circ T^X(\tau)p$ is a homeomorphism of a nbd. of 0 in $\mathbb{R}^2$ onto a nbd. of p. The function $h(t,\tau)$ describes how the solutions $T^Y(\cdot)q$ and $T^{[X,Y]}(\cdot)q$ "cross" or are "tangent" etc.; eg., $h(t,\tau) = 0$ means the vectors

$Y(T^Y(t) \circ T^X(\tau)p)$ and $[X,Y](T^Y(t) \circ T^X(\tau)p)$ are linearly dependent. By direct computation, one may verify that $h(t,\tau) = \Delta(T^Y(t) \circ T^X(\tau)p)w(T^Y(t) \circ T^X(\tau)p)$. We know from [7] that $w \not\equiv 0$ is necessary and sufficient that $\mathcal{A}(\varepsilon,p)$ have nonempty interior for all $\varepsilon > 0$; by theorem 3.2 of [1], it follows that $\dim \mathcal{J}_0(\mathcal{D}) = 2$ is equivalent to $h(t,\tau) \not\equiv 0$. With this assumption we may show that a necessary and sufficient condition for the system (6) to be locally controllable along T^X at p is that there exist $\varepsilon > 0$ such that $h(t,\tau)$ does not change sign as a function of t, at $t = 0$, for $0 < \tau \leq \varepsilon$. (This is shown as in the argument in example 1.1.)

To check the sign of h, we expand it in a power series in t and τ having coefficients depending on the elements of $\mathcal{J}_0(\mathcal{D})_p$. Briefly, using superscript T to denote transpose and subscript x to denote partial derivative with respect to x, $d/d\tau\, h(0,\tau) = X^T Z_x^T [X,Y] + Z^T [X,Y]_x X$ with the right side evaluated at $T^X(\tau)p$. From (8), $-X^T(x)Z_x(x) - Z^T(x)X_x(x) = 0$ hence $d/d\tau\, h(0,\tau) = Z^T[[X,Y],X] + X^T(Z_x^T - Z_x)[X,Y]$. But $(Z_x^T(x) - Z_x(x)) = \begin{pmatrix} 0 & -w(x) \\ w(x) & 0 \end{pmatrix}$ so $X^T(T^X(\tau)p)(Z_x^T(T^X(\tau)p) - Z_x(T^X(\tau)p))[X,Y](T^X(\tau)p)$ $= \left(h(0,\tau) / \Delta(T^X(\tau)p)\right) (X^T(T^X(\tau)p) \begin{pmatrix} 0 & -1 \\ 1 & 0 \end{pmatrix} [X,Y](T^X(\tau)p))$. If $h(0,0) = 0$ we see $d/d\tau\, h(0,0) = (Z(p),[[X,Y],X](p))$; in a similar fashion one obtains $(d/d\tau)^m h(0,0) = (-1)^m(Z(p),(\mathrm{ad}^{m+1}X,Y)(p))$ if $(d/d\tau)^j h(0,0) = 0$ for $j = 0,\ldots,(m-1)$. Also, $d/dt\, h(t,\tau) = Z^T[[X,Y],Y] + Y^T(Z_x^T - Z_x)[X,Y]$ with the right side evaluated at $T^Y(t) \circ T^X(\tau)p$. Now $Y^T(x)(Z_x^T(x) - Z_x(x))[X,Y](x) = \Delta(x)w(x)Z^T(x)[X,Y](x) = (Z^T(x)[X,Y](x))^2$ so $d/dt\, h(t,\tau) = Z^T(T^Y(t) \circ T^X(\tau)p)[[X,Y],Y](T^Y(t) \circ T^X(\tau)p) + h^2(t,\tau)$. If $h(0,0) = 0$, $d/dt\, h(0,0) = (Z(p),[[X,Y],Y](p))$. In general

$$\frac{d^{k+m}}{dt^k d\tau^m} h(0,0) = (-1)^{k+m+1}(Z(p),(\mathrm{ad}^m X,(\mathrm{ad}^{k+1}Y,X))(p))$$

if all derivatives of lower order vanish.

Substitute these in the taylor series expansion

$$h(t,\tau) = \sum_{k=0}^{\infty} \tau^k/k!\ (d/d\tau)^k h(0,0) + t \sum_{k=0}^{\infty} \tau^k/k!(d/dt)^1(d/d\tau)^k h(0,0)$$
$$+ t^2/2! \sum_{k=0}^{\infty} \tau^k/k!\ (d/dt)^2(d/d\tau)^k h(0,0) + \ldots$$

If an element of the form $(Z(p),(ad^{k+1}X,Y)(p)) \neq 0$ for some $k = 0,1,\ldots$, then this element and $Y(p)$ are linearly independent; $h(0,\tau) \neq 0$ in some nbd. $0 < \tau \leq \varepsilon$, $\varepsilon > 0$ hence $h(t,\tau)$ does not change sign as a function of t, at $t = 0$, for $0 < \tau \leq \varepsilon$. This gives sufficient condition (4).

If $(Z(p),(ad^{k+1}X,Y)(p)) = 0$ for all k, dim span $\mathscr{S}_p^0 = 1$ and we continue to the coefficients of t in the expansion, specifically $(Z(p),(ad^k X,(ad^2 Y,X))(p))$, $k = 0,\ldots$. If, for some k, such a coefficient is not zero, dim span $\mathscr{S}_p^1 = 2$ and h is an "odd function" of t for sufficiently small τ; thus changes sign at $t = 0$. Inductively, if the smallest m such that dim span $\mathscr{S}_p^m = 2$ is even, h does not change sign as a function of t, at $t = 0$, for small $\tau > 0$; if such an m is odd, h does change sign. This gives the result.

References.

1. Sussmann, H.J. and Jurdjevic, V.; Controllability of Nonlinear Systems, J. Diff. Eqs. 12(1972), 95-116.

2. Lobry, C.; Contrôlabilité des systèmes non linèaires, SIAM J. Control, 8(1970), 573-605.

3. Hermes, H., LaSalle, J.P.; Functional Analysis and Time Optimal Control, Academic Press, N.Y. (1969).

4. Krener, A.J.; A Generalization of Chow's Theorem and the Bang-Bang Theorem to Nonlinear Control Systems (to appear), SIAM J. Control.

5. Hermes, H.; On Local and Global Controllability,

6. Hermes, H.; Controllability and the Singular Problem, SIAM J. Control, 2(1965), 241-260.

7. Hermes, H. and Haynes, G.W.; On the Nonlinear Control Problem with Control Appearing Linearly, J. SIAM Control, 1(1963), 85-108.

THE HIGH ORDER MAXIMAL PRINCIPLE

Arthur J. Krener

University of California, Davis

1. INTRODUCTION

First order necessary conditions for optimality are the Pontryagin Maximal Principle (PMP) or the Euler-Lagrange and Legendre-Clebsch conditions. In many problems of practical importance, particularly those when the control enters linearly, these conditions are inconclusive for determining the optimal control. In 1964 Kelley [1] discovered a generalization of the Legendre-Clebsch condition and this was extended by Robbins [2,3], Tait [4], Goh [5,6], Kopp and Moyer [7], and Kelley, Kopp and Moyer [8]. An excellent survey of these papers and related results is Gabasov and Kirillova [9].

This condition, now known as the Generalized Legendre-Clebsch Condition (GLC), was developed by studying the high order effect of special control variations on the cost functional of the system. In 1969 Jacobson [10] used similar techniques to develop a different necessary condition. Many of the derivations of the GLC explicitly or implicitly assume that the problem is normal, i.e., there exists sufficient control variations to enable one to cancel out undesirable lower order effects of the special control variations. When there are terminal constraints, this assumption is necessary to insure they can be satisfied, and even without constraints the assumption is necessary to insure the correct form of the test. There generally has been no discussion of when such an assumption is valid or what to do when it is invalid.

In the PMP no assumption of normality is necessary; it is taken care of by the transversality condition on the adjoint variable and the additivity of first order variations. The purpose of this paper is to develop a High Order Maximal Principle (HMP) which extends the PMP, includes the GLC and Jacobson's condition when they apply and also includes numerous other conditions which can be specifically constructed for the problem at hand without assuming normality. We shall only sketch the proof to indicate why the hypotheses are necessary; for a rigorous proof see Krener [12].

2. FIRST ORDER VARIATIONS

The problem we wish to consider is to minimize

$$y_o(x(t_e)) \tag{2.1}$$

where the system equation is

$$\dot{x} = f(x,u) \tag{2.2}$$

subject to the constraints

$$x(t_o) = x^o \tag{2.3}$$

$$y_i(x(t_e)) = 0, \quad i = 1,\dots,m \tag{2.4}$$

$$u(t) \in \Omega, \quad t \in [t_o,t_e]. \tag{2.5}$$

For convenience $x = (x_o,x_1,\dots,x_n)$ with $x_o = t$ and $u = (u_1,\dots,u_n)$. We assume that f in C^∞ with respect to $x_1,\dots,x_n$, $u_1,\dots,u_k$, and a piecewise C^∞ with respect to x^o. Infinite differentiability is not essential, it is only to avoid counting the degree of differentiability needed in a particular setting. Piecewise differentiability means that left and right limits always exist and there are only a finite number of jumps in any compact interval. To speed the exposition we shall ignore the jumps, they are easily handled using standard techniques. We restrict to piecewise C^∞ controls, $u(t)$. The functions $y_o(x),\dots,y_m(x)$ are assumed to be C^∞ and the matrix $(\partial y_i/\partial x_j(x))$ $i = 0,\dots,m$, $j = 0,\dots,n$ is assumed to be of rank $m + 1$. Suppose u^o satisfies (2.5) and the corresponding solution $t \to x(t)$ of (2.2) satisfies (2.3) and (2.4). To develop necessary conditions for u^o we choose some point $t_1 \in (t_o,t_e)$ and modify u^o by replacing it with another admissible control, u^1, in a neighborhood of t_1.

To be more precise, let $s \to \gamma_i(s)x$ be the family of solutions of the differential equation $d/ds\ \gamma_i(s)x = f(\gamma_i(s)x,u^i)$ with initial conditions $\gamma_i(0)x = x$ for $i = 0,1$. Consider the family of trajectories whose locus of endpoints is

(2.6) $\gamma_o(t_e-t_1)\gamma_1(s)x(t_1-s) = \gamma_o(t_e-t_1)\gamma_1(s)\gamma_o(-s)x(t_1)$.

This locus is differentiable with respect to s and its effect on y_o and y_i studied, yielding first order necessary conditions.

The above control variation was made before t_1; it could have been made after t_1 or around t_1 as follows:

(2.7) $\gamma_o(t_e-t_1-s)\gamma_1(s)x(t_1) = \gamma_o(t_e-t_1)\gamma_o(-s)\gamma_1(s)x(t_1)$

or

(2.8) $\gamma_o(t_e-t_1-s/2)\gamma_1(s)x(t_1-s/2)$
$= \gamma(t_e-t_1)\gamma_o(-s/2)\gamma_1(s)\gamma_o(-s/2)x(t_1)$

and the same necessary conditions result.

Another type of control variation (2.9) is to stop short of t_e or to continue past t_e using the same control. This is equivalent to abbreviating or lengthening the trajectory at any intermediate point, t_1.

(2.9) $\gamma_o(\pm s)x(t_e) = \gamma_o(t_e-t_1)\gamma_o(\pm s)x(t_1)$.

A standard proof of the Maximal Principle is to consider the convex cone K^1 of vectors generated by the derivatives with respect to s of all expressions of the form (2.6) through (2.9). This cone is a measure of the controllability at x^e. This is true because the magnitude of control variation can be changed by multiplying s by a nonnegative constant, the derivative of two control variations made jointly at two different times is the sum of their individual derivatives, and for two variations of type (2.6) made at the same time with controls u^1 and u^2, the derivative of (2.10) is the sum of their individual derivatives.

(2.10) $\gamma_o(t_e-t_1)\gamma_2(s)\gamma_1(s)\gamma_o(-2s)x(t_1)$.

Since we are interested in the effect that these variations have on $y_o,\dots,y_m$ we define another convex cone L^1 of $m + 1$ dimensional vectors by

(2.11) $$L^1 = \left[\frac{\partial y_i}{\partial x_j}(x^e)\right] K^1.$$

Using a fixed point argument (Halkin [11]) it can be shown that if

u^o is optimal then L^1 can be separated from the m + 1 vector (-1,0,...,0) by linear functional $\nu = (\nu_o,\ldots,\nu_m)$ where $\nu_o \leq 0$. This linear functional on m + 1 vectors defines a linear functional $\lambda(t_e) = \nu[\partial y_i/\partial x_j(x^e)]$ on n + 1 vectors which can be pulled back along x(t) using the adjoint differential equation and the result is the Pontryagin Maximal Principle:

If u^o is optimal then there is nontrivial $\lambda(t)$ such that if $H(\lambda,x,u) \triangleq \lambda f(x,u)$ then

(2.12) $$\dot{\lambda}(t) = -\frac{\partial H}{\partial x}(\lambda(t),x(t),u^o(t)),$$

(2.13) $$\lambda(t_e) = \nu\left[\frac{\partial y_i}{\partial x_j}(x^e)\right] \quad \text{with } \nu_o \leq 0,$$

(2.14) $$0 = H(\lambda(t),x(t),u^o(t)) \geq H(\lambda(t),x(t),u) \quad \forall\, t \in [t_o,t_e] \text{ and } \forall\, u \in \Omega.$$

3. THE HIGH ORDER MAXIMAL PRINCIPLE WITH TERMINAL CONSTRAINTS

A first order control variation is inconclusive if

(3.1) $$\frac{d}{ds}\gamma_o(t_e-t_1)\gamma_1(s)\gamma_o(-s)x(t_1) = 0 \quad \text{at } s = 0.$$

Let

(3.2) $$\beta(s)x = \gamma_1(s)\gamma_o(-s)x.$$

We consider the effect that the second derivative has on the cost and terminal constraints. In fact if we could include the vector

(3.3) $$\frac{d^2}{ds^2}\gamma_o(t_e-t_1)\beta(0)x(t_1)$$

in the cone K^1 we would obtain a new necessary condition. This is not possible since the vectors of K^1 are first order effects. However if we replace the parameter s by $s^2/2$ in any one of the variations (2.6) through (2.10) then the first derivative of the new variation vanishes and the second derivative is the old first derivative. One might hope to construct a cone of second derivatives since the magnitude of a second order control can be changed by multiplying s by a nonnegative constant. However in general the joint second derivative of two variations satisfying (3.1) made at different times is not equal to the sum of their individual second derivatives. That is, if $\beta_i(s)x$ is a control variation with u^i of type (3.2) satisfying (3.1) at t_i for $i = 1,2$ then

$$(3.4)\quad \frac{d^2}{ds^2}\,(\gamma_o(t_e-t_2)\beta_2(0)\gamma_o(t_2-t_1)\beta_1(0)x^1)$$
$$= \frac{d^2}{ds^2}\,\gamma_o(t_e-t_2)\beta_2(0)x^2 + \frac{d^2}{ds^2}\,\gamma_o(t_e-t_1)\beta_1(0)x^1$$
$$+ \frac{\partial}{\partial x}\left(\frac{d}{ds}\,\gamma_o(t_e-t_2)\beta_2(0)\gamma_o(t_2-t_1)x^1\right)\frac{d}{ds}\,\beta_1(0)x^1 .$$

where $x^i \triangleq x(t_i)$ and $\frac{\partial}{\partial x}\left(\frac{d}{ds}\,\gamma_o(t_e-t_2)\beta_2(0)\gamma_o(t_2-t_1)x^1\right)$ is the $(n+1) \times (n+1)$ matrix of partial derivatives of $x \to d/ds\ \gamma_o(t_e-t_2)\beta_2(0)\gamma_o(t_2-t_1)x$ evaluated at x^1. However, if we restrict to variations satisfying

$$(3.5)\qquad \frac{d}{ds}\,\beta(0)x^1 = 0$$

then the last term is zero and the second derivatives do add.

Given two control variations (2.6) made at the same time, it was possible to construct a variation (2.10) whose first derivative was the sum of their individual first derivatives. This also fails in general for second derivatives. To see why let $\beta_i(s)x = \gamma_i(s)\gamma_o(-s)x$ for $i = 1,2$ and suppose they both satisfy (3.5).

Assume f and u^i are C^∞ at x^1. Then

$$(3.6)\qquad \frac{d}{ds}\,\beta_i(0)x = f_i(x^1) - f_o(x^1) = 0$$

$$(3.7)\quad \frac{d^2}{ds^2}\,\beta_i(0)x = \frac{\partial f_i}{\partial x}(x^1)f_i(x^1) - 2\,\frac{\partial f_i}{\partial x}(x^1)f_o(x^1) + \frac{\partial f_o}{\partial x}(x^1)f_o(x^1)$$

where $f_i(x) \triangleq f(x,u^i(x_1))$ for $i = 0,1,2$. On the other hand, let $\beta_3(s)x = \gamma_2(s)\gamma_1(s)\gamma_o(-2s)x$. Then

$$(3.8)\qquad \frac{d}{ds}\,\beta_3(0)x = f_2(x^1) + f_1(x^1) - 2f_o(x^1) = 0$$

$$(3.9)\quad \frac{d^2}{ds^2}\,\beta_3(0)x^1 = \frac{\partial f_2}{\partial x}(x^1)f_2(x^1) + 2\,\frac{\partial f_2}{\partial x}(x^1)(f_1(x^1)-2f_o(x^1))$$
$$+ \frac{\partial f_1}{\partial x}(x^1)f_1(x^1) - 4\,\frac{\partial f_1}{\partial x}(x^1)f_o(x^1) + 4\,\frac{\partial f_o}{\partial x}(x^1)f_o(x^1).$$

We need a stronger condition than (3.5) to insure additivity. If

$$(3.10)\qquad \frac{d}{ds}\,\beta_i(0)x(t) = 0 \quad \text{for} \quad t \in (t_1-\varepsilon,\ t_1+\varepsilon),$$

then

$$\frac{\partial f_i}{\partial x}(x^1)f_o(x^1) = \frac{\partial f_o}{\partial x}(x^1)f_o(x^1) \tag{3.11}$$

and this along with (3.6) implies

$$\frac{d^2}{ds^2}\beta_3(0)x^1 = \frac{d^2}{ds^2}\beta_1(0)x^1 + \frac{d^2}{ds^2}\beta_2(0)x^1 . \tag{3.12}$$

It may appear unnecessary to require (3.10); after all if we wish to add two variations satisfying (3.5) at the same time, why not move one to an arbitrarily close time so that it changes very little and then add them. The reason this cannot be done is because the variation that is moved might not satisfy (3.5) at the new time. Actually, the stronger condition will prove very helpful in constructing new variations from old ones as will be seen in Section 4.

Generalizing, we define a <u>control variation</u> to $\underline{u^o}$ as

$$\beta(s)x = \gamma_o(p_2(s))\gamma_n(q_n(s))\cdots\gamma_1(q_1(s))\gamma_o(p_1(s))x \tag{3.13}$$

where each γ_i corresponds to an admissible u^i and $p_i(s)$, $q_i(s)$ are polynomials satisfying $p_i(0) = q_i(0) = 0$ and $q_i(s) \geq 0$ for small $s > 0$. This includes all previous variations before, after, or around t_1 and also (2.9). A control variation is of order k at x^1 if for some $\epsilon > 0$

$$\frac{d^j}{ds^j}\beta(s)x(t) = 0 \quad \text{for} \quad j = 1,\dots,k-1 \text{ and } t \in (t_1-\epsilon,\ t_1+\epsilon). \tag{3.14}$$

If a variation is of order k at x^1 and if and all the controls involved are C^∞ at x^1 then it can be shown that the kth derivative of (3.13) does not depend on whether it is done before, after, or around x^1, i.e., it depends only on the sum, $p_1(s) + p_2(s)$, rather than on the individual $p_i(s)$. Furthermore the kth derivative of two k^{th} order variations made jointly at different times is just the sum of the individual k^{th} derivatives. Finally, for two k^{th} order variations, β_i, made at the same time, where $\beta_2(s)x = \gamma_n(q_n(s))\cdots\gamma_1(q_1(s))\gamma_o(p(s))x$, it can be shown that the variation β_3 defined below is of kth order and its kth derivative is the sum of the kth derivatives of β_1 and β_2.

$$\beta_3(s)x = \gamma_n(q_n(s))\cdots\gamma_1(q_1(s))\beta_1(s)\gamma_o(p(s)x. \tag{3.15}$$

We define K^k as the convex cone generated by all vectors of the form

$$\frac{d^k}{ds^k} \gamma_o(t_e - t_1)\beta(s)x^1 \tag{3.16}$$

where β is of order k at x^1. By replacing s by s^h a control variation of order k is shifted to one of order $k \cdot h$ therefore $K^k \subseteq K^{k\cdot h}$. This allows us to define the convex cones $K = \bigcup_{k\geq 1} K^k$ and $L = \left[\frac{\partial y_i}{\partial x_j}(x^e)\right]K$. The rest of the development of the HMP proceeds as before using a fixed point argument. The result is the High Order Maximal Principle with terminal constraints.

If u^o is optimal then there is a nontrivial $\lambda(t)$ such that

$$\dot{\lambda}(t) = -\frac{\partial H}{\partial x}(\lambda(t), x(t), u^o(t)), \tag{3.17}$$

$$\lambda(t_e) = \nu\left[\frac{\partial y_i}{\partial x_j}(x^e)\right] \quad \text{with} \quad \nu_o \leq 0, \tag{3.18}$$

$$0 = H(\lambda(t), x(t), u^o(t)) \geq H(\lambda(t), x(t), u) \quad \forall\, t \in [t_o, t_e] \text{ and } \forall\, u \in \Omega, \tag{3.19}$$

(3.20) if $\beta(s)x$ is a control variation of order k at x^1 then

$$\lambda(t_1)\frac{d^k}{ds^k}\beta(0)x^1 \leq 0.$$

If the terminal constraints are absent then additivity of variations is not needed and the situation is much simpler. Condition (3.18) reduces to $\lambda(t_e) = -dy_o(x^e)$ and $\lambda(t)$ is defined by (3.17). Condition (3.19) remain the same but (3.20) changes. The first nonzero derivative of the form

$$\frac{d^k}{ds^k} y_o(\gamma_o(t_e - t_1)\beta(0)x^1) \tag{3.21}$$

must be nonpositive.

4. EXAMPLES

Consider the problem of minimizing $y(x(t_e))$ subject to

$$\dot{x} = a_o(x) + u_1(t)a_1(x), \tag{4.1}$$

$$|u| \leq 1,\ x(t_o) = x^o,\ y_i(x(t_e)) = 0, \quad i = 1,\dots,m. \tag{4.2}$$

Let $u^o(t)$ be the candidate for the optimal control and $\lambda(t)$ satisfy (3.17) and (3.18).

Using a first order variation, $\beta_o(s)x = \gamma_o(\pm s)x$, of type (2.9) and condition (3.20) we obtain

$$H(\lambda(t),x(t),u^o(t)) = 0 \tag{4.3}$$

and using a different variation, $\beta_1(s)x = \gamma_1(s)\gamma_o(-s)x$ of type (2.6) where u^1 = constant and (3.20) we obtain

$$H(\lambda(t),x(t),u^o(t)) \geq H(\lambda(t),x(t),u^1). \tag{4.4}$$

We see that (3.20) implies (3.19).

If there are no terminal constraints, by (4.3) we can apply (3.21) to β_o at t_e,

$$\lambda(t_e)\frac{\partial a_o}{\partial x}(x^e) + a_o(x^e)\frac{\partial^2 y_o}{\partial x^2}\, a_o(x^e) \leq 0.$$

If $|u^o(t)| < 1\ \forall\ t \in [t_o,t_e]$, then (4.4) implies that

$$\frac{\partial H}{\partial u}(\lambda(t),x(t),u^o(t)) = \lambda(t)a_1(x(t)) = 0,\ \forall\ t \in [t_o,t_e]. \tag{4.5}$$

To differentiate (4.5) with respect to t we note that if $b(x)$ is a vector valued function of x then

$$\frac{d}{dt}\lambda(t)b(x(t)) = \lambda(t)[a_o,b](x(t)) + u^o(t)\lambda(t)[a_1,b](x(t)) \tag{4.6}$$

where the Lie bracket is defined by

$$[b_1,b_2](x) = \frac{\partial b_2}{\partial x}(x)b_1(x) - \frac{\partial b_1}{\partial x}(x)b_2(x) \tag{4.7}$$

and satisfies

$$[b_1,b_2] = -[b_2,b_1] \tag{4.8}$$

$$[b_1[b_2,b_3]] = [[b_1,b_2]b_3] + [b_2[b_1,b_3]]. \tag{4.9}$$

Differentiating (4.5) we obtain

$$\frac{d}{dt}\frac{\partial H}{\partial u} = \lambda(t)[a_o,a_1](x(t)) = 0 \tag{4.10}$$

$$\frac{d^2}{dt^2}\frac{\partial H}{\partial u} = \lambda(t)[a_o[a_o,a_1]](x(t)) + u^o(t)\lambda(t)[a_1[a_o,a_1]](x(t)) = 0. \tag{4.11}$$

If the coefficient of u^o is not zero then u^o is determined by (4.11). If it is zero then we can continue differentiating until, perhaps, some higher derivative determines u. It can be shown using (4.8) and (4.9) that the first k, such that $\partial/\partial u(d^k/dt^k(\partial H/\partial u))$ is not zero for all t, will be even.

Henceforth for convenience we assume that $u^o \equiv 0$. Jacobson's condition [10] is based on the fact that if there are no terminal constraints then (4.5) implies (3.21) is zero for β_1 with $k = 1$. Applying (3.21) with $u^1 = \pm 1$ and $k = 2$ we obtain

$$(4.12) \qquad \lambda(t)\frac{\partial a_1}{\partial x}(x(t))a_1(x(t)) + a_1(x(t))\frac{\partial^2}{\partial x^2}y_o(\gamma_o(t_e-t)x(t))a_1(x(t)) \leq 0.$$

Kelley constructed a control variation of order 2 from a pair of variations of order one which cancel each other out. Let $\beta_2(s)x = \gamma_2(s)\gamma_1(s)\gamma_o(-2s)x$ where $u^1 = 1$ and $u^2 = -1$ then $d/ds\ \beta_2(0)x = 0$ and $d^2/ds^2\ \beta_2(0)x = [a_o,a_1](x)$. Reversing u^1 and u^2 gives another variation of order 2 with the opposite second derivative, but together they yield (4.10). By the technique (3.15) of adding control variations made at the same time we obtain $\beta_3(s)x = \gamma_1(s)\gamma_2(2s)\gamma_1(s)\gamma_o(-4s)x$ a variation of order 3 and the condition

$$(4.13) \qquad \lambda(t)\frac{d^3}{ds^3}\beta_3(0)x(t) = 12\lambda(t)[a_o[a_o,a_1]](x(t)) - 4\lambda(t)[a_1[a_o,a_1]](x(t)) \leq 0.$$

By (4.11) and $u^o \equiv 0$ this becomes Kelley's condition

$$(4.14) \qquad \frac{\partial}{\partial u}\frac{d^2}{dt^2}\frac{\partial H}{\partial u} = \lambda(t)[a_1[a_o,a_1]](x(t)) \geq 0.$$

If $d^3/ds^3\ \beta_3(0)(x(t)) = 0$, then β_3 is a control variation of order 4 and its fourth derivative yields a new condition.

If u^1 and u^2 are reversed in β_3 then another variation of order 3 is obtained with third derivative, $-12[a_o[a_o,a_1]](x(t)) - 4[a_1[a_o,a_1]](x(t))$. If $[a_1[a_o,a_1]](x(t)) = 0$, then by adding these two variations of order 3 a variation, β_4, of order 4 is obtained. The fourth derivative of this new variation can be applied to (3.20) to yield a new test, if the fourth derivative is zero then β_4 is a variation of order 5 and its fifth derivative can be applied to (3.20). Another possible way to construct a variation of order 5 is to reverse u^1 and u^2 in β_4 to obtain a different variation of order 4 and then add them. Kelley, Kopp and Moyer [8] used this technique to obtain the GLC for this problem. The first nonzero derivative of the form

$$(-1)^h \frac{\partial}{\partial u} \frac{d^{2h}}{dt^{2h}} \frac{\partial H}{\partial u} (\lambda(t), x(t), u^o(t)) \tag{4.15}$$

must be nonpositive. This appears with the opposite inequality if $\nu_o \geq 0$. To derive this test an assumption of normality is necessary even if there are no terminal constraints. No explicit definition of normality was given by Kelley, Kopp and Moyer; therefore the usefulness of this test is severely limited.

An alternate approach is possible if $[a_1[a_o,a_1]](x(t))$ is a linear combination of $a_o(x(t))$, $a_1(x(t))$ and $[a_o,a_1](x(t))$ it is possible to reparametrize β_3 by s^2, β_2 by s^3, β_1 and β_o by s^6, and combine them into a variation of order > 6 where $[a_1[a_o,a_1]](x(t))$ is cancelled out. The general method for constructing new variations is to take two (or more) variations which cancel each other, adjust their parameters so they are of the same order and combine them. It is shown in [9] that Goh's [6] generalization of (4.15) can be obtained in this way for the problem with several control variables entering linearly.

REFERENCES

1. H. J. Kelley, A second variation test for singular extremals, AIAA J., 2(1964), pp. 1380-1382.

2. H. M. Robbins, Optimality of intermediate-thrust arcs of rocket trajectories, AIAA J. 3(1965), pp. 1094-1098.

3. H. M. Robbins, A generalized Legendre-Clebsch condition for the singular cases of optimal control, IBM J. Res. Develop., 11(1967), pp. 361-372.

4. K. Tait, Singular problems in optimal control, Ph.D. thesis, Harvard, 1965.

5. B. S. Goh, The second variation for the singular Bolza problem, SIAM J. Control, 4(1966), pp. 309-325.

6. B. S. Goh, Necessary conditions for singular extremals involving multiple control variables, SIAM J. Control, 4(1966), pp. 716-731.

7. R. E. Kopp and H. G. Moyer, Necessary conditions for singular extremals, AIAA J., 3(1965), pp. 1439-1444.

8. H. J. Kelley, R. E. Kopp and H. G. Moyer, Singular Extremals, Topics in Optimization, Academic Press, N.Y., 1967, pp. 63-101.

9. R. Gabasov and F. M. Kirillova, High order necessary conditions for optimality, SIAM J. Control, 10(1972), pp. 127-168.

10. D. H. Jacobson, A new necessary condition of optimality for singular control problems, SIAM J. Control, 7(1969), pp. 578-595.

11. H. Halkin, A maximum principle of the Pontryagin type for systems described by nonlinear difference equations, SIAM J. Control, 4(1966), pp. 90-111.

12. A. J. Krener, The high order maximum principle for singular extremals, to appear.

OPTIMAL CONTROL ON MANIFOLDS

M.L.J.Hautus

Department of Mathematics

Technological University Eindhoven

1. INTRODUCTION

Recently, a number of general methods for obtaining necessary conditions for optimality have been derived (see [2],[3],[5]). In [4], the author has given a general method, starting from the following basic problem:

BP(S,f): "Given a set S in R^n and a function $f: R^n \to R^1$, determine $x \in S$ such that f(x) is maximal".

The method is particularly useful in the case, where S is given as an intersection $S = S_1 \cap \ldots \cap S_k$ of a finite number of sets. The key tool is the concept *derived cone*, which is a kind of linear approximation of the set S. Explicitly:

<u>Definition</u>. If $b \in S \subseteq R^n$, then a closed convex cone C is called a *derived cone* of S at b, if for every k-tuple $p_1,\ldots,p_k$ of vectors in riC (the relative interior of C), there exists a $\mathcal{C}'$-map $\xi: \mathcal{N} \cap R^k_+ \to S$ satisfying:

$$\xi(\tau) = b + \sum_{i=1}^{k} p_i\tau_i + o(\tau) \qquad (\tau \to 0) \tag{1}$$

Here $\mathcal{N}$ is a neighborhood of the origin in R^k and $R_+^k := \{\tau \in R^k | \tau_i \geq 0$ $(i = 1,\ldots,k)\}$ is the positive orthant of R^k.

Definition. If C is a convex cone, then the *polar cone* of C is defined by:

$$C^0 := \{\psi \in R_n \mid \psi p \leq 0 \quad \text{for all } p \in C\}. \tag{2}$$

(We denote by R^n the set of column vectors and by R_n the set of row vectors, so that ψp is the matrix product and hence the inner product of ψ and p.)

The relevance of derived cones for BP(S,f) is clear from the following (elementary) result:

Lemma 1. Let b be a solution of BP(S,f) and C a derived cone of S at b. If f is $\mathcal{C}'$ at b , then

$$\nabla f(b) \in C^0$$

($\nabla f(b)$ denotes the gradient of f at b and is considered a row vector).

In the literature on optimization theory, a great number of different types of approximating cones have been introduced (see, e.g. [7]). For all types, Lemma 1 is valid. The distinguishing property of derived cones is the following result:

Lemma 2. Let $b \in S := S_1 \cap S_2$ and let C_1, C_2 be derived cones of S_1 and S_2, respectively. If C_1 and C_2 are not separated (that is, if there does not exist a nonzero $\varphi \in R_n$ such that $\varphi x \leq 0 \leq \varphi y$ for all $x \in C_1$, $y \in C_2$), then $C_1 \cap C_2$ is a derived cone of $S_1 \cap S_2$.

It is because of Lemma 2 that derived cones are useful for obtaining necessary conditions in the case that S is given as an intersection of a number of sets. In order to show this, let us first recall the following property of polar cones (see [6]):

Lemma 3. If C_1 and C_2 are nonseparated convex cones, then

$$(C_1 \cap C_2)^0 = C_1^0 + C_2^0 .$$

Then we obtain the following theorem, which is the main result of [4]:

Theorem 1. Let $b \in S := S_1 \cap \ldots \cap S_k$ be a solution of BP(S,f). If f is $\mathcal{C}'$ at b and if $C_1,\ldots,C_k$ are derived cones of $S_1,\ldots,S_k$ at b, then there exist a number $\rho \geq 0$, and vectors $\psi_i \in C_i^0$ $(i = 1,\ldots,k)$ such that

i) $(\rho,\psi_1,\ldots,\psi_k) \neq 0$ (nontriviality condition)

ii) $\rho\nabla f(b) = \psi_1 + \ldots + \psi_k$ (polar rule).

It will be illustrative to give a proof of this theorem for the case k = 2 (the general result follows by an induction argument). If C_1 and C_2 are not separated, then $C_1 \cap C_2$ is a derived cone of $S_1 \cap S_2$. According to Lemma 1 it follows that

$$\nabla f(b) \in (C_1 \cap C_2)^0 = C_1^0 + C_2^0 .$$

If C_1 and C_2 are separated, then the polar rule holds with $\rho = 0$, $\psi_1 = -\psi_2 = \varphi$ (the separating vector). □

In order to apply Theorem 1 in a given situation one has to construct derived cones of sets occurring in the problem. We give two examples:

Example 1. If $r : R^n \to R^1$ is $\mathcal{C}'$ at b and $r(b) = 0$, $\nabla r(b) \neq 0$, then $C := \{p \in R^n \mid \nabla r(b)p = 0\}$ is a derived cone of the set $S := \{x \in R^n \mid r(x) = 0\}$ at b.

Example 2. If $g : R^n \to R^1$ is $\mathcal{C}'$ at b and $g(b) = 0$, $\nabla g(b) \neq 0$, then $C := \{p \in R^n \mid \nabla g(b)p \leq 0\}$ is a derived cone of $S := \{x \in R^n \mid g(x) \leq 0\}$ at b. If $g(b) < 0$, then R^n is a derived cone of S.

Using these examples one can apply Theorem 1 to obtain well-known results in mathematical programming, discrete-time and continuous-time optimal control (see [4]).

2. OPTIMIZATION ON MANIFOLDS

Theorem 1 can easily be extended to manifolds. We have to adapt the definitions and results of the introduction slightly:

BP'(S,f): "given a set S in a manifold M and a function $f : M \to R^1$, determine $x \in S$ such that f(x) is maximal".

Definition. If $b \in S \subseteq M$, then a closed convex cone $C \subseteq M_b$ (the tangent manifold of M at b) is called a derived cone of S if for any k-tuple of vectors $p_1, \ldots, p_k$ in riC, there exists a $\mathscr{C}'$-map $\xi: R_+^k \to S$, satisfying $\frac{\partial \xi}{\partial \tau_i} = p_i$ (the tangent vector to $\tau_i \mapsto \xi(\tau)$ at b).

The polar C^0 of a derived cone C is a cone in M_b^*, the dual space. Hence, elements of C^0 are covariant vectors. Now, Lemma 1 and 2 and Theorem 1 are easily seen to hold (for example, by using local coordinates). Note that $\nabla f(b) = df(b)$ is the differential of f at b, and hence an element of M_b^*.

Let us consider optimal control problems on a manifold M.

Definition. We call a continuous function $f : M \times U \to T(M)$, where U is a topological space, a $\mathscr{C}'$ *vector field depending continuously on* $u \in U$, if $x \mapsto f(x,u)$ is a vector field for every $u \in U$, such that $(x,u) \mapsto D_x f(x,u)$ is continuous ($D_x f(x,u)$ is the derivative of f with respect to x).

We assume that we have

01) A topological space U.

02) A $\mathscr{C}'$ *vector field* $f: M \times U \to T(M)$, depending continuously on $u \in U$.

03) A $\mathscr{C}'$ function $h: M \to R^1$.

04) A point $c \in M$, a set $X \subseteq M$, and a number $T > 0$.

Given a piecewise continuous control function $u: [0,T] \to U$, we consider the differential equation

$$\dot{x}(t) = f(x,u), \quad x(0) = c \; . \tag{3}$$

The solution of this equation will be denoted by x_u.
A piecewise continuous control function $u: [0,T] \to U$ is called *admissible* if x_u exists on [0,T] and satisfies $x_u(T) \in X$. Then we can

formulate the optimal control problem:

OCP: "Determine an admissible control u such that $h(x_u(T))$ is maximal".

The extension of the maximum principle to this situation is straightforward. However, the condition that the control constraint set be independent of x is rather restrictive here.

Example. Let $M := S^{n-1}$ (the unit sphere in R^n). Let the system be given by $\dot{x} = f(x,u)$, $u \in U$, $x(0) = c \in M$. In order that $x(t) \in M$ for all t, the following condition must be satisfied: $x'f(x,u) = 0$ (where the prime denotes transposition). If $x \mapsto f(x,u)$ is a vector-field (as is assumed for all $u \in U$ in OCP), then this condition is satisfied. In general, however one has to consider the condition $x'f(x,u) = 0$ as an additional restriction on u depending on x.

In order to formulate such optimal control systems, we add the following assumption to 01,...,4: We are given:

05) A set valued map $x \mapsto U(x)$, where $x \in M$, $U(x) \subseteq U$.

Now we call a piecewise continuous control function *admissible*, if x_u exists on [0,T] and satisfies $x_u(T) \in X$, and if $u(t) \in U(x_u(t))$ for $0 \le t \le T$.

The optimal control problems reads as follows:

OCP': "Find an admissible control that maximizes $h(x_u(T))$".

We reformulate this problem in a more abstract fashion, replacing the differential equation by a contingent equation. Let us assume that we have:

A1) A manifold M, a point $c \in M$, a subset $X \subseteq M$, and a number $T > 0$.
A2) A $\mathcal{C}^1$ function $h: M \to R^1$.
A3) A set valued map $x \mapsto V(x)$, where $x \in M$ and $V(x) \subseteq M_x$ for each x.

We denote by Λ the set of piecewise continuously differentiable functions $x: [0,T] \to M$, with $x(0) = c$, $\dot{x}(t) \in V(x(t))$ for all t at which x is differentiable. The optimal control problem reads:

OCP": "Find $x \in \Lambda$ satisfying $x(T) \in X$ such that $h(x(T))$ is maximal".

Using the *reachable set*, defined by $W := \{x(T) \mid x \in \Lambda\}$ we observe that OCP" is equivalent to $BP(X \cap W, h)$ in the following sense: If x is a solution of OCP" then x(T) is a solution of BP. If b is a solution of BP, then any function $x \in \Lambda$ with $x(T) = b$ is a solution of OCP" (note that such a function exists). Thus we have reduced the optimal control problem to the basic problem. The major problem now is to construct a derived cone for W (we assume that a derived cone C of X is given). For that aim we make two additional assumptions (see also [1,4.4]). Let $\bar{x}$ be an optimal trajectory.

A4) If $x_0 \in M$, then there exists for every $y \in V(x_0)$ a $\mathcal{C}'$ function $v: \mathcal{N} \to T(M)$, where $\mathcal{N}$ is a neighborhood of x_0, such that $v(x) \in V(x)$ and $v(x_0) = y$. (Note that v is a vector field defined on $\mathcal{N}$.)

A5) For some $\delta > 0$, there exists a $\mathcal{C}'$ vector field $w: \mathcal{D} \to T(M)$ where $\mathcal{D} := \{(x,t) \mid 0 \le t \le T, |x-\bar{x}(t)| < \delta\}$, depending piecewise continuously on t (that is, there exist numbers $t_0, t_1, \ldots, t_n$, satisfying $0 = t_0 < t_1 < \ldots < t_n = T$, such that for $i = 1, \ldots, n$ there exist $\mathcal{C}'$ vectorfields f_i: $[t_{i-1}, t_i] \to T(M)$ depending continuously on $t \in [t_{i-1}, t_i]$, and satisfying $f_i = f$ for $t_{i-1} < t < t_i$). The vectorfield w has the property $w(x,t) \in V(x)$ for every $(x,t) \in \mathcal{D}$.

Using these assumptions we can construct perturbations of $\bar{x}$. An *elementary perturbation* of $\bar{x}$ is of the form $\pi = (t_0, y_0)$, where $t_0 \in (0,T)$ is a point where $\bar{x}$ is continuous and $y_0 \in V(\bar{x}(t_0))$. To an elementary perturbation corresponds a perturbed trajectory $x_\pi(\cdot,\varepsilon) \in \Lambda$, defined by

$$\begin{cases} x_\pi(t,\varepsilon) = \bar{x}(t) & (0 \le t \le t_0 - \varepsilon) \\ \dot{x}_\pi(t,\varepsilon) = v(x_\pi(t,\varepsilon)) & (t_0 - \varepsilon < t < t_0) \\ \dot{x}_\pi(t,\varepsilon) = w(x_\pi(t,\varepsilon),t) & (t_0 < t < T) \\ x_\pi \text{ is continuous.} & \end{cases}$$

It is a consequence of the theory of ordinary differential equations, that for small $\varepsilon > 0$, such a function exists, is unique and that x_π depends $\mathcal{C}^1$ on ε. The tangent vector of $\varepsilon \mapsto x_\pi(T,\varepsilon)$ is given by

$$p_\pi := \Phi(T,t_0)\ (y_0 - \dot{\bar{x}}(t_0))$$

where Φ is the transition matrix of the variational equation:

$$\dot{y}(t) = A(t)y(t)$$

where $A(t) := D_x w(\bar{x}(t),t)$.

Using elementary (and combined) perturbations one can show (see [4]):

<u>Lemma 4</u>. The perturbation cone, that is, the closed convex cone P generated by the perturbation vectors p_π, is a derived cone of W at $b := \bar{x}(T)$.

Therefore we can apply Theorem 1: There exists $\rho \geq 0$, $\varphi \in C^0$, $\psi \in P^0$ such that $\rho\nabla h(b) = \varphi + \psi$, and $(\rho,\varphi,\psi) \neq 0$.

Defining $\psi(t) := \psi\Phi(T,t)$ we obtain the maximum principle:

<u>Theorem 2</u>. Let $\bar{x}$ be a solution of OCP" and let the conditions A1) ... A5) be satisfied. Let C be a derived cone of Λ at $b := \bar{x}(T)$. Then there exist a number $\rho \geq 0$ and a vector $\varphi \in C^0$, not both zero, such that the solution of the adjoint equation

$$\dot{\psi}(t) = -\psi(t)A(t)$$

with boundary condition

$$\psi(T) = \rho\nabla h(b) - \varphi$$

where $A(t) := D_x w(\bar{x}(t),t)$, satisfies

$$\psi(t)\bar{x}(t) = \max_{y \in V(\bar{x}(t))} \psi(t)y$$

for all t at which $\bar{x}$ is $\mathcal{C}^1$.

If we go back to OCP', we see that with $V(x) := f(x,U(x))$, conditions A1,2,3) are satisfied. In order to obtain that conditions A4,5) are satisfied we assume that U is a manifold and that

$$U(x) = \{u \in U \mid r(x,u) = 0\}$$

where $r: M \times U \to R^{\ell}$ is a $\mathcal{C}'$ map. Then we can show

Lemma 5. Let $\bar{u}$ be a solution of OCP' and $\bar{x} := x_{\bar{u}}$. If $x \mapsto r(x,u)$ is regular (that is $d_x r(x,u)$ has rank ℓ at all points $(\bar{x}(t),u)$, $(0 \le t \le T,\ u \in U(\bar{x}))$, then conditions A4) and A5) are satisfied.

The proof depends on an application of the implicit function theorem (compare also [1]). Applying Theorem 2 to this problem we find

Theorem 3. Let $\bar{u}$ be a solution of OCP', let the conditions of Lemma 5 be satisfied and let $f: M \times U \to T(M)$ be $\mathcal{C}'$. Let C be a derived cone of X at $\bar{x}(T)$. Then there exist a number $\rho \ge 0$, a vector $\varphi \in C^0$, a piecewise continuous function $\lambda: [0,T] \to R^{\ell}$, such that the solution of the adjoint system

$$\dot{\psi} = -D_x H(\psi,\bar{x},\bar{u},\lambda)$$

$$\psi(T) = \rho \nabla h(\bar{x}(T)) - \varphi$$

satisfies the maximum principle

$$H(\psi,\bar{x},\bar{u},\lambda) = \max\ \{H(\psi,\bar{x},v,\lambda) \mid r(\bar{x}(t),v) = 0\}$$

and

$$D_u H(\psi,\bar{x},\bar{u},\lambda) = 0.$$

Here $H(\psi,x,u,\lambda) := \psi f(x,u) - \lambda r(x,u)$.

Note that the functions v and w do not appear explicitly in Theorem 3. The classical maximum principle, Theorem 2 and Theorem 3 can be viewed as special cases of a general formulation. Define the relation

$$R := \{(x,\dot{x}) \in M \times T(M) \mid \dot{x} \in V(x)\}.$$

Let $\bar{x}$ be a solution of OCP, (or OCP', OCP") and let for every $(x,y) \in R$, $D_{x,y}$ be a derived cone of R at (x,y). Then there exists $\rho \geq 0$, $\varphi \in C^0$ and a function ψ: $[0,T] \to T(M)^*$, satisfying

$$(\dot{\psi},\psi) \in D^0_{\bar{x},\dot{\bar{x}}}, \quad \psi(T) \in \rho\nabla h(\bar{x}(t)) - \varphi.$$

It would be interesting to find out under what conditions this principle is generally valid.

REFERENCES

[1] W.G. Boltjanski, *Mathematische Methoden der optimalen Steuerung*, Geest & Portig K.G., Leipzig 1971.

[2] M. Canon, C. Cullum and E. Polak, *Constrained minimization problems in finite dimensional spaces*, SIAM J. Control,4 (1966), pp. 528-547.

[3] H. Halkin and L.W. Neustadt, *General necessary conditions for optimization problems*, Proc. Nat. Acad. Sci. (1966), pp. 1066-1071.

[4] M.L.J. Hautus, *Necessary conditions for multiple constraint optimization problems*, to appear in SIAM J. Control, 11, November.

[5] L.W. Neustadt, *A General theory of extremals*, J. Comput. System. Sci, 3 (1969), pp. 57-92.

[6] R.T. Rockafellar, *Convex analysis*, Princeton University Press, N.J. 1970.

[7] A.P. Wierzbicki, *Maximum principle for semiconvex performance functionals*, SIAM J. Control 10, pp. 444-459.

PROBLEMS IN GEODESIC CONTROL.

JOHN GROTE

RESEARCH FELLOW AT THE CONTROL THEORY CENTRE,

UNIVERSITY OF WARWICK.

Introduction.

In recent papers [2,4,10,11] we have seen control systems defined on Lie groups. That is, if G is a Lie group, L(G) its Lie algebra of right invariant vector fields, then systems of the form

$$\frac{dx}{dt}(t) = X_o(x(t)) + \sum_{i=1}^{m} u_i(t)\, X_i\, (x(t)) \quad \text{where}$$

$X_o, \ldots, X_m$ are in L(G) and x(t) is in G, have been studied. The problems of realisation, observability, [2], to some extent optimality, [10], and in particular controllability, [2,4,11] have been investigated.

The controllability results obtained depend on the one to one relationship between the subalgebras of L(G) and the Lie subgroups of G. This relationship is obtained by observing that any subalgebra of L(G) defines a distribution on G whose maximal integral manifold through e, the identity of G, is a Lie subgroup of G. Thus, one way of viewing the theory is that of the action of the Lie subgroup, corresponding to the subalgebra generated by (X_i), on the group G.

A natural generalisation of the above is to study systems of the form $\dot{x} = f(x,u)$, where $f_u : M \to T(M)$ defined by $f_u(x) = f(x,u)$ is a suitably differentiable section of the tangent bundle, T(M), of a differential manifold M. The problem of controllability for such systems has been studied in [6,7,8,11]. However, in [11] is suggested another generalisation, that of the action of a Lie group on a differential manifold. It is this generalisation I wish to pursue a little here.

1. Fibre Bundles [1,3,5,9]

Let G be a Lie group and P a differentiable manifold. G acts differentiably on P to the right if there is a map $f:P \times G \to P$, and we write $f(p,g) = pg$, satisfying

1) for each $g \in G$ the map $g: P \to P$ defined by $g(p) = pg$, is a diffeomorphism.
2) for all $g, h \in G$ and $p \in P$, $p(gh) = (pg)h$.

G is said to act <u>freely</u> if the only element of G having a fixed point on P is the identity.

<u>Definition</u>. A differentiable principal fibre bundle is a triple (P,M,G), where P and M are differential manifolds and G is a Lie group such that

1) G acts freely and differentiably on P to the right.
2) M is the quotient space of P by equivalence under G, and the projection $\pi: P \to M$ is differentiable. (Thus G is transitive only on $\pi^{-1}(x)$, $x \in M$).
3) P is locally trivial, that is, every point $x \in M$ has a neighbourhood U such that there exists a diffeomorphism $\psi:\pi^{-1}(U) \to U \times G$, $\psi(p) = (\pi(p), \phi(p))$ where $\phi:\pi^{-1}(U) \to G$ satisfies $\phi(pg) = \phi(p)\, g$ for all $p \in \pi^{-1}(U)$ and $g \in G$.

We call P the <u>total space</u> or <u>bundle space</u>, M the <u>base space</u> and G the <u>structure group</u>. For each $x \in M$, $\pi^{-1}(x)$ is a closed submanifold of P, called the <u>fibre</u> over x and diffeomorphic to G.

As remarked in [11], there is a canonical homomorphism, λ, from L(G) into the Lie algebra of vector fields on P, defined as follows:-

Given any $p \in P$ we have a map, also denoted $p: G \to P$, defined by $p(g) = pg$. Let $dp:T_eG \to T_pP$ denote the induced map on the tangent spaces, then identifying T_eG with L(G) we define

$\lambda:L(G) \to F'(P)$ by

$\lambda(X)(p)=dp\,(X(e))$, and denote $\lambda(X)$ by X^*.

(F'(P) is the set of differentiable vector fields on P)

<u>Definition</u>. A vector field Y on P is called <u>fundamental</u> if $Y = X^* = \lambda(X)$ for some $X \in L(G)$.

<u>Lemma</u> If X^* is a fundamental vector field on P, then

1) X^* is differentiable.
2) $X^*(p)$ is tangential to the fibre through p, for all $p \in P$.
3) $dR_g X^* = \lambda(\mathrm{ad}\, g^{-1}X)$, where $R_g:G \to G$ is right multiplication in G.

<u>Lemma</u> $[X^*, Y^*] = \lambda[X,Y]$, and thus the fundamental vector fields on P form a subalgebra of F'(P).

<u>Example</u>. Let us consider P is a Lie group, G, and examine the earlier work in the principal bundle setting.

Given the system

$$\frac{dx}{dt}(t) = \sum_{i=1}^{m} u_i(t)X_i(x(t)) ,$$

where the $X_i \in L(G)$, let $L(H)$ denote the subalgebra generated by the X_i and H the corresponding subgroup of G. Then, we consider the action of H on G as the principal fibre bundle $(G,G/H,H)$, the fibre through each g being isomorphic to H. The vector fields X_i are fundamental on G and the set of attainability from g is precisely the fibre through g. Of course, this is the simpler symmetric case, but the non-symmetric case could be dealt with, just as in [4].

Thus a simple generalisation of the Lie group work is to consider the system

$$\frac{dx}{dt}(t) = X_o(x(t)) + \sum_{i=1}^{m} u_i(t)\, X_i(x(t)) , \text{ where}$$

$X_o, \ldots, X_m$ are fundamental vector fields on a principal bundle (P, M, G) and $x(t) \in P$. However, one can expect no new results, only a slightly more general setting.

A second problem is to consider vector fields on P "transverse" to the fibres, in some sense, and in this way induce a motion on the base space M. We shall see later that this too can be thought of as a generalisation of the Lie group work. To affect this generalisation we give the idea of a connection.

2. Connections in Principal Bundles [1,5]

Let (P,M,G) be a principal bundle. For each $p \in P$, let T_pP denote the tangent space to P at p. Denote by $V(p)$ the subspace of T_pP consisting of vectors tangential to the fibre through p. $V(p)$ is called the vertical subspace and $V:P \to T(P)$ is a (dim G) - dimensional differentiable distribution on P.

<u>Definition</u>. A <u>connection</u> on the principal bundle (P,M,G) is a (dim M) - dimensional distribution $H:P \to TP$ such that

1) H is differentiable.
2) $H(p) + V(p) = T_pP$ (i.e. $H(p)$ is a linear complement of $V(p)$ for all $p \in P$).
3) $dR_g\, H(p) = H(pg)$ for all $p \in P$, $g \in G$.

If $v \in T_pP$ is such that $v \in H(p)$, then v is called <u>horizontal</u>. The projection $\pi:P \to M$ induces a linear map $d\pi: T_pP \to T_{\pi(p)}M$. When a connection is given, $d\pi$ maps the horizontal subspace $H(p)$ isomorphically onto $T_{\pi(p)}M$. The horizontal lift of a vector field , X , on M is a unique vector field $\bar{X}$ on P which is horizontal and which satisfies $d\pi\,(\bar{X}(p)) = X(\pi(p))$, for all $p \in P$.

Given the connection H on the principal fibre bundle we can define the concept of parallel displacement along any curve c in the base space, M. Let c be a piecewise differentiable curve in M, then a horizontal lift of c, is a curve c^* in P such that

1) $\pi c^* = c$ and
2) c^* is horizontal, in the sense that its tangent vectors are horizontal.

Now let c be any differentiable curve on M and p an arbitrary point of P such that $\pi(p) = c(o)$. Let c^* be the unique horizontal lift of c such that $c^*(o) = p$. Then $c^*(t)$ is in $\pi^{-1}(c(t))$ and thus by varying the point p in $\pi^{-1}(c(o))$ we obtain a map from $\pi^{-1}(c(o))$ to $\pi^{-1}(c(t))$. We call this map parallel displacement along the curve c.

Given a connection H on (P, M, G) we have a canonical L(G) - valued 1-form, ω, on P defined as follows:-

Given a vector field Y on P , let VY denote the vertical component of Y, then

$$\omega(Y) = X \text{ , where } X \in L(G) \text{ is such that}$$
$$\lambda(X) = VY \text{ where } \lambda: L(G) \to F'(P) \text{ is the homomorphism of 1.}$$

Lemma 1) $\omega(Y) = 0$ iff Y is horizontal
2) $\omega(X^*) = X$ for all $X \in L(G)$
3) $\omega(dR_g X) = \operatorname{ad} g^{-1}\, \omega(X)$ for all $g \in G$ and $X \in F'(P)$

Alternatively, if ω is an L(G) - valued 1-form on P satisfying 2) and 3) then ω defines a connection on (P,M,G), with kernel ω defining the horizontal distribution on P. The form ω is called the connection form of the given connection, H.

Definition. The curvature form of a connection H is the L(G) - valued 2-form $D\omega$, where ω is the connection form of H, and $D\omega$ is defined by

$$D\omega(X,Y) = d\omega(HX,HY) ,$$

where HX and HY are the horizontal components of X and Y respectively.

We now restrict our attention to a particular principal bundle over an arbitrary base space, M, the so called bundle of bases. This allows us some additional structure, and in particular the definition of a geodesic.

3. Affine Connections [1,5]

Given a differentiable manifold, M, we define a principal bundle with base space M, called the bundle of bases as follows:-

Let B(M) be the set of (n+1) - tuples $(x, e_1,\ldots,e_n)$ where $x \in M$ and $e_1,\ldots,e_n$ is a basis for T_xM. Then B(M) is a differentiable manifold. Moreover, $GL(n,R)$ acts to the right on $B(M)$. For, if $g = (g_{ij}) \varepsilon GL(n,R)$ then

$$(x,e_1,\ldots,e_n)g = (x, \sum_i g_{i1}e_i,\ldots, \sum_i g_{in}e_i)$$

Thus $(B(M), M, GL(n,R))$ is a principal fibre bundle, called the bundle of bases, with $\pi:B(M) \to M$ defined by

$$\pi(x, e_1,\ldots, e_n) = x.$$

<u>Definition</u>. A connection in $(B(M), M, GL(n,R))$ is called an <u>affine connection</u>.

Suppose now that we have an affine connection in $(B(M), M, GL(n,R))$, then we have canonically defined horizontal vector fields. For, given $b = (x,e_1,\ldots e_n) \varepsilon B(M)$ we have a map, also denoted $b:R^n \to T_xM$ defined as follows

$$b(y) = (x , \sum_i y_i e_i) \quad , \quad y = (y_1,\ldots, y_n) \varepsilon R^n.$$

Now recall that $d\pi|H(b)$ is an isomorphism onto T_xM and define the vector field $E(y)$ on $B(M)$ by

$$E(y)(b) = (d\pi|H(b))^{-1}(by).$$

The vector field, $E(y)$, on $B(M)$,

for a fixed $y \varepsilon R^n$, is called a basic vector field.

<u>Lemma</u>
1) $E(y)$ is differentiable.
2) $E(y)$ is horizontal.
3) $dR_g\ E(y) = E(g^{-1}y)$.

<u>Definition</u>. The projection onto M of any integral curve of a basic vector field of $B(M)$ is called a <u>geodesic</u>.

<u>Theorem</u>. If $c:(o,t_1)\to M$ is a geodesic then the vector field $\frac{dc}{dt}(t)$ is parallel along c, in the following sense:-

Choose any $b \varepsilon B(M)$ such that $\pi(b) = c(o)$ and let c^* be the horizontal lift of c through b. Then we have from 2. the idea of parallel translation of $\pi^{-1}(c(o))$ along c. Suppose $b = (c(o), e_1,\ldots e_n)$ then let the parallel translate of b along c^* be $(c(t), e_1(t),\ldots e_n(t))$. Now $\frac{dc(o)}{dt} = \sum_i z_i e_i$ and

$\frac{dc}{dt}(t) = \sum_i z_i(t)\ e_i(t)$. Then $\frac{dc}{dt}$ is parallel along c if the $z_i : (o,t_1) \to R$ are constant functions.

Consider the system defined on $B(M)$

$$\frac{dx}{dt}(t) = \sum_{i=1}^{m} u_i(t)\, E(y_i) \quad \text{——}(*).$$

Let α be an integral curve of this system on $B(M)$ in the sense of [11], such that $\alpha(o) = (x, e_1,\ldots,e_n) = b$.

Definition. We say $y \in M$ is geodesically accessible from $x \in M$ for the system (*), if there exists an integral curve α of (*) such that $\alpha(o) = (x,e_1,\ldots,e_n)$ and $\pi_0\alpha(t) = y$ for some $t > o$.

Denote by $gA(x)$ the set of all such points, geodesically accessible from x. (Note we choose $b \in \pi^{-1}(x)$, arbitrarily, as the initial point of the system (*), but it is then considered fixed.)

Problem. (Geodesic Controllability) What is the structure of $gA(x)$?

By definition, $gA(x)$ is the projection into M of the maximal integral manifold of the involutive distribution on $B(M)$ generated by $(E(y_i))_i$. Thus we first examine the involutive distribution generated by the $(E(y_i))_i$. To consider the vector field $[E(y_1), E(y_2)]$ on $B(M)$ we first introduce the standard (or Solder) 1-form of an affine connection.

The R-valued Solder 1-forms, θ_i, are defined as follows:-

Given $v \in T_bB(M)$, $b = (x,e_1,\ldots,e_n)$ then $d\pi(v) \in T_xM$ can be expressed in terms of the basis $(e_1,\ldots,e_n)$ as $d\pi(v) = \Sigma\, z_ie_i$

Definition. $\theta_i\colon T_b\, B(M) \to R$ is defined by $\theta_i(v) = z_i$.

We define an R^n-valued solder 1-form, $\theta : T_bB(M) \to R^n$ by $\theta(v) = (\theta_1(v),\ldots,\theta_n(v))$

Corollary. If $E(y)$ is a basic vector field then $\theta(E(y))(b) = y$.

Definition. The torsion form of an affine connection is the 2-form $D\theta = d\theta \circ H$. (i.e. $D\theta(v_1, v_2) = d\theta(HV_1, HV_2)$, where HV_i denotes the horizontal component of V_i).

Theorem. $[E(y_1), E(y_2)] = -\lambda\, D\omega(E(y_1), E(y_2)) - E(D\theta(E(y_1),E(y_2)))$.

That is the curvature and torsion give the vertical and horizontal components of the brackets of two basic vector fields.

Thus, rather than tackle the problem in its full generality we shall restrict our attention to manifolds plus affine connection for which the curvature and/or torsion have a particularly simple form.

For example, if the curvature form is identically zero (i.e. the manifold is flat) then we see that the Lie bracket of basic vector fields is horizontal. To obtain that the basic vector fields form a subalgebra we need a further regularity condition on the torsion (that the covariant differential of the torsion is zero). This is just the case for a Lie group, G, with the direct connection [1] (or + connection [5]). In this case the geodesics are just the solutions curves of the the right invariant vector fields and $-E(D\theta(X,Y)) = [X,Y]$ for X and $Y \in L(G)$. Thus $gA(e)$ is the subgroup corresponding to the Lie subalgebra generated by the $E(y_i)$, as expected, and furthermore is a <u>totally geodesic submanifold</u> in the following sense.

<u>Definition</u>. A submanifold N of a manifold M with an affine connection is said to be totally geodesic at a point x of N, if for every $X \in T_x N$, the geodesic c of M such that $c(o) = x$ and $\frac{dc}{dt}(o) = X$ lies in N for small values of the parameter t. If N is totally geodesic at every point $x \in N$, then it is called a totally geodesic submanifold.

If the torsion form is identically zero, for example Riemannion manifolds, we see that the Lie bracket of basic vector fields is vertical. If, in addition, we have that the covariant differential of the curvature is zero then $[E(y_1), E(y_2)]$ is a fundamental vector field on $B(M)$. If we join these two observations we have trivially

<u>Theorem</u>. On a flat Riemannian manifold $[E(y_1), E(y_2)] = 0$ and thus 1) the integral manifold of the system (*) is of dimension m.

2) the set $gA(x)$ has dimension m.

<u>Proof</u>. 1) The Lie algebra generated by $(E(y_i))_i$ is of dimension m.

2) π is a diffeomorphism when restricted transverse to the fibres.

Finally we turn to a less trivial example of manifolds plus affine connection for which the torsion form is zero, and the covariant differential of the curvature is zero, namely symmetric spaces with the canonical affine connection.

4. <u>Symmetric Spaces</u> [3,5]

Let G be a connected Lie group with an involutive automorphism δ ($\delta^2 = 1$, $\delta \neq 1$). Let K be a closed subgroup which lies between the closed subgroup of all fixed points of δ and its identity component. G/K is then called a symmetric space defined by δ. Clearly $(G, G/_K, K)$ is a principal fibre bundle and δ induces a connection, called the <u>canonical connection</u>, on $(G, G/K, K)$ as follows:-

Let $L(G)$ be the Lie algebra of G and $d\delta : L(G) \to L(G)$ denote the map induced by δ. Let

$$V = \{X \in L(G) \mid d\delta(X) = X\} \text{ and}$$
$$H = \{X \in L(G) \mid d\delta(X) = -X\}.$$

<u>Lemma</u> 1) $L(G) = V + H$ and 2) V coincides with $L(K)$

<u>Proof</u>. 1) Given $X \varepsilon L(G)$ suppose $d\delta(X) = Y$
then $d\delta(X) = X + (Y-X)$.
and $d\delta(Y-X) = d\delta(Y) - d\delta(X) = d\delta . d\delta X - Y = X - Y = -(Y-X)$.
2) trivial since $\delta|K$ is just the identity.

Thus $V = L(K)$ is the vertical subspace of $L(G)$ for the principal bundle $(G, G/_K, K)$ and H defines a connection on $(G, G/_K, K)$. We obtain an affine connection on $G/_K$, by defining a bundle homomorphism, [5], $(G, G/_K, K) \to (B(G/_K), G/_K, GL(n,R))$ and defining the <u>canonical affine connection</u> as the image of the canonical connection in $(G, G/_K, K)$.

<u>Theorem</u> [5]. Given the canonical affine connection on a symmetric space $G/_K$, then for any $X \varepsilon H$ let $g(t) = \exp tX$ and $c(t) = \pi_0 g(t)$, then

1) c is a geodesic
2) The torsion is zero
3) The covariant differential of the curvature is zero.

We now return to the control problem. We see, by 1), that for this case our basic vector fields are $\{X | X \varepsilon H\}$ and that by 2) and 3) $[H,H] \subset V$. In fact this is immediate from the definitions since

$$d\delta[X,Y] = [-X,-Y] = [X,Y] \quad \text{for} \quad X,Y \varepsilon H.$$

Moreover $d\delta[X,Y] = [X,-Y] = -[X,Y]$ for $X \varepsilon V$, $Y \varepsilon H$ and thus $[V,H] \subset H$. (In fact the Lie bracket of a fundamental vector field with a basic vector field is a basic vector field for all affine connections.)

<u>Corollary</u>. $[[H,H],H] \subset H$

<u>Definition. A Lie triple of</u> a Lie algebra is a subspace t such that $[[t,t],t] \subset t$.

Thus, for the present case, we may write the system (*) as

$$\frac{dx(t)}{dt} = \sum_{i=1}^{m} u_i(t)\, X_i, (**), \text{ where } X_i \varepsilon H.$$

Denote the subspace of H spanned by the X_i by H_0, We look first for the maximal integral manifold of the distribution H_0 on $B(G/_K)$. Let $L(G_1) \subset L(G)$ denote the Lie algebra generated by H_0, and $H_1 = L(G_1) \cap H$.

<u>Theorem</u> H_1 is a Lie triple, called the Lie triple generated by H_0.

<u>Proof</u>. Given $X,Y,Z \varepsilon H_1$ then $X,Y,Z \varepsilon H$ and thus $[[X,Y],Z] \varepsilon H$. But $[[X,Y],Z] \varepsilon L(G_1)$ and thus $[[X,Y],Z] \varepsilon L(G_1) \cap H = H_1$.

<u>Theorem</u> [3,5] Let $G/_K$ be a symmetric space and $L(G) = V + H$

the canonical decomposition. Then there is a natural one to one correspondence between the Lie triples $H_1 \subset H$ of $L(G)$ and the set of complete totally geodesic submanifolds M_1 through o, the equivalence class of e in $G/_K$. The correspondence is given by $T_o M_1 = H_1$, under the identification $T_o(G/_K) = H$.

Theorem. Given the system (**) defined on the symmetric space $G/_K$ with the canonical affine connection, then the geodesically accessible set gA(o) is the totally geodesic submanifold of $G/_K$ through o corresponding to the Lie triple generated by $(X_i)_i$.

Proof Let H_1 denote the Lie triple generated by $(X_i)_i$ and set $V_1 = [H_1, H_1]$. If $L(G_1)$ is the Lie algebra generated by the $(X_i)_i$ then we see by definition that $L(G_1) = H_1 + V_1$. Let G_1 be the subgroup corresponding to the subalgebra $L(G_1)$ and K_1 be the subgroup corresponding to the subalgebra V_1. Then clearly $K_1 = K \cap G_1$ and thus $\pi(G_1) = G_1/_K = G_1/_{K_1} = gA(o)$.

But $G_1/_{K_1}$ is precisely the totally geodesic manifold corresponding to H_1 of the previous theorem.

Note that the restriction to accessibility from o is not serious, at least for the case K compact. In this case we can put a Riemann metric on $G/_K$ such that the canonical affine connection is precisely the Riemannian connection and thus $G/_K$ is a Riemannian symmetric space. In this case we have the theorem:-

Theorem [3] Let M be a Riemannian symmetric space and $x \in M$. If G is the group of isometries of M and K the subgroup of G which leaves x fixed, then $G/_K$ is diffeomorphic to M, and clearly $o = x$.

Example. Let $SO(n+1)$ denote the $(n+1)\times(n+1)$ orthogonal matrices. We define $\delta: SO(n+1) \to SO(n+1)$ by
$\delta(A) = SAS^{-1}$, where $S = \begin{pmatrix} -1 & 0 \\ 0 & I_n \end{pmatrix}$. The identity component of the subgroup of fixed points is the set of matrices of the form $\begin{pmatrix} 0 & 0 \\ 0 & B \end{pmatrix}$, where $B \in SO(n)$.

We write $SO(n)$ for this subgroup and then $SO(n+1)/_{SO(n)}$ is clearly a symmetric space.

The Lie algebra $so(n+1)$ of $SO(n+1)$ is the set of skew-symmetric $(n+1)\times(n+1)$ matrices. The subalgebra V is just the set $\begin{pmatrix} 0 & 0 \\ 0 & so(n) \end{pmatrix}$ and the subspace H, the set of matrices of the form $\begin{pmatrix} 0 & -\xi^t \\ \xi & 0 \end{pmatrix}$ where ξ is a column vector in R^n.

Now $SO(n+1)/SO(n)$ is diffeomorphic to S^n, the unit sphere in R^{n+1}. The diffeomorphism is defined by

$$f(a_{i,j}) = (a_{1,1}, a_{1,2}, \dots, a_{1,n+1}) .$$

Moreover, the canonical affine connection on $SO(n+1)/SO(n)$ coincides with the Riemannian connection of S^n as an imbedded submanifold of R^{n+1}. Geodesic controllability from o, in this case coincides with geodesic controllability from (1,0,...,0) on S^n. However, by defining $S = \begin{pmatrix} 1 & 0 & 0 \\ 0 & -1 & 0 \\ 0 & 0 & I_{n-1} \end{pmatrix}$

we obtain a different imbedding of $SO(n)$ in $SO(n+1)$ and controllability from o becomes controllability from the point (0,1,0,...,0) in S^n, etc.

Let us now consider the set $gA(o)$ on $SO(n+1)/SO(n) = S^n$.

Suppose we are given the point, $x \in S^n$, corresponding to $o \in SO(n+1)/SO(n)$, and a subspace of $T_x S^n$ as admissible directions. We can lift the subspace to a subset, H_o, of H under the identification of H with $T_x S^n$ and then find the Lie triple generated by H_o. In fact H_o is itself a Lie triple and we obtain that if H_o has dimension m, then $G_1 = SO(m+1)$, $V_1 = SO(m)$ and the corresponding totally geodesic submanifold $SO(m+1)/SO(m)$ through o, is the sphere S^m, such that $T_x S^m$ is the given subspace.

References

[1] R.L. Bishop and R.L. Crittenden, Geometry of Manifolds, Academic Press, New York, 1964.

[2] R.W. Brockett, System Theory on Group Manifolds and Coset Spaces, SIAM J. on Control, 1972.

[3] S. Helgason, Differential Geometry and Symmetric Spaces, Academic Press, New York, 1962.

[4] V.J. Jurdjevic and H.J. Sussmann, Control Systems on Lie Groups, J. Differential Equations, to appear.

[5] S. Kobayashi and K. Nomizu, Foundations of Differential Geometry, vols. I and II, Interscience, New York, 1963/69.

[6] C. Lobry, Controllabilité des Systems Non Lineaire, SIAM J. on Control, 1970.

[7] C. Lobry. Une Proprieté Generique des Couples de Vecteurs, Czech. Math. J. 22, 1972.

[8] C. Lobry, Geometrical Structure of Orbits of Dynamical Polysystems, Control Theory Centre Report No. 19, University of Warwick.

[9] S. Sternberg, Lectures on Differential Geometry, Prentice

Hall, New Jersey, 1964.

[10] H.J. Sussmann, The Bang-Bang Problem for Linear Control Systems on Lie Groups, SIAM J. on Control, to appear.

[11] H.J. Sussmann and V.J. Jurdjevic, Controllability of Non Linear Systems, J. Differential Equations, to appear.

CONTROLLABILITY IN NONLINEAR SYSTEMS*

Ronald M. Hirschorn

Department of Mathematics, Queen's University
Kingston, Ontario
Canada

ABSTRACT

In this paper we obtain an explicit expression for the reachable set for a class of nonlinear systems. This class is described by a chain condition on the Lie algebra of vector fields associated with each nonlinear system. These ideas are used to obtain a generalization of a controllability result for linear systems in the case where multiplicative controls are present.

1. INTRODUCTION

The purpose of this paper is to study the controllability of nonlinear systems of the form $dx/dt = f(x, u)$ where the state space is a real analytic manifold M. Our goal is to identify a large class of nonlinear systems for which there is an explicit expression for the reachable set from any initial state x_0 at time t. Recent work by Sussman and Jurdjevic [2] has established some of the basic properties of the reachable set but, except for two special classes of systems (Brockett's Theorem 7 of [1] and symmetric systems i.e. $f(\cdot, u) = -f(\cdot, -u)$) the problem of obtaining an explicit expression for the reachable set from x at time t remains open.

* This work was performed while the author was at Harvard University, Division of Engineering and Applied Physics, Cambridge, Massachusetts and was supported in part by the U.S. Office of Naval Research under Joint Services Electronic Program by Contract N00014-67-A-0298-0006 and by the National Aeronautics and Space Administration under Grant NGR 22-007-172.

We shall assume, as in [2], that the vector field on M, $f(\cdot, u)$, corresponding to a constant control u, is complete. One nice consequence of this assumption is the existence of a 1-parameter group, X_t^u, for each vector field $f(\cdot, u)$ where u is a constant control. The collection of all such 1-parameter groups and their products (under composition) is a group, G, of diffeomorphisms of M. For the systems which we shall be considering the reachable set from a state x at time t can be expressed as the orbit of the point x under a certain subset of the group G. Thus one can study the properties of the reachable set from x by examining the structure of the orbits of x under G and certain subsets of G (see for example, Lobry [3]). The orbit of x under G and the Lie algebra $\mathscr{L}$ of vector fields generated by the vector fields $f(\cdot, u)$ corresponding to constant control are closely related since the orbit of x under G is the integral submanifold of $\mathscr{L}$ through x (cf. [3]). Thus the structure of the reachable set from x can be studied by examining the structure of the Lie algebra of vector fields $\mathscr{L}$ associated with a system.

Our approach will be to study the structure of the group of diffeomorphisms G directly. In general this global approach is ineffective because of the lack of structure in G, but in the case where $\mathscr{L}$ is finite dimensional G can be given the structure of a Lie group with Lie algebra isomorphic to $\mathscr{L}$. This interesting result, which is due to Palais [7], enables us to reformulate the original control system as a nonlinear system on the Lie group G for which the vector fields corresponding to constant controls are right-invariant. For these reasons we will restrict our attention to nonlinear systems for which the associated Lie algebra of vector fields $\mathscr{L}$ is finite dimensional. For a subclass of these systems the associated Lie algebra $\mathscr{L}$ of vector fields satisfies a certain chain condition. For these systems we are able to obtain an explicit description of the reachable set at time t from any initial state x, subject to one additional constraint on the associated transformation group G. For nonlinear systems of the form

$$\frac{dx}{dt} = A(x) + \sum_{i=1}^{m} u_i B_i(x)$$

on M, where $A, B_1, \ldots, B_m$ are complete analytic vector fields, this additional constraint on G can be dropped.

This paper is organized as follows: In Section 2 we introduce notations and present some of the known controllability results. In Section 3 we describe some of Palais' results on Lie transformation groups, prove our main result, and present an example. In Section 4 these results are applied to obtain a generalization of Brockett's Theorem 7 of [1] and a generalization of the exact time controllability result for linear systems where multiplication controls are present.

2. PRELIMINARIES

We shall assume that the reader is familiar with the basic notions of differential geometry and Lie theory (cf. [4]). The following notation will be used:

M - a real analytic Hausdorff differentiable manifold
$V(M)$ - the set of smooth analytic vector fields on M
$\mathrm{diff}(M)$ - the group of diffeomorphisms of M
M_x - the tangent space to M at $x \in M$
TM - the tangent bundle of M
$\underline{G}$ - a Lie group
$\mathscr{L}(\underline{G})$ - the Lie algebra of right-invariant vector fields on $\underline{G}$
$\exp: \mathscr{L}(\underline{G}) \to \underline{G}$ is the standard exp map (cf. [4])
$\underline{R}$ - the real numbers
$\underline{R}^m$ - m-dimensional Euclidean space
$\underline{R}^+$ - the open interval $(0,\infty)$, a semigroup under addition

Suppose $h \subset \underline{G}$ and $\mathscr{h} \subset \mathscr{L}(\underline{G})$. Then $\{h\}_G$ is the group generated by h in $\underline{G}$; $\{\mathscr{h}\}_{LA}$ is the Lie algebra generated by $\mathscr{h}$ in $\mathscr{L}(\underline{G})$; $\mathrm{cl}\, h$ is the closure of h in $\underline{G}$; and $\{\mathscr{h}\}_{LS}$ is the linear span of $\mathscr{h}$.

We regard $V(M)$ as a Lie algebra over $\underline{R}$. For all $X, Y \in V(M)$, we define the Lie bracket of X and Y by $[X, Y](m) = X(m)Y - Y(m)X$ (cf. [4]), and set $\mathrm{ad}_X Y = [X, Y]$. If $M = \underline{R}^n$ we identify M_x with $\underline{R}^n$ for all $x \in \underline{R}^n$ and $[X, Y](m) = (dY)_m X(m) - (dX)_m Y(m)$ where $(dY)_m$ is the Jacobian for the map $Y: \underline{R}^n \to \underline{R}^n$.

We shall consider nonlinear systems of the form

$$\frac{dx}{dt}(t) = f(x(t), \underline{u}(t)) \tag{†}$$

where: (i) the functions $\underline{u}(\cdot)$ are contained in $\mathscr{U}_p^{(m)}$, the class of piecewise constant functions from $[0,\infty)$ into $\underline{R}^m$.

(ii) $f: M \times \underline{R}^m \to TM$ is jointly continuously differentiable and for each $\underline{u} \in \underline{R}^m$, $f(\cdot, \underline{u})$ is a complete analytic vector field. Thus for all $\underline{u}(\cdot) \in \mathscr{U}_p^{(m)}$ the corresponding solution to (†) with $x(0) = x$, $\pi(x, \underline{u}(t), t)$, exists for all $t \in \underline{R}^+$.

(iii) The Lie algebra generated by the collection of vector fields $D \triangleq \{f(\cdot, \underline{u}) : \underline{u} \in \underline{R}^m\}$ is finite dimensional.

<u>Remark</u>: Property (ii) implies that for each vector field $f(\cdot, \underline{u}) \in D$ there is a 1-parameter group of $f(\cdot, \underline{u})$, $X_t^{\underline{u}}$. That is $(d/dt) X_t^{\underline{u}} \cdot p = f(X_t^{\underline{u}} \cdot p, \underline{u})$ and $X_0^{\underline{u}} \cdot p = p$ for all $p \in M$.

<u>Definition</u>: A point $y \in M$ is <u>reachable from</u> $x \in M$ <u>at time</u> t iff there exists $\underline{u}(\cdot) \in \mathscr{U}_p^{(m)}$ such that $y = \pi(x, \underline{u}(t), t)$. We define the <u>reachable set from</u> x <u>at time</u> t, $\mathscr{R}_t(x)$, as set of points in M reachable from x at time t.

We are interested in obtaining global results and will consider the problem of determining the following subsets of $\mathrm{diff}(M)$:

$$G = \left\{ X_{t_1}^{\underline{u}_1} \circ X_{t_2}^{\underline{u}_2} \circ \cdots \circ X_{t_n}^{\underline{u}_n} : \underline{u}_i \in \underline{R}^m,\ t_i \in \underline{R},\ n=1,2,\ldots \right\}$$

$$G^+ = \left\{ X_{t_1}^{\underline{u}_1} \circ \cdots \circ X_{t_n}^{\underline{u}_n} : \underline{u}_i \in \underline{R}^m,\ t_i \in \underline{R}^+,\ n=1,2,\ldots \right\}$$

$$G_t = \left\{ X_{t_1}^{\underline{u}_1} \circ \cdots \circ X_{t_n}^{\underline{u}_n} : \underline{u}_i \in \underline{R}^m,\ t_i \in \underline{R}^+,\ \sum_{i=1}^{n} t_i = t,\ n=1,2,\ldots \right\}$$

For all $x \in M$, $G \cdot x$, $G^+ \cdot x$, $G_t \cdot x$ will denote the orbits through x of G, G^+, G_t respectively i.e. $G_t \cdot x = \{g(x) : g \in G_t\}$. We will be interested in obtaining an explicit expression for G_t. This will result in an explicit expression for $\mathscr{R}_t(x)$, since for all $x \in M$, $t \in \underline{R}^+$, $\mathscr{R}_t(x) = G_t \cdot x$. The orbits of G_t and the structure of the finite dimensional Lie algebra $\mathscr{L}$ generated by D in V(M) are closely related. In particular the following decomposition of $\mathscr{L}$ reveals some of the basic properties of the orbits of G_t (cf. [2]):

$$\mathscr{L} = \{X : X \in D\}_{LA}$$
$$\mathscr{B} = \{r_1 X_1 + \cdots + r_n X_n : r_i \in \underline{R},\ r_1 + \cdots + r_n = 0,\ X_i \in D,\ n=0,1,\ldots\}_{LA}$$
$$\mathscr{L}_0 = \text{the ideal generated by } \mathscr{B} \text{ in } \mathscr{L}$$

For all $x \in M$, $I(\mathscr{L}, x)$, $I(\mathscr{L}_0, x)$ and $I(\mathscr{B}, x)$ will denote the maximal integral submanifolds through x corresponding to $\mathscr{L}$, $\mathscr{L}_0$ and $\mathscr{B}$ respectively i.e. $I(\mathscr{L}, x)$ is the largest connected submanifold N of M such that $N_y = \{X(y) : X \in \mathscr{L}\}$ for all $y \in N$. Its existence is a consequence of the global version of Frobenius' theorem [5].
Definition: For all $x \in M$, $t \in \underline{R}^+$, $I^t(\mathscr{L}_0, x) \stackrel{\Delta}{=} X_t^{\underline{u}}(I(\mathscr{L}_0, x))$ where $X_t^{\underline{u}}$ is the 1-parameter group associated with $f(\cdot, \underline{u}) \in D$. Lemma 3.6 of [2] states that $I^t(\mathscr{L}_0, x)$ is well defined, i.e. independent of the choice of $\underline{u} \in \underline{R}^m$.
Theorem 2.1: (Sussman and Jurdjevic [2]) Consider the nonlinear system (†) with the associated triple of Lie algebras $(\mathscr{L}, \mathscr{L}_0, \mathscr{B})$. Then for all $t \in \underline{R}^+$, $x \in M$, $G_t \cdot x \subset I^t(\mathscr{L}_0, x)$ and $G_t \cdot x$ has a nonempty dense interior in $I^t(\mathscr{L}_0, x)$.

This result holds even when $\mathscr{L}$ is infinite dimensional. In the case where M is a Lie group, and D is a collection of right-invariant vector fields on M, there is an explicit expression for G_t if the Lie algebra $\mathscr{L}$ satisfies a certain chain condition [6]. We will now describe this algebraic condition where M is the Lie group $\underline{G}$.

Suppose $\mathscr{Q}$ is a Lie subalgebra of $\mathscr{L}$ and $A \in D$ is a right-invariant vector field on $\underline{G}$. Consider the chain of Lie subalgebras of $\mathscr{L}$,

$$\mathscr{Q} \subset \widetilde{\mathscr{Q}} \subset \{\mathscr{Q}, A\}_{LA} \qquad (*)$$

where $\widetilde{\mathscr{Q}}$ is the ideal generated by $\mathscr{Q}$ in $\{\mathscr{Q}, A\}_{LA}$.
Definition: The chain (*) is called an A-chain if $\mathscr{Q}$ is an ideal in $\widetilde{\mathscr{Q}}$. If $\mathscr{Q}$ is contained in h, a Lie subalgebra of $\mathscr{L}$, and (*) is an

A-chain, we will call (*) an A-chain from $\mathcal{h}$.
Definition: Suppose that $\mathcal{h}$ is a Lie subalgebra of $\mathscr{L}$ and (*) is an A-chain from $\mathcal{h}$. Then the Lie algebra $\{\mathcal{h}, \tilde{\mathscr{Q}}\}_{LA}$ is said to be A-generated from $\mathcal{h}$.
Definition: Suppose $\mathscr{B}_0, \mathscr{B}_n$ are Lie subalgebras of $\mathscr{L}_0$ and $\mathscr{B}_0 \subset \mathscr{B}_n$. A chain of Lie subalgebras

$$\mathscr{B}_0 \subset \mathscr{B}_1 \subset \cdots \subset \mathscr{B}_{n-1} \subset \mathscr{B}_n \subset \mathscr{L}_0$$

is called an A-series for $\mathscr{B}_0$ terminating at $\mathscr{B}_n$ if $\mathscr{B}_{i+1}$ is A-generated from $\mathscr{B}_i$ for $i = 0, 1, \ldots, n-1$. The A-radical for $\mathscr{B}_0$, $\mathscr{R}(A; \mathscr{B}_0)$, is the largest Lie subalgebra, $\mathcal{h}$, of $\mathscr{L}_0$ with the property that there exists an A-series for $\mathscr{B}_0$ terminating at $\mathcal{h}$. $\mathscr{R}(A;\mathscr{B})$ is well defined and unique as a consequence of the definitions and the finite dimensionality of $\mathscr{L}$ (see [6]).
Theorem 2.2: (Hirschorm [6]) Consider the nonlinear system (†) defined on the Lie group $\underline{G}$ with the associated collection of right-invariant vector fields D and triple of Lie algebras $(\mathscr{L}, \mathscr{L}_0, \mathscr{B})$. Let $A(\cdot) = f(\cdot, 0) \in D$. Suppose that the A-radical for $\mathscr{B}$ is $\mathscr{L}_0$ and suppose that for all $t \in \underline{R}^+$, $\exp tA\{\exp \mathscr{B}\}_G \subset \mathrm{cl}\ G_t$. Then for all $t \in \underline{R}^+$, $x \in \underline{G}$, $G_t \cdot x = I^t(\mathscr{L}_0, x)$.

For the nonlinear system (†) on M the assumption that $\mathscr{L} = \{D\}_{LA}$ is finite dimensional results in many simplifications. In particular, the subgroup $G \subset \mathrm{diff}(M)$ which contains G_t and G^+ can be given the structure of a Lie group with Lie algebra isomorphic to $\mathscr{L}$. This result, due to Palais, enables us to reformulate (†) as a nonlinear system on the Lie group $\underline{G}$ and determine G_t explicitly for a large class of nonlinear systems.

3. DYNAMICAL SYSTEMS WITH LIE TRANSFORMATION GROUPS

Suppose $X, Y \in V(M)$ are complete vector fields. If M is not compact then neither $[X, Y]$ nor $X+Y$ need be complete (see Palais [7]). We have the following result:
Theorem 3.1: (Palais [7]) Let $\mathscr{L}$ be a finite dimensional Lie algebra of vector fields on the Hausdorff differentiable manifold M. Then the following conditions are equivalent:

(A) Every $L \in \mathscr{L}$ is complete

(B) The set of $L \in \mathscr{L}$ which are complete generate the Lie algebra $\mathscr{L}$.

Definition: (Palais) A finite dimensional Lie algebra of vector fields on a Hausdorff differentiable manifold M will be called an infinitesimal group of M if it satisfies conditions (A) or (B) of Theorem 3.1.

Thus if D is the set of complete vector fields associated with the nonlinear system (†) and $\mathscr{L} = \{D\}_{LA}$ then $\mathscr{L}$ is an infinitesimal group of M as a consequence of property (iii) of (†).

Definition: (Palais) Let $\underline{H}$ be a connected Lie group whose underlying group is a subgroup of diff(M). We shall call $\underline{H}$ a connected Lie transformation group of M if the mapping $\varphi: (h,p) \mapsto h(p)$ of $\underline{H} \times M \to M$ is smooth. We call φ the natural global $\underline{H}$-transformation group. For all $p \in M$ we define the mapping $\varphi_p: \underline{H} \to M$ by setting $\varphi_p(h) = h(p)$ and define the mapping $\varphi^+: \mathscr{L}(\underline{H}) \to V(M)$ by setting $\varphi^+(L)(p) = (d\varphi_p)_e(L(e))$ for all $L \in \mathscr{L}(\underline{H})$, $p \in M$. The range of φ^+ is called the infinitesimal group of $\underline{H}$.

Remark: Palais shows that φ^+ is an isomorphism from $\mathscr{L}(\underline{H})$ onto $\mathscr{L}$, the infinitesimal group of $\underline{H}$. Thus for all $L \in \mathscr{L} = \varphi^+(\mathscr{L}(\underline{H}))$ there exists a unique vector field $\widetilde{L} \in \mathscr{L}(\underline{H})$ with the property that $\varphi^+(\widetilde{L}) = L$ and $\exp t\widetilde{L}$ is the 1-parameter group of L.

The following interesting result is due to Palais [7]:

Theorem 3.2: Every infinitesimal group of M is the infinitesimal group of a unique connected Lie transformation group of M.

If D is the set of complete vector fields associated with a nonlinear system then $\mathscr{L} = \{D\}_{LA}$ is an infinitesimal group of M. Theorem 3.2 implies the existence of a unique connected Lie transformation group $\underline{G}$ of M with infinitesimal group $\mathscr{L}$. As a consequence of the above remark, $\mathscr{L}$ is isomorphic to $\mathscr{L}(\underline{G})$ i.e. $(\varphi^+): \mathscr{L} \to \mathscr{L}(\underline{G})$ is an isomorphism. For all $L \in \mathscr{L}$ we set $\widetilde{L} = (\varphi^+)^{-1}(L)$. Then

$$G = \underline{G} = \{\exp t_1\widetilde{L}_1 \cdots \exp t_n\widetilde{L}_n : t_i \in \underline{R},\ \varphi^+(\widetilde{L}_i) = L_i \in \mathscr{L},\ n=1,2,\ldots\}$$

and G^+ and G_t are subsets of $\underline{G}$. Thus associated with each nonlinear system (†) is a triple of Lie algebras $(\mathscr{L}, \mathscr{L}_0, \mathscr{B})$ and a triple of connected Lie transformation groups $(\underline{G}, \underline{G}_0, \underline{B})$ where $\mathscr{L}$, $\mathscr{L}_0$, and $\mathscr{B}$ are the infinitesimal groups of $\underline{G}$, $\underline{G}_0$ and $\underline{B}$ respectively. Here $\underline{G} \supset \underline{G}_0 \supset \underline{B}$ and $\underline{G}_0$ is a normal subgroup of $\underline{G}$ since $\mathscr{L}_0$ is an ideal in $\mathscr{L}$ (cf. [4]). We now use these ideas to obtain an explicit expression for G_t and $\mathscr{R}_t(x)$ for a class of nonlinear systems.

Theorem 3.3: Consider the nonlinear system (†) on M with the associated triple of Lie algebras $(\mathscr{L}, \mathscr{L}_0, \mathscr{B})$ and connected Lie transformation group $(\underline{G}, \underline{G}_0, \underline{B})$. Let $A \in \mathscr{L}$ be the vector field $A(\cdot) = f(\cdot, \underline{0})$. Suppose that the A-radical for $\mathscr{B}$ is $\mathscr{L}_0$ and suppose that for all $t \in \underline{R}^+$, $(\exp t\widetilde{A})\underline{B} \subset \mathrm{cl}\, G_t$. Then for all $t \in \underline{R}^+$, $x \in M$, $G_t = (\exp t\widetilde{A})\underline{G}_0$ and

$$\mathscr{R}_t(x) = (\exp t\widetilde{A})\underline{G}_0 \cdot x = I^t(\mathscr{L}_0, x) \quad .$$

Proof: Since $\mathscr{L}$ is isomorphic to $\mathscr{L}(\underline{G})$ we can define a control system on $\underline{G}$. This system satisfies the conditions of Theorem 2.2, hence $G_t = (\exp t\widetilde{A})\underline{G}_0$. It is well known that $(\exp t\widetilde{A})\underline{G}_0 \cdot x = I^t(\mathscr{L}_0, x)$ (cf. [3]). Q.E.D.

Corollary: Consider the nonlinear system on M,

$$\frac{dx}{dt}(t) = A(x(t)) + \sum_{i=1}^{m} u_i(t) B_i(x(t))$$

where $u_i(\cdot) \in \mathscr{U}_p^{(1)}$ and $A, B_1, \ldots, B_m$ are complete vector fields on M. Associated with this system is the triple of finite dimensional Lie algebras $(\mathscr{L}, \mathscr{L}_0, \mathscr{B})$ and connected Lie transformation groups $(\underline{G}, \underline{G}_0, \underline{B})$. Then $\mathscr{B} = \{B_1, \ldots, B_m\}_{LA}$, $\mathscr{L} = \{A, \mathscr{B}\}_{LA}$, and $\mathscr{L}_0$ is the ideal generated by $\mathscr{B}$ in $\mathscr{L}$. If $\mathscr{R}(A;\mathscr{B}) = \mathscr{L}_0$ then

$$G_t \cdot x = (\exp t\widetilde{A}) \underline{G}_0 \cdot x \qquad \text{for all } x \in M, \ t \in R^+ .$$

Proof: The first assertion is easily verified, and the second follows from the observation that if $\mathscr{R}(A;\mathscr{B}) = \mathscr{L}_0$ then this system satisfies the conditions of Theorem 3. 3. Q. E. D.

Example: Consider the nonlinear system on $\underline{R}^3$,

$$\frac{dx}{dt}(t) = A(x(t)) + u_i(t) B_1(x(t)) + u_2(t) B_2(x(t)) \qquad \text{(N)}$$

where for all $\underline{x} = (x_1, x_2, x_3) \in \underline{R}^3$, $A(\underline{x}) = (x_2^2, 1, x_1)$, $B_1(\underline{x}) = (x_2^2, -1, 1)$ and $B_2(\underline{x}) = (4, 0, -2x_2)$. It is easy to verify that $ad_A B_1(\underline{x}) = (4x_2, 0, -x_2^2)$, $ad_A^2 B_1(\underline{x}) = (4, 0, -6x_2)$, $B_3(\underline{x}) \triangleq [B_1, B_2](\underline{x}) = (0, 0, 2)$, $ad_A B_3 = [ad_A B_1, B_3] = [ad_A B_1, B_2] = [B_1, B_3] = [B_2, B_3] = 0$, $ad_A B_2 = [ad_A^2 B_1, B_1] = -3B_3$, $[ad_A B_1, B_1] = B_2$ and $ad_A^3 B_1 = -5B_3$. Thus A, B_1, and B_2 are complete vector fields on $\underline{R}^3$ and $\mathscr{L}$ has a basis $\{A, B_1, B_2, B_3, ad_A B_1, ad_A^2 B_1\}$, $\mathscr{B}$ has a basis $\{B_1, B_2, B_3\}$ and $\mathscr{L}_0$ has a basis $\{B_1, B_2, B_3, ad_A B_1, ad_A^2 B_1\}$. The bracket structure of $\mathscr{L}$ which is displayed above implies that $\mathscr{B}$ is an ideal in $\mathscr{L}_0$. Thus $\mathscr{R}(A;\mathscr{B}) = \mathscr{L}_0$ and since $\mathscr{L}$ is finite dimensional, the system (N) satisfies the conditions of the Corollary to Theorem 3. 3. It is not difficult to verify that $\underline{G}_0 \cdot x = I(\mathscr{L}_0, \underline{x}) = \underline{R}^3$ for all $\underline{x} \in \underline{R}^3$: for all $\underline{x} = (x_1, x_2, x_3)$, $\underline{y} = (y_1, y_2, y_3) \in \underline{R}^3$, the integral curve through x for B_1 at time $(x_2 - y_2)$ is $\underline{z} = (z_1, y_2, z_2)$. The integral curve through $\underline{z}$ for B_2 at time $(y_1 - z_1)/4$ is $\underline{q} = (y_1, y_2, q_3)$. Finally, the integral curve through $\underline{q}$ for B_3 at time $(y_3 - q_3)/2$ is (y_1, y_2, y_3). Thus for all $\underline{x} \in \underline{R}^3$, $t \in \underline{R}^+$, $\mathscr{R}_t(x) = G_t \cdot x = \underline{R}^3$.

4. LINEAR SYSTEMS WITH MULTIPLICATIVE CONTROLS

In this section we will consider an application of Theorem 3. 3 to the controllability problem for a class of linear systems with multiplicative controls. The following result is due to Brockett [9]. Consider the system on $\underline{R}^n$:

$$\frac{dx}{dt}(t) = (A + u_1(t) B_1 + \cdots + u_m(t) B_m) x(t) + F\underline{v}(t) \ ; \ x(0) = \underline{0}$$

where $A, B_1, \ldots, B_m$ are n by n matrices over $\underline{R}$, F is an n by ℓ matrix over $\underline{R}$ and $u_i(\cdot) \in \mathscr{U}_p^{(1)}$, $\underline{v}(\cdot) = (v_1(\cdot), \ldots, v_\ell(\cdot)) \in \mathscr{U}_p^{(\ell)}$. Let $\underline{f}_i \in \underline{R}^n$ be the ith column of F and set $\mathscr{L} = \{A, B_1, \ldots, B_m\}_{LA}$ and

$$\mathscr{L}^i = \left\{L_{k_1} L_{k_2} \cdots L_{k_i} : L_{k_j} \in \mathscr{L} \text{ for } j = 1, \ldots, \ell\right\}_{LS} .$$

Then $\mathscr{R}_t(\underline{0}) = \{\mathscr{L}^i f_j : i = 0, 1, \ldots \text{ and } j = 1, \ldots, \ell\}_{LS}$.

For the case where $B_1 = \cdots = B_m = 0$, $\mathscr{R}_t(\underline{0}) = \text{range}(F, AF, \ldots, A^{n-1}F)$ which is the usual result for linear system (cf. [8]). This result, which can be proved using Theorem 3. 3, depends on the fact that $\mathscr{R}_t(\underline{0})$ is a vector space when $x(0) = \underline{0}$ (cf. [6]). The case where $x(0) \neq \underline{0}$ is treated by the corollary to the following theorem:

Theorem 4. 1: Consider the nonlinear system on M

$$\frac{dx}{dt} = A(x) + \sum_{i=1}^{m} u_i(t)\, B_i(x) + \sum_{i=1}^{\ell} v_i(t)\, F_i(x)$$

where $A, B_1, \ldots, B_m, F_1, \ldots, F_\ell$ are complete analytic vector fields on M and $u_i(\cdot), v_i(\cdot) \in \mathscr{U}_p^{(1)}$. Associated with this system is the triple of Lie algebras $(\mathscr{L}, \mathscr{L}_0, \mathscr{B})$ where $\mathscr{B} = \{B_1, \ldots, B_m, F_1, \ldots, F_\ell\}_{LA}$, $\mathscr{L} = \{A, \mathscr{B}\}_{LA}$ and $\mathscr{L}_0$ is the ideal generated by $\mathscr{B}$ in $\mathscr{L}$. Consider the Lie algebras $(\widetilde{\mathscr{L}}, \widetilde{\mathscr{L}}_0, \widetilde{\mathscr{B}})$ where $\widetilde{\mathscr{B}} = \{B_1, \ldots, B_m\}_{LA}$, $\widetilde{\mathscr{L}} = (A, \mathscr{B}\}_{LA}$ and $\widetilde{\mathscr{L}}_0$ is the ideal generated by $\widetilde{\mathscr{B}}$ in $\widetilde{\mathscr{L}}$. Suppose $\mathscr{L}$ is finite dimensional, $\mathscr{R}(A; \widetilde{\mathscr{B}}) = \widetilde{\mathscr{L}}_0$, and

$$\left[ad_{X_1} ad_{X_2} \cdots ad_{X_n} F_i, F_j\right] = 0 \quad \text{for } X_i \in \{A, B_1, \ldots, B_m\},$$
$$i, j = 1, 2, \ldots, \ell \text{ and } n = 0, 1, \ldots$$

Then for all $x \in M$, $t \in \underline{R}^+$,

$$\mathscr{R}_t(x) = G_t \cdot x = I^t(\mathscr{L}_0, x) .$$

Proof: We will show that $\mathscr{R}(A; \mathscr{B}) = \mathscr{L}_0$. This theorem then follows directly from Theorem 3. 3. Set

$$\mathscr{Q}_0 = \left\{ad_{X_1} ad_{X_2} \cdots ad_{X_n} F_m : n = 0, 1, \ldots;\ m = 1, \ldots, \ell;\ X_i \in \widetilde{\mathscr{B}}\right\}_{LS}$$

Then $\mathscr{Q}_0 \subset \mathscr{B}$ and if $\widetilde{\mathscr{Q}}_0$ is the ideal generated by $\mathscr{Q}_0$ in $\{\mathscr{Q}_0, A\}_{LA}$, $[\mathscr{Q}_0, \widetilde{\mathscr{Q}}_0] = 0$. Thus $\mathscr{Q}_0 \subset \widetilde{\mathscr{Q}}_0 \subset \{\mathscr{Q}_0, A\}_{LA}$ is an A-chain from $\mathscr{B}$. Set $\mathscr{B}_1 = \{\mathscr{B}, \widetilde{\mathscr{Q}}_0\}_{LA}$ and

$$\mathscr{Q}_1 = \left\{ad_{X_1} \cdots ad_{X_n} ad_A^k ad_{Y_1} \cdots ad_{Y_m} F_j : n, m, k = 0, 1, \ldots;\right.$$
$$\left. X_i, Y_i \in \widetilde{\mathscr{B}}, \text{ and } j = 1, 2, \ldots, \ell\right\}_{LS} .$$

If $\widetilde{\mathscr{Q}}_1$ is the ideal generated by $\mathscr{Q}_1$ in $\{\mathscr{Q}_1, A\}_{LA}$, then $\widetilde{\mathscr{Q}}_1$ is an abelian Lie algebra as a consequence of the hypothesis on the bracket structure of $\mathscr{L}$, and $\mathscr{Q}_1 \subset \widetilde{\mathscr{Q}}_1 \subset \{\mathscr{Q}_1, A\}_{LA}$ is an A-chain from $\mathscr{B}_1$. We set $\mathscr{B}_2 = \{\mathscr{B}_1, \widetilde{\mathscr{Q}}_1\}_{LA}$ and continue this process until $\mathscr{B}_n = \mathscr{B}_{n+1}$ for some positive integer n, whose existence is guaranteed by the finite dimensionality of $\mathscr{L}$. There exists an A-series from $\widetilde{\mathscr{B}}$ terminating at $\mathscr{L}_0$, $\widetilde{\mathscr{B}} \subset \widetilde{\mathscr{B}}_1 \subset \widetilde{\mathscr{B}}_2 \subset \cdots \subset \widetilde{\mathscr{B}}_m = \mathscr{L}_0$, as $\mathscr{R}(A;\widetilde{\mathscr{B}}) = \mathscr{L}_0$ by assumption. Setting $\mathscr{B}_{n+i} = \{\widetilde{\mathscr{B}}_i, \mathscr{B}_n\}_{LS}$ it follows that $\mathscr{B} \subset \mathscr{B}_1 \subset \cdots \subset \mathscr{B}_n \subset \mathscr{B}_{n+1} \subset \cdots \subset \mathscr{B}_{n+m} = \mathscr{L}_0$ is an A-series from $\mathscr{B}$ terminating at $\mathscr{L}_0$, hence $\mathscr{R}(A;\mathscr{B}) = \mathscr{L}_0$ and the proof is complete. Q. E. D.

Corollary: Consider the system on $\underline{R}^n$

$$\frac{dx}{dt}(t) = (A + u_1(t) B_1 + \cdots + u_m(t) B_m) x(t) + F \underline{v}(t)$$

where $A, B_1, \ldots, B_m$ are n by n matrices over $\underline{R}$, $F = (\underline{f}_1, \ldots, \underline{f}_\ell)$ where $\underline{f}_i \in \underline{R}^n$ is the ith column of F, and $u_i(\cdot) \in \mathscr{U}_p^{(1)}$, $\underline{v}(\cdot) = (v_1(\cdot), \ldots, v_\ell(\cdot)) \in \mathscr{U}_p^{(\ell)}$. Associated with this system is the triple of Lie algebras $(\mathscr{L}, \mathscr{L}_0, \mathscr{B})$ where $\mathscr{B} = \{B_1, \ldots, B_m\}_{LA}$, $\mathscr{L} = \{\mathscr{B}, A\}_{LA}$ and $\mathscr{L}_0$ is the ideal generated by $\mathscr{B}$ in $\mathscr{L}$. Suppose $\mathscr{R}(A;\mathscr{B}) = \mathscr{L}_0$. Then for all $x \in \underline{R}^n$, $t \in \underline{R}^+$,

$$\mathscr{R}_t(x) = e^{tA}\left\{e^{\mathscr{L}_0}\right\}_G \underline{x} + \{\mathscr{L}^i \underline{f}_j : i=0, 1, \ldots \text{ and } j=1, \ldots, \ell\}_{LS}$$

Proof: Let

$$\widetilde{A} = \begin{pmatrix} A & 0 \\ 0 & 0 \end{pmatrix}, \quad \widetilde{B}_i = \begin{pmatrix} B_i & 0 \\ 0 & 0 \end{pmatrix}, \quad F_i = \begin{pmatrix} 0 & \underline{f}_i \\ 0 & 0 \end{pmatrix}$$

be a collection of n+1 matrices over $\underline{R}$. It is easily verified by direct computation that the system

$$\frac{dX}{dt}(t) = (\widetilde{A} + u_1(t)\widetilde{B}_1 + \cdots + u_m(t)\widetilde{B}_m + v_1(t)\widetilde{F}_1 + \cdots + v_\ell(t)\widetilde{F}_\ell) X(t)$$

satisfies the conditions of Theorem 4. 1. Thus for this bilinear system the reachable set from the identity matrix I_{n+1} at time t is

$$\widetilde{\mathscr{R}}_t(I_{n+1}) = e^{t\widetilde{A}}\left\{e^{\widetilde{\mathscr{L}}_0}\right\}_G = I^t(\widetilde{\mathscr{L}}_0, I_{n+1}) .$$

By direct computation

$$\widetilde{\mathscr{L}}_0 = \left\{\begin{pmatrix} \mathscr{L}_0 & \mathscr{L}^i \underline{f}_j \\ 0 & 0 \end{pmatrix} : i=0, 1, \ldots \text{ and } j=1, 2, \ldots, \ell\right\}_{LS} .$$

Thus if $\pi_n : (x_1, \ldots, x_{n+1}) \mapsto (x_1, \ldots, x_n)$ is the projection from $\underline{R}^{n+1}$ onto $\underline{R}^n$.

$$\mathscr{R}_t(\underline{x}_0) = \pi_n\left(\widetilde{\mathscr{R}}_t(I_{n+1})\begin{pmatrix} \underline{x}_0 \\ 1 \end{pmatrix}\right) = e^{tA}\left\{e^{\mathscr{L}_0}\right\}_G \underline{x}_0 + \{\mathscr{L}^i \underline{f}_j : i=0, 1, \ldots \text{ and } j=1, \ldots, \ell\}_{LS} .$$

Q. E. D.

Theorem 4.1 is a generalization of Brockett's Theorem 7 of [1]. For the case where $B_1 = \cdots = B_m = 0$ we have $\mathscr{L}_0 = \mathscr{B} = \{0\}$, $\mathscr{R}(A;\mathscr{B}) = \mathscr{L}_0$ and the corollary states that

$$\mathscr{R}_t(x) \doteq e^{tA}x + \{A^i \underline{f}_j : i=0,1,\ldots \text{ and } j=1,\ldots,\ell\}_{LS}$$

$$= e^{tA}x + \text{range}\,(F, AF, \ldots, A^{n-1}F)$$

which is the usual result for linear systems (cf. [8]).

ACKNOWLEDGEMENTS

The author would like to express his appreciation to Professor R.W. Brockett. His interest, encouragement and advice were invaluable. He also wishes to thank John Baillieul for his helpful comments on the original manuscript.

BIBLIOGRAPHY

1. Brockett, R.W., "System Theory on Group Manifolds and Coset Spaces", SIAM J. Control 10, No. 2 (May 1972).

2. Sussman, H.J. and V. Jurdjevic, "Controllability of Nonlinear Systems", J. of Diff. Equ. 12, No. 1 (July 1972).

3. Lobry, C., "Geometrical Structure of Orbits of Dynamical Polysystems", Control Theory Center, U. of Warwick, R 19 (July 1972).

4. Warner, R.W., Foundations of Differentiable Manifolds and Lie Groups, Scott, Foresman and Co. (1971).

5. Lobry, C., "Contrôlabilité des systèmes non linéaires", SIAM J. Control 8, 573-605 (1970).

6. Hirschorn, R.M., "Topological Semigroups and Controllability in Bilinear Systems", Ph.D. Thesis, Div. of Eng. and Appl. Phys., Harvard Univ. (Sept. 1973).

7. Palais, R.S., A Global Formulation of the Lie Theory of Transformation Groups, Mem. of the Amer. Math. Soc., 22 (1957).

8. Brockett, R.W., Finite Dimensional Linear Systems, John Wiley and Sons, Inc. (1970).

9. Brockett, R.W., "Control Theory on Lie Groups", Proc. NATO Advanced Study Instit., Imperial College, London (1973).

CONTROL THEORY IN TRANSFORMATION SEMIGROUPS

Gerald S. Goodman

Institute of Mathematics

University of Florence

The idea of using control-theoretic methods in the study of semigroups of transformations goes back fifty years to a paper by the Czech mathematician Karl Loewner devoted to the theory of conformal mapping [11]. There he showed how extremal problems for univalent conformal mappings on the unit disk could be treated as problems of optimal control, with the infinitesimal generators of the semigroup entering as control variables. Despite the vast importance which this work has had in the theory of conformal mapping, it has remained practically unknown outside of that field, and the semigroup ideas upon which it is based have been, largely, neglected, even though, as Loewner foresaw, they can be utilized elsewhere.

Now that current thinking in control theory is beginning to move in this direction, it seems timely to give an account of Loewner's viewpoint in a way which emphasizes its general applicability. In so doing, we shall see that the function-theoretic aspects of his work are relatively incidental. Our main emphasis will be on the behavior of two-parameter families of transformations which possess a functorial character with respect to time akin to the transition functions of a non-stationary flow. Such families occur in Loewner's work in a subsidiary, technical rôle. By elevating them to objects of fundamental concern and learning how to treat them, we can get more easily to the basis of Loewner's thinking, and then carry it over to the treatment of other semigroups, such as the stochastic matrices which occur in the theory of Markov processes.

In this way, we hope to alert the reader to the existence

of rich areas of pure mathematics where control-theoretic methods can apply.

1.

By the second decade of this century, when Loewner began his research in conformal mapping, the Riemann Mapping Theorem, which says that, apart from the obvious exceptions, every simply-connected schlicht (i.e., non-overlapping) subdomain of the complex plane is the conformal image of the unit disk under a mapping of the form $w = f(z)$, where f is analytic and univalent in $|z| < 1$, had been placed on a firm foundation by Koebe and others. The study of the special class of analytic functions with this property of assuming different values at different points of the unit disk--the so-called schlicht or univalent functions-- thus began to attract interest, and there grew up a "geometric" theory of functions, concerned with the interplay between the geometric properties of the image domain and the analytic properties of the mapping function. Already in the most general case, where it is only assumed that the image domain is schlicht, the univalency of the mapping exercises some influence on the magnitude of the Taylor coefficients of the mapping function f in its expansion about the origin, and one problem that was posed, by Bieberbach in 1916, was to describe the precise region of variability S_n of the initial coefficients $a_2, \ldots, a_n$, under the assumption that $a_0 = f(0) = 0$ and $a_1 = f'(0) = 1$. The class S of univalent functions satisfying this normalization enjoys the fundamental property of being compact in the topology of local uniform convergence. In order to arrive at inequalities that must be satisfied by normalized univalent functions, one can thus seek to determine the extreme values of continuous functionals defined on S. The problem of maximizing $|a_n|$ is an example of such an extremal problem, still unsolved for general n, despite a host of partial results, all supporting the appealing conjecture that the maximal value is n.

Similar problems can be posed within the subclass of univalent functions whose image domains are uniformly bounded, and, since we have homothetic transformations at our disposal, we may renormalize these mappings and suppose that their image domains belong to the unit disk. We are thus led to consider extremal problems in the class $\overline{S}$ of univalent self-mappings of the unit disk, i.e., of univalent functions h on $|z| < 1$ that satisfy the inequality $|h(z)| < 1$ for $|z| < 1$ and are so normalized that $h(0) = 0$ and $0 < h'(0) \leq 1$. In this subclass $\overline{S}$, the problem of determining the coefficient regions $\overline{S}_n$ (now beginning with a_1) is transformed by Loewner into a reachability or controllability problem for a certain control system, and the coefficient problem of maximizing $|a_n|$ becomes a problem of optimal control.

2.

To achieve this reduction, Loewner started off by regarding the functions in $\bar{S}$ as the elements of a transformation semigroup, closed under composition. Considered in this way, the elements of $\bar{S}$ were found to have a remarkable decomposition property: <u>every one can be written as the composite of elements which lie arbitrarily close to the identity</u>. This he proved by function-theoretic means. Then, following the practice of Sophus Lie, he suggested replacement of these finite transformations that lie close to the identity by "infinitesimal transformations", i.e., by their linearizations about the identity, the "composition" of which amounts to the integration of a differential equation of the form

$$\frac{dh}{dt} = V_t[h],$$

where, for each $t \geq 0$, V_t is an <u>infinitesimal generator</u> of the semigroup $\bar{S}$.

By an adroit use of Schwarz's Lemma, he established that the infinitesimal generators V were of the form $-h\,p(h)$, where $p(h)$ is any analytic function of positive real part on the disk $|h| < 1$, with $p(0) \geq 0$. Such functions form a complete convex cone with apex at the origin, generated by the section in which $p(0) = 1$. Their coefficient regions were completely characterized by Carathéodory in 1907, well prior to the work of Loewner.

<u>Loewner's Equation</u> thus has the form

$$\frac{dh}{dt} = -h\,p_t(h), \qquad t \geq 0,$$

with $h = h_t(z)$ taking the initial values $h_o(z) = z$, $|z| < 1$. It is clearly a control equation, with the functions $p_t(h)$, or the coefficients in their Taylor development about the origin, entering as control variables. When the latter are measurable functions of t, the solutions exist in the Carathéodory sense for all time, and the uniqueness of the solutions in their dependence upon the initial values z, together with the fact that $|h_t(z)|$ is monotone decreasing, assures that the resulting transformation h_t belongs to $\bar{S}$ for each fixed t. Under the usual normalization $p_t(0) \equiv 1$, the limit of $e^t h_t$ exists as $t \to \infty$ and belongs to S. Thus, extremal problems in S can also be treated by means of Loewner's equation.

Loewner's decision to allow the infinitesimal generators to vary with t was not gratuitous. If the right side of the equation is independent of t, the variables are separable, and the solution can be expressed by quadratures. Loewner showed how the resulting formulas could be interpreted geometrically and was thus able to conclude that the mappings generated would be confined to a subclass of $\bar{S}$ so narrow that its properties would not be typical for $\bar{S}$ as a whole. He thus rejected as inadequate the idea of investigating $\bar{S}$ by means of its one-parameter subsemigroups and, instead, raised the question as to whether, by allowing the infinitesimal generators to vary with t, the entire semigroup $\bar{S}$ could be generated.

An—affirmative—response to this question had to wait some decades, when the compactness properties of Carathéodory solutions of the control equation came to be appreciated [3], [13]. This response was prepared for, while, at the same time, it was rendered less urgent, by Loewner's own main contribution: his theory of slit mappings.

3.

A <u>slit mapping</u> is a function in $\bar{S}$ whose image domain is bounded by the periphery of the unit circle together with a Jordan arc. Using standard methods of conformal mapping, Loewner showed that the class of slit mappings is <u>dense</u> in $\bar{S}$. He then established that every slit mapping could be generated by his differential equation using a restricted class of controls, viz., continuous controls taking values at the <u>extreme points</u> of the section $p(0) = 1$ of the control region. Such functions are of the form

$$p_t(h) = \frac{1+\varkappa(t)h}{1-\varkappa(t)h} = 1 + 2\varkappa(t)h + 2\varkappa(t)^2 h^2 + \cdots,$$

where $|\varkappa(t)| \equiv 1$. A slit mapping is therefore characterized by one single function $\varkappa(t)$.

Loewner was drawn to slit mappings because of controllability considerations. He emphasized that the general solution h_{st}, $s \leq t$, of his equation could be regarded as the transition function of a non-stationary flow, whose momentary velocity field is V_t: $h_{st}(z)$ is the position at time t of a particle that starts at the point z at time s. It thus satisfies the <u>functorial conditions</u>

$$h_{st} = h_{ut} \circ h_{su}, \quad s \leq u \leq t, \qquad \text{and } h_{ss} = \text{the identity}.$$

For s and t fixed, h_{st} is a univalent mapping of the unit disk into itself which satisfies the inequality $|h_{st}(z)| \leq |z|$ for

$|z| < 1$. The functoriality thus implies that for $s \leq u$ the mapping h_{st} will be "subordinate" to the mapping h_{ut} in the sense that the image of the unit disk under h_{st} will be contained in the image of the unit disk under h_{ut}. In other words, for t fixed, say at some terminal value t_o, the image domains must expand continuously with increasing s. Accordingly, before a mapping in S can be generated, it must be possible to embed its image domain in a continuum of domains which expand with increasing s until they eventually exhaust the unit disk. Conversely, if g_s, $0 \leq s \leq t_o$, is any such family of mappings, then the function $h_{st} = g_t^{-1} \circ g_s$ is well-defined and satisfies the foregoing functoriality condition.

In the case of slit mappings, these general considerations tell us that if such a mapping is to be generated at all, it can be generated in only one way. For the only possibility of embedding a slit domain in a continuum of expanding domains is just to shorten the slit. The controllability problem is thus soluble in a unique way for slit mappings, and the kinematics arise naturally from the parametrization of the slit itself.

4.

The derivation of the differential equation for slit mappings is the central result of Loewner's paper, but we shall not pursue this. Instead, we shall indicate how, by general considerations, we can arrive at the differentiability a.e. with respect to t of the functor h_{st} by adopting an opportune change of time scale (already used by Loewner).

Our first observation is that the functoriality of h_{st} is presrved under any continuous, order preserving map of the parameter interval, i.e., under any continuous, monotone increasing change of parameters. Since the differentiability will clearly depend upon the parametrization chosen, the problem is really to select a parametrization which will confer differentiability. The problem is thus akin to the celebrated Fifth Problem of Hilbert, concerning the introduction of differentiable coordinates in continuous groups.

In this connection, we note, with Loewner, that the mappings in S are so normalized that, by the chain rule, the values of $\gamma = h'(0)$ --the so-called "interior mapping radius" of h-- will multiply when the mappings h are composed. Thus γ defines a homomorphism of S onto the semigroup $(0, 1]$ under multiplication. Elementary function theory shows that γ is continuous and takes the value 1 only at the identity. To prove that the inverse of γ, which is multivalued in general, is continuous in a neighborhood of the identity, Loewner established a displacement estimate, which can be written in the form

$$|h(z) - z| \le [1 - \gamma]\, |z| \left|\frac{1+|z|}{1-|z|}\right| ,$$

for $|z| < 1$. Thus the size of γ governs the nearness of h to the identity. (Loewner used this fact in arriving at his decomposition of univalent functions.)

Suppose now that h_{st} is any continuous functor from the ordered interval $[0, t_0]$ into $\overline{S}$. Then $\gamma[s,t] = h'_{st}(0)$ is multiplicative over adjacent intervals, from which it follows that $\gamma[0,t]$ is a decreasing function of t. Accordingly, Loewner could introduce $-\log \gamma[0,\cdot]$ as his new time scale, and we do the same. In this new parametrization, we have $\gamma[s,t] = e^{s-t}$, so that $1 - \gamma[s,t] \le t - s$. The displacement inequality thus yields the estimate

$$|h_{st}(z) - h_{ss}(z)| \le t - s \times \text{locally bounded terms};$$

in other words, for fixed z, h_{st} satisfies a Lipschitz condition in the second variable when the first is held fixed at s. Replacing z by h_{os} and using the functoriality then gives us a Lipschitz condition on the entire trajectory $h_{ot}(z)$ and, therefore, its differentiability a.e. There is enough uniformity in the estimates to show that the exceptional set is independendent of z, and in this way we can arrive at infinitesimal generators and Loewner's equation in the Carathéodory sense. For details, see [3] or [13].

This use of a reparametrization to achieve Lipschitzianity is akin to Lebesgue's introduction of arc length in order to prove the differentiability a.e. of functions of bounded variation.

Once expressed in the new parameters, the functors h_{st} become equiabsolutely continuous, and the corresponding control functions satisfy the normalization $p_t(0) \equiv 1$. This fact allows us to pass from the dense set of slit mappings to the rest of $\overline{S}$ by applying a selection theorem and thus to establish that every mapping in S can be generated by Loewner's differential equation when the control functions are allowed to be measurable in t and the solution is understood in the Carathéodory sense. Renormalizing and passing with t to ∞ then gives a means of generating every function in S.

5.

One may speculate as to why Loewner himself did not arrive at this result. The reason appears to be threefold. In the first place, for him, the integration of a differential equation corresponded to the intuitive idea of composing infinitesimal transformations--the kinematic counterpart being the reconstruction of a flow from its velocity field. This idea goes back to Euler and runs through the work of Lie, but is apparently lost in the Carathéodory approach, where the exceptional set at which the differential equation fails to hold seems to vary with the trajectory. It was only much later, when Scorza-Dragoni [15] proved that the exceptional set does not really depend upon the trajectory, that the notion of velocity field, and thus of infinitesimal transformations, could be reinstated.

A second reason appears to be this. If we regard the coefficients in the Taylor expansion about the origin of $p_t(h)$ as control parameters, then the number necessary to generate a given mapping in $\bar{S}$ is countable. Loewner considered his prime accomplishment to be the reduction of this number to just one--the function $\varkappa(t)$. He knew, furthermore, that, even with continuous $\varkappa$'s, it was possible to generate mappings which are not slit mappings. If $\varkappa$ is allowed to have jumps, it may be that the whole of $\bar{S}$ can be generated. He was working in this direction, and no counterexample has ever been found, despite serious attempts [14].

Finally, Loewner regarded his method as essentially a continuity method: he expressed functionals on $\bar{S}$ as integrals involving $\varkappa$ as a parameter, which he proceeded to estimate directly. The resulting estimates then carried over to $\bar{S}$ by continuity, and to S by renormalization. In this way, he established the sharp inequality $|a_3| \leq 3$ for functions in class S. There remains the problem of determining the extremal mappings, but these turn out to be slit mappings, for reasons that only became clear later on, in the work of Schiffer [16].

6.

Loewner's paper bears a "I" after the title, added by the editor (Bieberbach) in anticipation of further results. Loewner was attempting to establish further coefficient inequalities by varying the control, but his efforts were inconclusive and "II" never appeared. In 1962, he assigned me the task of exploring what can be done by use of the Pontryagin Maximum Principle, and in the next year I arrived at the following result concerning the problem of maximizing a smooth functional of the initial coefficients of a function in class $\bar{S}$ [3]. The Hamiltonian that arises in the Maximum Principle, when evaluated along an optimal trajectory, serves as an integrating factor for Loewner's equation for the corresponding extremal mapping. Now, applying a classical result, if $M[h,t]$ is an integrating factor for Loewner's equation, then $M[h_{ot}(z),t]\,h'_{ot}(z)$ is a differential invariant. Evaluating it at $t=0$, when h_{ot} reduces to the identity, and then at the terminal time, when it goes over into the extremal mapping, and equating gives then a differential equation for the extremal mapping in which no trace of the control functions appears. A similar result holds in class S. This differential equation, first derived by Schiffer in 1938 by an entirely different method [16], is the key to a detailed study of the coefficient bodies S_n. From it one can conclude that the extremal mappings transform the unit disk onto domains bounded by piecewise analytic slits, and the corresponding κ is piecewise analytic. Since Schiffer's equation is one of the deepest results in conformal mapping, its derivation in this way is a testimony to the remarkable power of the Pontryagin Maximum Principle.

Years before the Maximum Principle had been formulated in any generality, Schiffer had proved, conversely, that whenever Loewner's differential equation generates a solution of Schiffer's differential equation, the Maximum Principle holds along the trajectories of the initial coefficients [17]: thus the two criteria for extremality are equivalent. This is important because deep investigations of Teichmüller [18] and others have shown—using differential geometric methods—that any solution of Schiffer's equation has extremal properties with regard to the n-th coefficient when the previous coefficients are held fixed. Consequently, the Pontryagin Maximum Principle is a sufficient condition for optimality in coefficient problems when terminal constraints are introduced. We thus have a control-theoretic result proved by function-theoretic means. Perhaps, by eliminating the function theory, the result can be extended to other control systems.

The Pontryagin Maximum Principle itself can be derived very easily, by calculus, in a semigroup setting. For lack of space, we cannot describe the procedure here. A hint of what is involved will be found in [5]. Nor can we go into the following application of the above sufficiency result: any function in S whose coefficients form a probability distribution can be embedded in a functorial family h_{st} possessing the same properties. This result characterizes the probability generating functions of continuous parameter Markov branching processes in the "supercritical" case [8].

7.

We have seen in sec. 4 that the derivation of Loewner's differential equation can be based upon his displacement inequality and the fact that $\gamma = h'(0)$ is a homomorphism of $\overline{S}$ onto $(0, 1]$. Both of these facts remain valid when $\overline{S}$ is replaced by the larger semigroup H that arises when the assumption of univalence of the power series in $\overline{S}$ is dropped. Moreover, both semigroups have the same infinitesimal generators. Thus, Loewner's equation likewise governs the behavior of continuous functors h_{st} that take values in H. Since the solutions of that equation necessarily lie in $\bar{S}$, we see that univalency imposes itself automatically.

This suggests a characterization of $\bar{S}$ in H in terms of finite, rather than infinitesimal, decompositions, viz., that $\bar{S}$ consists of precisely those elements of H which can be written as the composite of elements of H that lie arbitrarily close to the identity. We recall, from sec. 2, that Loewner found that every element in $\bar{S}$ enjoys this property. It remains to prove the converse, but something more general is true [4].

We introduce a triangular array by assigning to each $n = 1, 2, \ldots$ and each $k = 1, \ldots, k_n$ a function h_{kn} in H in such a way that $\gamma_{kn} = h'_{kn}(0)$ tends to 1 uniformly in k as $n \to \infty$. Suppose that the marginal products

$$h_n = h_{k_n n} \circ \cdots \circ h_{1n}$$

converge to a function h when $n \to \infty$ (this can always be achieved by passing to a subsequence). Then, either $h'(0) > 0$ and h belongs to $\bar{S}$, or $h'(0) = 0$ and h vanishes identically. Thus, "approximate" decomposability is enough to characterize $\bar{S}$.

To prove this, we make use of an argument which occurs in [13]. According to a theorem of Landau, any function in h with leading coefficient γ will necessarily be univalent in the disk $|z| < r(\gamma)$, where $r(\gamma) = [1-(1-\gamma^2)^{1/2}]/\gamma$. By Schwarz's Lemma, univalency in a disk about the origin is not lost under composition, so the composite of k functions in H will have radius of univalence about the origin no smaller than the minimal Landau radius among the lot. Since $r(\gamma)$ tends to 1 with γ, this means that the radius of univalence of the marginal products h_n approches 1 when $n \to \infty$. On any fixed subdomain of the unit disk, h thus agrees with the limit of univalent functions and therefore is either univalent on the entire disk or else is a constant equal to $h(0)$.

When the limit function h is univalent, it can be embedded in a continuous family h_{st} in $\bar{S}$ which is functorial and has its parameters normalized. Pommerenke has observed that this embedding can be arrived at directly from the triangular array by replacing the indices k for each fixed n by the values $t_{kn} = -\log[\gamma_{kn} \cdots \gamma_{1n}]$ and then using a compactness argument; cf. [13].

8.

In the foregoing sections, we have tried to formulate Loewner's ideas in such a way as to detach them from specifically function-theoretic considerations. To gauge how much we have succeeded, let us now replace the semigroup H by another semigroup—the class $\underline{S}$ of $n \times n$ stochastic matrices (i.e., matrices with non-negative entries whose row-sums equal 1) under multiplication. Consider a continuous function P_{st} from the ordered interval $[0,t_o]$ into $\underline{S}$ which satisfies the functoriality conditions

$$P_{st} = P_{su} P_{ut} \quad \text{for } s \leq u \leq t, \quad \text{and } P_{ss} = I;$$

such functors arise as transition probabilities of non-stationary, continuous parameter Markov chains. Can we introduce a time scale that will allow us to exhibit such functors as the solutions of Kolmogorov's differential equation

$$\frac{dP}{dt} = P Q_t \quad \text{a.e.}$$

where, for each fixed t, Q_t is an infinitesimal generator of $\underline{S}$?

Our experience teaches us to seek a continuous homomorphism γ from $\underline{S}$ into $[0,1]$ which is 1-1 at the identity and whose deviation from 1 governs the deviation of the corresponding elements of $\underline{S}$ from I. An obvious choice is $|\det P| = \gamma$, where P ranges over $\underline{S}$, but this equals 1 whenever P is a permutation matrix, so no displacement estimate of the type we are seeking can hold. Nevertheless, if P is a value taken by P_{st}, it necessarily can be written as the product of matrices in $\underline{S}$ that lie near I. Now, we showed in [7] that stochastic matrices $P = [p_{ij}]$ near I satisfy $\min_i [p_{ii}] \geq \det P > 0$, and the class of all matrices in $\underline{S}$ which satisfy this inequality is closed under multiplication. Hence, the values taken by P_{st}, while nominally in $\underline{S}$, are actually confined to a smaller semigroup $\underline{M}$ in which the following displacement estimate holds:

$$1 - \det P \geq 1 - p_{ii} \geq p_{ij}, \qquad i, j = 1, \dots, n, \; i \neq j.$$

We can therefore use $-\log \det P_{0\cdot}$ as a new time scale and derive the Kolmogorov differential equations for P_{st}: v. [7].

Probabilistically speaking, $-\log |\det P|$ is like an entropy: if P is regarded as the channel matrix of a discrete, noisy, memoryless channel, then it works as a measure of distortion or noise. In the case of continuous parameter Markov chains, $\det P$ is a probability, and $-\log \det P$ is the expectation of certain related Poisson processes.

Product integration of Kolmogorov's equation amounts to the generation of $\underline{M}$ by "composition of infinitesimal transformations", and the problem of describing which matrices in $\underline{S}$ can be embedded in a continuous parameter Markov chain now becomes a reachability question [7]. The differential equation implies that $\prod_{i=1}^{n} p_{ii} \geq \det P > 0$ holds throughout $\underline{M}$, but the conjecture in [7] that this condition is sufficient for reachability has since been proved false by D. Williams, who observed that equality cannot hold for an embeddable matrix unless one of its entries vanishes. This same observation can be used to show that M is not convex when $n > 2$. From general findings about non-negative matrices [1], it is known that M is contractible to certain of its boundary points; S. Johansen has recently shown how controllability considerations allow the contractions to be performed along rays [10]. The theory of chattering controls implies that M is the closure relative to GL(n) of finite products of matrices generated by constant controls Q taking values on the extreme rays of the control region. Finite products alone suffice to generate int M [10]; hence M is the relative closure of its interior.

The case $n = 2$ is very special. M here consists of all stochastic matrices with $\det P = \operatorname{tr} P - 1 > 0$: M is thus convex. Every element can be generated by time-independent Q's, and the bang-bang principle holds. When $n > 2$, convex hull M is unknown, and the elements generated by constant controls do not even form a semi-group.

The elements of M can also be characterized as the non-singular limits of triangular arrays P_{kn} with $\gamma_{kn} \to 1$ uniformly in k as $n \to \infty$. Since no determination of the matrices in M by means of of their mapping properties has been found, we must fall back on Pommerenke's compactness argument. This decription of M has served as the point of departure for an interesting study by Johansen [9]. By introducing the notion of an "accompanying array", he is able to give an independent proof and to derive the differentiability results of [7] in a new way. This approach seems of value in justifying Loewner's "Schritt ins Infinitesimale" by which he would pass from finite to infinitesimal transformations, and can probably be used to arrive directly at differentiability results in the semigroup of infinitely decomposable elements in the class of $n \times n$ non-negative matrices without reducing them--as can be done--to the stochastic case by change of dependent variable.

In closing, we call attention to Loewner's paper [12], where, under a symmetry condition on the Q-matrices, the reachable set from any unit vector in $\mathbb{R}^n$ is explicitly determined.

Collaboration with S. Johansen has been made possible by a grant from the Scientific Affairs Division of NATO, and the preparation of this paper has been supported by the Consiglio Nazionale delle Ricerche.

1. Brown, D.R., On clans of non-negative matrices, Proc. Amer. Math. Soc. 15 (1964), 671-674.
2. Goodman, G.S., On the determination of univalent functions with prescribed initial coefficients, Arch. Rat. Mech. & Anal. 24 (1967), 78-81.
3. Goodman, G.S., Univalent functions and optimal control, Thesis, Stanford Univ., Stanford, Calif., 1968.

4. Goodman, G.S., Foundations of Loewner's theory of schlicht functions, lecture given at the Conference on Math. Analysis, Otaniemi, Finland, 1966.

5. Goodman, G.S., A method for comparing univalent functions, Bull. Amer. Math. Soc. 75 (1969), 517-521.

6. Goodman, G.S., On a theorem of Scorza-Dragoni and its application to optimal control, in "Math. Theory of Control", Balakrishnan & Neustadt eds., Academic Press, N. Y., 222-233.

7. Goodman, G.S., An intrinsic time for non-stationary finite Markov chains, Zeit. f. Warsch. 16 (1970), 165-180.

8. Goodman, G.S., to appear.

9. Johansen, S., A central limit theorem for finite semigroups and its application to the imbedding problem for finite state Markov chains, Zeit. f. Warsch. 26 (1973), 171-190.

10. Johansen, S., The bang-bang problem for stochastic matrices, Zeit. f. Warsch. 26 (1973), 191-195.

11. Loewner, K., Untersuchungen über schlichte konforme Abbildungen des Einheitskreises, I, Math. Ann. 89 (1923), 103-121.

12. Loewner, K. = C., A theorem on the partial order derived from a certain transformation semigroup, Math. Zeit. 79 (1959), 53-60.

13. Pommerenke, Chr., Über die Subordination analytischer Funktionen, J. Reine u. Angew. Math. 218 (1965), 159-173.

14. Pommerenke, Chr., On the Loewner differential equation, Mich. Math. J. 13 (1966), 435-443.

15. Scorza-Dragoni, G., Una applicazione della quasicontinuità semiregolare delle funzioni misurabili rispetto ad una e continue rispetto ad un'altra variabile, Atti Accad. Naz. Lincei, Rend. Cl. Sci. Fis. Mat. Nat. (8) 12 (1952), 55-61.

16. Schiffer, M., A method of variation within the family of simple functions, Proc. Lon. Math. Soc. 44 (2nd ser.) (1938), 432-449.

17. Schiffer, M., Sur l'équation différentielle de M. Löwner, C. R. Acad. Sci. Paris 221 (1945), 369-371.

18. Teichmüller, O., Ungleichungen zwischen den Koeffizienten schlichter Funktionen, S.-B. Preuss. Akad. Wiss., Phys.- Math. Kl. (1938), 363-375.

An account of Loewner's equation and Schiffer's (variational) method as seen by a function-theorist can be found in

L.V. Ahlfors, "Conformal Invariants", McGraw-Hill, 1973.

For more details concerning the approach of this paper one may consult my lectures on "Semigroups and Control Theory", available from the International Center for the Mechanical Sciences, Udine, Italy.

THE IMBEDDING PROBLEM FOR FINITE MARKOV CHAINS

Søren Johansen

University of Copenhagen

1. INTRODUCTION

The problem of characterizing the stochastic matrices which can occur in a continuous time Markov chain was first formulated by Elfving in 1937, see [7] and [8]. The problem was mentioned by Chung in 1960, [1] p 203, and in the last 10 years a number of papers have appeared.

In this note we shall present a brief outline of the methods and results in the papers [12] - [18] and relate them to the other work in the area.

We first introduce the basic notation and definitions. Let P denote an n × n stochastic matrix with elements p_{ij}, i.e.

$$p_{ij} \geqq 0, \quad \Sigma_j\, p_{ij} = 1.$$

An intensity matrix Q is a matrix with elements q_{ij}, such that

$$q_{ij} \geqq 0,\ i \neq j,\ \Sigma_j\, q_{ij} = 0.$$

A Markov chain is a continuous family

(1.1) $$\{P(s,t),\ 0 \leqq s \leqq t < t_0\}$$

of stochastic matrices satisfying the Chapman-Kolmogorov equations:

(1.2) $$P(s,t) = P(s,u)P(u,t),\ 0 \leqq s \leqq u \leqq t < t_0,$$

and

(1.3) $$P(s,s) = I,\quad 0 \leqq s < t_0.$$

The stochastic matrix P is called imbeddable if there exists a Markov chain such that

(1.4) $$P(0,1) = P.$$

We say that P is imbeddable in a homogeneous chain if the family (1.1) depends only on t-s.

An important idea in the discussion of the equations (1.2) is the following: The equations are clearly invariant under a homeomorphic change of time scale and it was observed by Goodman [10] that if one chooses

(1.5) $$\varphi(t) = -\ln \operatorname{Det} P(0,t)$$

as the new time scale, then the functions $P(\cdot,t)$ and $P(s,\cdot)$ become absolutely continuous and can be characterized as the unique solution to the Kolmogorov differential equations:

(1.6) $$\partial_t P(s,t) = P(s,t)Q(t),\quad t \notin N,$$

(1.7) $$\partial_s P(s,t) = -Q(t)P(s,t),\quad s \notin N,$$

(1.8) $$P(s,s) = I,$$

where N is a null set for Lebesgue measure, and $Q(\cdot)$ is a bounded measurable function with values in the set of intensity matrices.

The solution to these equations can be constructed as a product integral

$$P(s,t) = \prod_s^t (I + Q(u)du),$$

see Schlesinger [34] and Dobrušin [6].

In view of these results the imbedding problem can be formulated by means of the theory of differential equations. More specifically, consider the controlsystem (1.6) for s = 0 and P(t) = P(0,t):

(1.9) $$DP(t) = P(t)Q(t),\ t \notin N,$$

where the controller $Q(\cdot)$ is chosen in the convex cone of intensity matrices. In this language a matrix P can be imbedded if it can be reached using a bounded measurable controller in a finite time, and the imbedding problem is that of characterizing the reachable set.

In this formulation one naturally asks for a Bang-Bang representation of an imbeddable matrix, i.e. a representation of P as a finite product of matrices generated by the extremal intensity matrices [15] and [18], since this corresponds to reaching P by switching the controller a finite number of times between the extremal controllers.

One can think of the above as a semigroup approach, since clearly the set of stochastic matrices as well as the set of imbeddable matrices form a semigroup.

It is, however, also possible to apply a convex analysis to the set of stochastic matrices, which clearly form a convex compact set. The extreme points are easily identified with the matrices with entries 0 and 1 and they form a semigroup under multiplication. A stochastic matrix can then be represented as a convex combination of the extreme points or as a probability measure on a finite semigroup. Conversely given a probability measure on a finite semigroup one can construct the corresponding random walk which is a Markov chain with discrete time.

It is easily seen that convolution of the probability measures correspond to multiplication of the stochastic matrices, see [14] or [26].

Thus the stochastic matrices can be thougth of as a representation of the measures on finite semigroups, and there is a close connection between stochastic processes with independent increments and imbeddable stochastic matrices, [16]. This relation is used to suggest the definitions and results for Markov chains as well as a central limit theorem for random variables on finite semigroups and the definition of infinite factorizability.

There is also a relation between the Lévy-Khinchin representation of infinitely divisible distributions [12] and the Bang-Bang representation [15] and [18] in that they both clarify the rôle of the extreme intensities as generating the "building blocks" of the semigroup.

The basic structure of the stochastic matrices that is used is that they form a convex semigroup. The set has many other properties, like the extreme points form a semigroup, the multiplication is bilinear and there exists a homomorphism onto [0,1] with multiplication, namely the determinant. Furthermore there is a norm and a natural topology on the vectorspace generated by the stochastic matrices.

There are many other examples of convex semigroups, the most obvious is that the set of probability measures on a semigroup is itself a semigroup under convolution. The set of doubly stochastic matrices with multiplication,the set of characteristic functions of probability measures on R again with multiplication, but also the set of generating functions of probability measures on the positive integers with composition as the semigroup operation form convex semigroups.

Many of the results derived here for finite stochastic matrices can be proved for convex semigroups with some extra structure as indicated above, but we shall only be concerned here with results that have a direct interpretation in terms of Markov chains.

2. NON-HOMOGENEOUS MARKOV CHAINS

In this section we shall give some of the results obtained on the imbedding problem for finite state non-homogeneous Markov chains and in particular for processes with independent increments and values in a finite semigroup.

2.1. Definition. A stochastic matrix P is called infinitely factorizable if for all $\varepsilon > 0$ there exists $P_1, \ldots, P_n$ such that

$$P = P_1 \cdot \ldots \cdot P_n \tag{2.1}$$

$$||I - P_i|| \leqq \varepsilon, \; i = 1, \ldots, n, \tag{2.2}$$

where

$$||I - P|| = \sup_i \Sigma_j \, |\delta_{ij} - p_{ij}| = 2\sup_i (1 - p_{ii}).$$

This concept was first used by Loewner [25] who studied the semigroup of schlicht mappings of the unit disc into itself but has also been used by Maksimov [28] and [29] in the discussion of probability measures on groups.

2.2. Definition. A triangular array is a family $\{P_{m,k}, \; k = 1, \ldots, m, \; m = 1, 2, \ldots\}$ of stochastic matrices. The marginal products are

$$P_m = \prod_{k=1}^{m} P_{m,k}$$

and the limit is $\lim_m P_m$. The array is a null array if

$$\lim_m \sup_k ||I = P_{m,k}|| = 0.$$

The concept of a triangular array is well known in probability, see Grenander [11], Kendall [20], Davidson [5] and Gnedenko and Kolmogorov [9].

The relation between probability measures on semigroups and stochastic matrices suggests that the following result holds:

2.3. Theorem. Let P be a nonsingular stochastic matrix, then the following conditions are equivalent:

(2.3) P is imbeddable.
(2.4) P is infinitely factorizable.
(2.5) P is the limit of a triangular null array.

The proof of this can be found in [16].

If we consider the problem as a control problem we have to discuss the extremal controllers. The intensity matrices form a convex cone and an extremal element has at most one positive off-diagonal element. The stochastic matrix generated by an extremal intensity is called a Poisson matrix and is characterized by having at most one positive off-diagonal element.

By means of this we can prove [16].

2.4. Theorem. Let P be a nonsingular stochastic matrix,then P is imbeddable if and only if P can be approximated by a finite product of Poisson matrices.

Using this result and the techniques from control theory [23] we get the following Bang-Bang representation [15]:

2.5. Theorem. Let P be in the interior of the imbeddable matrices, then P has a representation as a finite product of Poisson matrices.

Finally one would like to extend this result to hold for all imbeddable matrices.

This has been proved in [18] for 3×3 matrices, and in fact if Det $P \geqq \frac{1}{2}$ we need only use 6 Poisson matrices to represent P.

Kingman and Williams [22] have studied the zero configuration of the imbeddable matrices and shown that it can be represented as a finite product of zero configurations which are reflexive and transitive.

The set of imbeddable matrices is not convex, but in [15] it is proved that it is starshaped around the stochastic matrix with equal entries, and it is quite easily seen that the convex hull is the set of all stochastic matrices.

In [18] another semigroup is considered namely the set of matrices that can be imbedded using symmetric intensity matrices. This problem continues an investigation by Loewner [24] who considered the control problem

$$DX(t) = - Q(t), \; X(t) \in R^n,$$

where Q(t) is a symmetric intensity matrix.

It can easily be seen that the symmetric stochastic matrices with two off-diagonal elements equal to $\frac{1}{2}$ and the rest equal to 0 are on the boundary of the imbeddable set. There are $\frac{1}{2}n(n - 1)$ such matrices, and it is easily seen that the convex hull of the set we get by taking products of not more than $\frac{1}{2}n(n - 1)$ of these equals the closed convex hull of the imbeddable matrices. Thus we have found the smallest convex semigroup containing the matrices imbeddable by symmetric intensities. The result is proved for n = 3 in [17], where also the supporting hyperplanes of the set are found. This gives thus a lot of necessary conditions for imbeddability.

The semigroup of probability measures on a finite semigroup is studied to give results about the special type of Markov chains formed by processes with independent increments.

The semigroup has all the properties of the set of stochastic matrices in particular we have a homomorphism defined by

$$h(\nu) = |\mathrm{Det}\ P(\nu)|$$

where $P(\nu)$ denote the transition probability matrix for the random walk determined by the probability measure ν.

Martin-Löf [30] used results about Markov chains to derive results about random walks and infinitely divisible distributions. We can prove [16] a central limit theorem and characterize the marginal distributions in processes with independent increments in a manner analogous to Theorem 2.3. This generalizes some results by Maksimov [27] and [28].

Finally we can turn these results around and obtain a criterion for imbeddability:

2.6. Theorem. A stochastic matrix is imbeddable if and only if it can be represented by a nonsingular infinitely factorizable probability measure on the extreme stochastic matrices.

3. HOMOGENEOUS CHAINS

For homogeneous chains the imbedding problem is that of finding a continuous one parameter semigroup of stochastic matrices that contains P. Since any continuous one parameter semigroup is of the form $\{P(t) = \exp tQ, 0 \leqq t < \infty\}$ for some intensity matrix Q, this problem is closely related to finding the logarithm of P. In fact Elfving [7] assumed that P had distinct eigenvalues, which then have to be positive or come in complex conjugate pairs, and then determined the various logarithms by diagonalizing the matrix.

He proved that only finitely many logarithms were admissible and the solution to the imbedding problem is then to search among these logarithms for an intensity matrix.

Cuthbert [3] and [4] discusses the logarithm function by means of the Jordan form and finds criteria for a unique imbedding. He also gives a criterion for imbeddability in terms of a series expansion for the logarithm. Speakman [35] gave an example of a stochastic matrix which could be imbedded in two different ways.

Elfving [7] obtained some simple inequalities for the eigenvalues of an imbeddable matrix and Runnenburg [32] described the region of the complex plane where the eigenvalues can be found, using results of Karpelewitch [19].

In [14] a criterion for imbeddability is given in terms of a powerseries expression for the logarithm. For n = 3 this gives manageable conditions for imbeddability in terms of P, P^2 and the eigenvalues of P.

Conditions of a different kind can be found using infinite divisibility.

3.1. Definition. The stochastic matrix P is called infinitely divisible if for all n there exists P_n such that $P = (P_n)^n$.

In this way we can obtain a result similar to Theorem 2.3.

3.2. Theorem. Let P be a nonsingular stochastic matrix, then the following conditions are equivalent:

(3.1) P is imbeddable in a homogeneous chain.
(3.2) P is infinitely divisible.
(3.3) P is the limit of a triangular null array with commuting elements in each row.

These results can be found in [16] but (3.2) was proved by Kingman [21] in 1962, and it was this result that revived the interest in the imbedding problem and it has inspired almost all the subsequent work in the area. See also Ott [31].

Vere-Jones [36] considered another commutative convex semigroup, namely the set of stochastic matrices that reduce to diagonal form by a fixed nonsingular transformation. He then proved the equivalence of (3.1) and (3.2) for this semigroup.

We can also obtain results for processes with independent increments similar to the results in the non-homogeneous case and then apply them to the imbedding problem as follows:

3.3. Theorem. A stochastic matrix P can be imbedded in a homogeneous Markov chain if and only if it can be represented by an infinitely divisible probability measure on the extreme stochastic matrices.

This result is found in [14]. It should be emphasized that the notions of infinite factorizability and infinite divisibility coincide if the semigroup is commutative.

Finally Cohen [2] has defined a class of semigroups obtained from a given semigroup and a kernel. This result is used to characterize a subclass of infinitely divisible matrices by means of a positive definiteness condition.

4. REFERENCES

1 Chung, K.L.: Markov Chains with Stationary Transition Probabilities. Berlin-Göttingen-Heidelberg: Springer 1960.

2 Cohen, J.W.: A Note on Skeleton Chains. Nieuw Arch. Wisk. 10, 180-186 (1962).

3 Cuthbert, J.R.: On Uniqueness of the Logarithm for Markov Semigroups. J. London Math. Soc., 4, 623-630 (1972).

4 Cuthbert, J.R.: The Logarithm Function for Finite-state Markov Semi-groups. J. London Math. Soc. 6, 524-532 (1973).

5 Davidson, R.: Arithmetic and Other Properties of Certain Delphic Semi-groups 1.Z. Wahrscheinlichkeitstheorie verw. Gebiete 10, 120-145 (1968).

6 Dobrušin, R.L.: Generalization of Kolmogorov's Equations for Markov Processes with a Finite Number of Possible States. Mat. Sb. N.S. 33, (75), 567-596 (1953).

7 Elfving, G.: Zur Theorie der Markoffschen Ketten. Acta Soc. Sci. Fennicae, n.Ser. A 2, No. 8, 1-17 (1937).

8 Elfving, G.: Über die Interpolation von Markoffschen Ketten. Soc. Sci. Fenn. Comment. Phys.-Math. 10, No.3, 1-8 (1939).

9 Gnedenko,B.W., Kolmogorov, A.N.: Grenzverteilungen von Summen unabhängiger Zufallsgrössen. Berlin:Akademie-Verlag 1959.

10 Goodman, G.: An Intrinsic Time for Non-Stationary Finite Markov Chains. Z. Wahrscheinlichkeitstheorie verw. Gebiete 16, 165-180 (1970).

11 Grenander, U.: Probabilities on Algebraic Structures. New York: Wiley 1963.

12. Johansen, S.: An Application of Extreme Point Methods to the Representation of Infinitely Divisible Distributions. Z Wahrscheinlichkeitstheorie verw. Geb. 5, 304-316 (1966).

13. Johansen, S. and Goodman, G.S.: Kolmogorov's Differential Equations for Non-stationary, Countable State Markov Processes with Uniformly Continuous Transition Probabilities. Proc. Camb. Phil. Soc. 73, 119-138 (1973).

14 Johansen, S.: Some Results on the Imbedding Problem for Finite Markov Chains. J. London Math. Soc.(1973).

15 Johansen, S.: The Bang-Bang Problem for Stochastic Matrices. Z. Wahrscheinlichkeitstheorie verw. Gebiete 26, 191-195 (1973).

16 Johansen, S.: A Central Limit Theorem for Finite Semigroups and Its Application to the Imbedding Problem for Finite State Markov Chains. Z. Wahrscheinlichkeitstheorie verw. Gebiete 26, 171-190 (1973).

17 Johansen, S.: The Stochastic Matrices Generated by Symmetric Intensities. Imperial College and University of Copenhagen 30 pp (1972).

18 Johansen, S. and Ramsey, F.: A Representation Theorem for 3 × 3 Stochastic Matrices. University of Copenhagen (1973).

19 Karpelewitch, F.I.: On the Characteristic Roots of a Matrix with Non Negative Elements. Isvestija Ser. Mat. 15, 361-383 (1951).

20 Kendall, D.G.: Delphic Semigroups, Infinitely Divisible Regenerative Phenomena, and the Arithmetic of p-functions. Z. Wahrscheinlichkeitstheorie verw. Gebiete 9, 163-195 (1968).

21 Kingman, J.F.C.: The Imbedding Problem for Finite Markov Chains. Z. Wahrscheinlichkeitstheorie verw. Gebiete 1, 14-24 (1962).

22 Kingman, J.F.C. and Williams, D.: The Combinatorial Structure of Non-homogeneous Markov Chains. Z. Wahrscheinlichkeitstheorie verw. Gebiete 26, 77-86 (1973).

23 Lee, E.B., Markus,L.: Foundations of Optimal Control Theory. New York: John Wiley(1968).

24 Loewner, C.: A Theorem on the Partial Order Derived from a Certain Transformation Semigroup. Math. Z. 72, 53-60 (1959).

25 Loewner, K.: Untersuchungen über schlichte konforme Abbildungen des Einheitskreises. I. Math. Ann. 89, 1o3-121 (1923).

26 Maksimov, B.M.: On the Relation Between Limit Theorems of Finite Groups and an Ergodic Theorem for Markov Chains with Doubly Stochastic Transition Matrices. Theor. Probability Appl. 10, 493-496 (1965).

27 Maksimov, B.M.: On the Convergence of Products of Independent Random Variables Taking Values from an Arbitrary Finite Group. Theor. Probability Appl. 12, 619-637 (1967).

28 Maksimov, B.M.: Random Processes with Independent Increments with Values in an Arbitrary Finite Group. Theor. Probability Appl. 15, 215-228 (1970).

29 Maksimov, B.M.: Divisible Distributions on Finite Groups. Theor. Probability Appl. 16, 306-318 (1971).

30 Martin-Löf, P.: Probability Theory on Discrete Semigroups. Z. Wahrscheinlichkeitstheorie verw. Gebiete 4, 78-101 (1965).

31 Ott, J.T.: Infinitely Divisible Stochastic Matrices. University of Rochester 33 pp (1972).

32 Runnenburg, J.Th.: On Elfving's Problem of Imbedding a Time-discrete Markov Chain in a Time-continuous One for Finitely many States I. Nederl. Akad. Wetensch. Proc. Ser. A. 65, 536-541 (1962).

33 Scheffer, C.L.: On Elfving's Problem of Imbedding a Time-discrete Markov Chein in a Time-continuous One for Finitely many States II. Nederl. Akad. Wetensch. Proc. Ser.A. 65, 542-548 (1962).

34 Schlesinger, L.: Neue Grundlagen für einen Infinitesimalkalkül der Matrisen. Math. Z. 33, 33-61 (1931).

35 Speakman, J.M.O.: Two Markov Chains with a Common Skeleton. Z. Wahrscheinlichkeitstheorie verw. Geb. 7, 224 (1967).

36 Vere-Jones,D.: Finite Bivariate Distributions and Semigroups of Non-negative Matrices. Quart. J. Math. Oxford Ser. 22, 247-270 (1971).

Some remarks on the geometry of systems

Robert Hermann

Mathematics Department, Rutgers

University, New Brunswick, New Jersey 08903

1. Introduction

The aim of this paper is to report briefly on work in progress concerning the applications of differential geometry to unify diverse areas of engineering and physics.

The immediate starting point for these observations is a paper by R. Kalman [11]. I will show that the differential geometry of fiber spaces is a natural setting for his ideas, and enables one to treat non-linear systems in a similar spirit.

2. Dynamical systems in terms of fiber spaces

In the modern systems and control theory literature [2,4] one finds a generalization of the classical notion of "dynamical system". In the most commonly accepted form, it is defined as a system of ordinary differential equations of the following form:

$$\frac{dx}{dt} = f(x,u,t). \qquad 2.1$$

Here, $x \in R^n$, $u \in R^m$, $t \in R$. x is the state vector, u the control vector, t the time parameter.

Let us first rewrite this system as a Pfaffian system. (I use the notations and ideas of [6,8]). Introduce indices and variables as follows:

$$1 \leq i,j \leq n$$

$$1 \leq a,b \leq n$$

$$x = (x_i);\ u = (u_a).$$

Let M be R^{n-m+1}, the space of variables (x_i, u_a, t). Let θ_i be the one form on M defined as follows

$$\theta_i = dx_i - f_i dt \qquad 2.2$$

Let I be the differential ideal of forms on I generated by the θ_i. Then, a curve $t \to (x(t),u(t))$ in $R^n \times R^m$ is a solution of 2.1 if and only if the curve $t \to (x(t),u(t),t)$ in M is an integral curve of I. This enables one to apply Chow's theorem on Pfaffian systems [3] to study controllability of the system defined by 2.1, [6]. This is the only method for studying controllability which works in the non-linear case.

A differential geometric version of these ideas may be described as follows. Let X be a manifold of dimension n, E a manifold of dimension n+m, and

$$\pi: E \to X$$

a mapping which defines E as a local-product fiber space. Let (x_i') be a coordinate system of functions on X. Set:

$$x_i = \pi^*(x_i'), \text{ functions on E.}$$

Consider a set (u_a) of functions on E such that:

$$(x_i, u_a)$$

form a coordinate system for E. We will call the (u_a) <u>fiber coordinates</u>, since they form a coordinate system when <u>restricted to the fibers of π</u>. We will call such coordinate systems for E <u>coordinate systems adapted to the fiber space (E,X,π)</u>.

Now let

$$M = E \times R \qquad 2.3$$

<u>Definition</u>. A <u>dynamical system</u> is defined by giving a fiber space (E,X,π), and a Pfaffian system I on $M=E\times R$, which, in the adapted coordinate systems (x_i,u_a), is generated by forms θ_i given by formula 2.2.

Such a dynamical system is said to be <u>linear</u> if X is a vector space, if $\pi: E \to X$ defines a vector bundle, i.e. if the fibers are vector spaces, and if, in adapted coordinate systems, with respect of the linearity of base and fiber, the functions $f_i(x,u,t)$ which define the ideal I are linear in x and u.

Let us consider the linear case, to see the relation to Kalman's work. Suppose (x_i') is a fixed linear coordinate system for X. Consider two sets (u_a), (u_a') of fiber coordinates, which are linear on the fibers.

For example, consider the following transformation law from the unprimed to the primed fiber coordinates:

$$u = u' - Lx \ , \qquad 2.4$$

where L is a $m\times n$ matrix. Suppose that:

$$(f_i) = f(x,u,t) = Fx + Gu, \qquad 2.5$$

where F is an $m\times n$ matrix, G an $n\times m$.

Then, if $f'_i(x,u,t)$ are the functions defining the ideal I in the coordinates (x,u',t), we have:

$$\begin{aligned} f' &= Fx+G(u'-Lx) \\ &= F'x+G'u' , \end{aligned} \qquad 2.6$$

with

$$\begin{aligned} F' &= F-GL \\ G' &= G . \end{aligned} \qquad 2.7$$

In the case that F and G are independent of t, Kalman finds the orbits under this group. The answer can be obtained from Kronecker's deep algebraic theory of "pencils" of linear maps.

For non-linear systems, one can look at things from the point of view of E. Cartan's theory of "G-structures" and their equivalence, [5]. Alternately, it can be said that one is studying the group of diffeomorphisms $\phi:M\rightarrow M$ which preserve the following geometric structures:

a) The ideal I

b) The adapted coordinate system, i.e. if (x_i,u_a) is a coordinate system adapted to the fiber space (E,X,π), so is $(\phi^*(x_i),\phi^*(u_a))$.

c) $\phi^*(t) = t$.

Kalman's theory essentially solves the "equivalence theorem" for subgroups of this group which preserve the linear, stationary-time systems.

Let us now turn to study the geometric formulation of the Pontrjagin minimal principle.

3. Singular characteristic curves and the Pontrjagin minimal principle.

It is known [1,6] that the calculus of variations problems arising from classical mechanics may be formulated differential-geometrically in terms of "symplectic manifolds" and "characteristic curves of closed 2-forms". This interpretation has led to insights into the relation between classical and quantum mechanics, quantum field theory, and even areas of pure mathematics, such as Lie group representation and automorphic function theory. There is also a relation between the Feynman path integral approach to quantum mechanics (which in turn is closely related to the theory of stochastic differential equations) and the symplectic manifold-characteristic curve approach to classical mechanics. (See [7], Chapter IV).

Now, the variational problems appearing in control theory are generalizations of those in classical mechanics. I believe that generalizing the ideas known from physics to control theory will lead to new interrelations between systems theory (particularly stochastic control theory) and physics.

Here is a sketch of the basic geometric formalism. Further details will appear in my work in preparation [9,10].

Consider a dynamical system, defined by differential equations 2.1. Add a "performance criterion", defined by a Lagrangian function

$$L(x,u,t). \qquad 3.1$$

Introduce new variables (λ_i), called the Lagrange multiplier variables. Let M' be the space of the variables (x,u,λ,t). Globally, this amounts to constructing M' as a vector bundle on M, with λ_i the linear coordinates on the fiber. However, for simplicity, we shall only work locally, in these coordinates.

Set:

$$H' = L-\lambda_i f_i \qquad 3.2$$

$$\omega = d(\lambda_i dx_i + H\, dt) \qquad 3.3$$

ω is a closed 2-form on M'. Its characteristic vectors are the tangent vectors $v \in T(M')$ such that:

$$v \lrcorner \omega = 0.$$

A point of M' is singular if the dimension of the space of characteristic vectors at that point is greater than at neighboring points. A curve in M' is a singular characteristic if its tangent vectors are characteristic, and if each point on the curve is singular.

For ω given by 2.3, the singular characteristic curves are readily calculated if the following non-degeneracy condition is satisfied:

The functions $(x_i, \lambda_i, \frac{\partial H}{\partial u_a}, t)$ 3.4

form a coordinate system for M'.

We shall suppose that 3.4 is satisfied. Let:

M'' = set of points of M' at which

$$\frac{\partial H}{\partial u_a} = 0. \qquad 3.5$$

In view of 3.4, M'' is a submanifold of M'. Let:

$$H' = H \text{ restricted to } M'' \qquad 3.6$$

$$\omega' = \omega \text{ restricted to } M''$$

$$= d\lambda_i \wedge dx_i + dH' \wedge dt' . \qquad 3.7$$

Here is the main result, which is proved by a standard calculation:

Theorem 3.1. Each singular characteristic curves of ω lies on the submanifold M''.

As a curve in M'', each such singular characteristic curve of ω is a non-singular characteristic curve of ω', given by the following Hamilton equations:

$$\frac{dx_i}{dt} = -\frac{\partial H'}{\partial \lambda_i} \qquad 3.8$$

$$\frac{d\lambda_i}{dt} = \frac{\partial H'}{\partial x_i}$$

Remark: If the usual "Legendre conditions" are assumed for the Lagrangian L, then H' can be defined as in the Pontrjagin Minimal Principle.

$$H'(x,\lambda,t) = \min_u H(x,u,\lambda,t) \qquad 3.9$$

Thus, we see that studying a triple (M',M'',ω) consisting of a manifold M', a closed-2 form ω on M', and a submanifold M'' consisting of the singular points of ω stands in the same relation to optimal control-systems theory as the study of symplectic manifolds does to classical mechanics.
I know of no recent work of a general mathematical nature on geometric structures of this type. I suggest that it would be extremely fruitful to pursue such work.

Bibliography

1. R. Abraham and J. Marsden, Foundations of mechanics, W.A. Benjamin, New York, 1967.

2. R. Brockett, Finite dimensional linear systems, J. Wiley, New York, 1970.

3. W.L. Chow, Uber systeme von linearen partiellen differential Gleichungen, Math. Ann. 117, 89-105 (1940).

4. C. Desoer, A second course on linear systems, Van Nostrand Reinhold, New York, 1970.

5. R. Hermann, Cartan connections and the equivalence problem, Contributions to Differential Equations, Vol. III, Interscience, New York, 1964.

6. ________, Differential geometry and the calculus of variations, Academic Press, New York, 1968.

7. ________, Lectures in Mathematical physics, Vol. II, W.A. Benjamin, Reading, Mass, 1972.

8. ________, Geometry, physics and systems, Marcel Dekker, New York, 1973.

9. ________, Algebra applied to systems theory and physics, to appear, Interdisciplinary Mathematics, New Brunswick, N.J.

10. ________, Differential geometric methods and ideas in engineering and physics, to appear, Interdisciplinary Mathematics, New Brunswick, N.J.

11. R. Kalman, Kronecker invariants and feedback, Ordinary Differential Equations, 1971 NRL-MRC Conference, 459-471, Edited by L. Weiss, Academic Press, New York.

MINIMAL REALIZATIONS OF NONLINEAR SYSTEMS

Héctor J. Sussmann

Rutgers University

In this paper we shall describe some recent results that we have obtained on the existence and uniqueness of minimal realizations of finite-dimensional autonomous nonlinear systems. A detailed presentation of these results, with complete proofs and a precise specification of technical assumptions will be given elsewhere (cf. Sussmann 4 and 5). Here we shall limit ourselves to a general outline, and we shall attempt to emphasize the importance of the differential-geometric method which, in our opinion, goes far beyond its mere use as a tool for proving theorems. We believe that, in addition to this important role, the use of Differential Geometry in Systems Theory has other advantages, namely, that it can provide a good framework for the theory, and that it can suggest what are the appropriate questions to be asked, and what kind of answers one should "naturally" expect. Finally, the interaction of Geometry and Systems Theory can also be fruitful by raising interesting mathematical questions and by leading to results in Geometry, as illustrated by Sussmann 1,2 and 3.

A good example of the use of Differential Geometry for the study of a specific problem of Systems Theory is provided by the work of Hermann, Brockett, Lobry, Krener, Haynes and Hermes, Jurdjevic and Sussmann, and others, on controllability of nonlinear systems. This line of research will be important here both for its successes and for its failures. The latter are, unfortunately, much easier to describe. The basic question that we would like to see answered remains unanswered. We have not been able to obtain a reasonable necessary and sufficient condition for controllability (in the sense that every state can be reached from every other state). Moreover, the question seems too hard and, perhaps, a complete answer should not be hoped for. However, the search for this answer has had some interesting by-products, of which two deserve to be mentioned here. The first one is that some properties other than controllability have been shown to be both important and easy to characterize. The so-called "accessibility property" (i.e. that reachable sets have nonempty interiors) turns out to be equivalent (at least in the real-analytic case) to an "algebraic" condition in terms of the rank of a Lie algebra of vector fields (Sussmann and Jurdjevic, Krener). The second by-product is soul-searching. If controllability is so hard to characterize, perhaps it is time to ask once again why we are trying to characterize it. In addition to all the obvious answers, there is one that leads directly into the direction of the present paper. Controllability is, for linear systems, one of the two basic ingredients that make up the concept of "minimal realization" (the other one being observability). If a satisfactory theory of minimal realizations of nonlinear systems is to be developed, then the study of nonlinear controllability seems to be a good place to start.

However, the problem that is really important is somewhat different. We must identify a property (C) of nonlinear systems which will play, in nonlinear realization theory, a role similar to that of controllability in the linear theory. Such a property will have to satisfy some obvious requirements. For instance, given a system $dx/dt = f(x,u)$, where the state variable x belongs to a manifold M, and given a state x_0 in M, we must be able to define the "controllable piece" through x_0, i.e. there must exist a submanifold S of M, which contains x_0 and is such that the "restriction" of our system to S makes sense (i.e. for each u and each s in S, $f(s,u)$ is tangent to S at s) and has property (C). This requirement suffices to eliminate controllability as a candidate for our property (C). Indeed, there are trivial examples in which M has no submanifold S such that the system, restricted to S, is controllable. Also, it is clear that, no matter what (C) is, we must not expect the "controllable piece" through x_0 to be simply the set of all states reachable from x_0 for, in general, this set is not a submanifold. If we now allow our differential-geometric intuition to give us a hint, it becomes clear that property (C) must be such that the "controllable piece through x_o" is the integral manifold through x_o of the involutive family of vector fields generated by the vector fields $x \longrightarrow f(x,u)$. These manifolds always exist in the real-analytic case. Moreover, if S is such a manifold, then S is connected, and the given system has a perfectly well defined restriction to S. In addition, this restriction has the accessibility property. We therefore agree to say that a system has property (C) if it has the accessibility property and its state space is connected. This is almost what we need, but for two objections. The first one is that, given a system with a state space M, and a state x_o, there are infinitely many integral manifolds of the given involutive family. They are all connected (by definition) and the restriction of our system to any of these manifolds is well defined and has the accessibility property. We would like to define S to be "minimal" in some sense (for otherwise there will be a still smaller "controllable piece" through x_o). However, it is clear that there is no smallest integral manifold through x_o. Even if we add the reasonable extra requirement that S should contain all the states that are reachable from x_o, there will still not exist, in general, a minimal S. The only "canonical" way of associating with each x_o in M an integral manifold S of the corresponding involutive family of vector fields seems to be to let S be the <u>maximal</u> integral manifold through x_o. To make sure that this manifold will be the "controllable piece" through x_o, we might try to modify our property (C) by adding to it a "maximality" condition. This, however, is clearly impossible. Indeed, the property that S is a maximal integral manifold of the associated involutive family does <u>not</u> depend only on the restriction of the system to S. It requires a knowledge of the system on <u>all</u> of M, as can be seen most clearly from the following equivalent characterization: the maximal integral manifold through x_o is the smallest set that contains x_o and that, whenever it contains a state x, it contains every state reachable from x and every state from which x is reachable. The second objection is that, in the non-analytic case, integral manifolds need not exist.

We dispose of the first objection by restricting ourselves to a smaller category of nonlinear systems, namely, to the class of <u>complete</u> systems. (We call a system complete if all the associated vector fields are complete; notice that this requires that their integral trajectories be defined for all times, positive and negative.) We stress that the completeness assumption is not just a simplification that could easily be done away with. Rather, it is absolutely essential for the theory. The second objection could be dealt with by using, in the C^∞ case, the property of "orbit-minimality" as a substitute for "accessibility plus connectednes" (Sussmann 5). However, later on it will become necessary, for different reasons, to restrict ourselves to the analytic case, so that we shall not pursue this question any further. We shall, therefore, work with analytic, complete systems (we call the system "analytic" if the state space is a real analytic manifold and all the associated vector fields are analytic). A system has property (C) if it has the accessibility property and the state space is connected. It is then true that, given a system with state space M, and a point x_0 in M, there exists a unique submanifold S through x_0 such that the restriction of the given system to S is a well defined system (i.e. is analytic and complete), and has property (C).

We now turn our attention to the second aspect of the concept of a minimal realization, namely, observability. It is remarkable that here we shall not encounter the difficulties that we experienced with controllability. We consider systems which, in addition to the properties of the preceding paragraph, have an output map g from the state space M into an output space N. We assume that N is a real analytic manifold, and that g is real analytic. Given an input $u(t)$, defined for t in $[0,T]$, and an initial state x in M, it is clear how to define the output $G_{x,u}$ that corresponds to x and u. Then $G_{x,u}$ is a function of t with values in N. If $G_{x,u}$ and $G_{y,u}$ are identically equal for all inputs u, we say the states x and y are *indistinguishable*. If there do not exist x and y that are indistinguishable but different, we say that our system is *observable*. The following theorem shows that this concept of observability has at least some of the properties that it should have. In the formulation of Theorem 1, we use the concept of *strong equivalence*, which we define as follows: two systems with the same set of inputs and the same output space are called strongly equivalent if every state of one of them is indistinguishable from some state in the other one. (The definition of "indistinguishability" clearly makes sense for states of different systems, as long as the inputs and outputs are the same.)

THEOREM 1. *Every system with property (C) is strongly equivalent to a system that has property (C) and is observable.*

If we put together Theorem 1 with our remarks on controllability, we can conclude that every input-output map $u \longrightarrow G_{x,u}$ that can be realized by a finite dimensional, analytic and complete system, and the choice of a fixed initial state of this system, can also be realized by a system which, in addition, is observable and "controllable" (i.e. satisfies (C)). It is natural to call such a system a <u>minimal</u> <u>realization</u> of the given input-output map.

The proof of Theorem 1 is too long to be given here. It appears in Sussmann 4, and its basic idea is simply to take the original state space M and divide by the equivalence relation R such that xRy if and only if x and y are indistinguishable. The main difficulty is to give the quotient state space M/R the structure of a manifold, for which it is necessary to show that R is "regular" as in Serre LG 3.26. The work of Sussmann 3 contains the necessary theorem. It is shown there that an equivalence relation R on a manifold M is regular, provided that it is closed and that it has "sufficiently many many symmetry vector fields". Here a symmetry vector field is a vector field whose corresponding local diffeomorphisms map equivalence classes to equivalence classes. The precise meaning of "sufficiently many" is that the set of such vector fields (which is necessarily a Lie algebra) be of maximal rank at each point. The application of this result to the existence of observable realizations is easy. First, it is clear that R is closed (because N is Hausdorff and g is continuous), Second, all the vector fields associated with our control system are symmetry vector fields for R .(That the local diffeomorphisms corresponding to positive times map classes into classes follows from the definition of indistinguishability; that this is also true for negative times follows by analyticity). Finally, let L denote the Lie algebra generated by these vector fields. Since the set of all symmetry vector fields is is a Lie algebra, it follows that all the vector fields in L are symmetry vector fields. The accessibility property then tells us precisely what we needed to know, namely, that L has sufficiently many elements. We remark that the real-analyticity is used in an essential way, and that there are trivial examples of C^{∞} systems for which Theorem 1 does not hold.

The second important question is that of uniqueness of minimal realizations. Given two systems with the same inputs and the same output space, and given states x and y , one in each system, assume that the input -output maps $u \longrightarrow G_{x,u}$ and $u \longrightarrow G_{y,u}$ are identical. Assume, moreover, that both systems are observable and have property (C). We would like to conclude that they are isomorphic in a very strong sense. As we now show, this turns out to be true in the real analytic case, and the analyticity assumption is essential.

Two systems $dx/dt = f_i(x,u)$, $y = g_i(x)$, $x \in M_i$, $y \in N$ ($i = 1,2$) are said to be isomorphic if there is a diffeomorphism F from M_1 onto M_2 such that, for each x in M_1 and each u in the control space, the differential of F at x maps $f_1(x,u)$ to $f_2(F(x),u)$ and that, in addition, $g_2F = g_1$. Such a mapping F is called an isomorphism between the given systems. There is another, "weaker", concept of isomorphism, defined as follows. A mapping $F:M_1 \longrightarrow M_2$ is a weak isomorphism if it is one-to-one and onto, it satisfies $g_2F = g_1$ and if, in addition, it maps trajectories of the first system into trajectories of the second one. It is easy to see that such a mapping will necessarily take the orbits of the family of vector fields associated with the first system into the corresponding orbits of the second system. Moreover, it is not hard to prove that the restriction of F to an orbit is a smooth mapping (the proof is given in Sussmann 5). From these two facts we conclude:

THEOREM 2. *A weak isomorphism between systems with property (C) is in fact an isomorphism.*

The concept of strong equivalence has been defined above. The following theorem is trivial:

THEOREM 3. *Two observable systems that are strongly equivalent are weakly isomorphic.*

We now state a theorem which will enable us to prove our desired result on systems with a fixed initial state. We shall give a brief sketch of the proof, to show the essential role played by analyticity.

THEOREM 4. *Suppose that there are states x_1 , x_2 of the first and second system, respectively, such that x_1 and x_2 are indististinguishable. Suppose, moreover, that both systems have property (C). Then the systems are strongly equivalent.*

Proof: Let us use the notation $t \longrightarrow X_t(x)$ to indicate the integral trajectory of the vector field X which goes through x when $t = 0$. If u belongs to the control space, let V^u , W^u denote the vector fields $x \longrightarrow f_1(x,u)$ (on M_1) and $x \longrightarrow f_2(x,u)$ (on M_2), respectively. Given states x_1', x_2' in M_1, M_2 it is esay to see that they will be indistinguishable if and only if the equality

$$(*) \quad g_1(V_{t_1}^{u_1} V_{t_2}^{u_2} \dots V_{t_k}^{u_k}(x_1')) = g_2(W_{t_1}^{u_1} W_{t_2}^{u_2} \dots W_{t_k}^{u_k}(x_2'))$$

holds for every positive integer k, every k-tuple $u_1, \dots, u_k$ of elements of the control space, and every k-tuple $t_1, \dots t_k$ of positive real numbers. Indeed, this is just another way of saying that the outputs corresponding to the initial states x_1' and x_2' are equal for any piecewise constant control $u(t)$. Since every control can be approximated by piecewise constant ones, the identity of the outputs for piecewise constant controls is equivalent to indistinguishability.

Since everything is real analytic, the preceding characterization of indistinguishability can be replaced by the condition that (*) holds for all $u_1, \dots u_k$ and all $t_1, \dots, t_k$, positive or not. From this it is clear that, if the state $x_1' \in M_1$ is given by

$$(**) \quad x_1' = V_{t_1}^{u_1} V_{t_2}^{u_2} \dots V_{t_k}^{u_k}(x_1)$$

then there is a state $x_2' \in M_2$ which is indistinguishable from x_1' (let x_2' be defined in the same way as x_1', with the V's replaced by W's and x_1 by x_2; notice that the completeness of the W's is used in this step). But every $x_1' \in M_1$ can be obtained in this way, because of property (C). It follows that for every state of the first system there is an indistinguishable state in the second system. Since the roles of the two systems can be reversed, it follows that they are strongly equivalent. Q.E.D.

If we now combine Theorems 1,2,3,4 and our remarks on property (C), we get an existence and uniqueness theorem for minimal realizations of input-output maps that arise from a nonlinear system with an output map and a fixed initial state.

THEOREM 5. *Let there be given a mapping which to each control* $u(t)$ *(defined for* t *in some interval* $[0,T]$ *) assigns a curve* $t \to y(t)$, $t \in [0,T]$ *in the real analytic manifold* N. *Assume that there exists a finite dimensional real-analytic complete system*

$$\dot{x} = f(x,u), \quad x \in M$$
$$y = g(x)$$

<u>and a state $x_o \in M$ which realize the given mapping, in the sense that this mapping is the input-output map $u \longrightarrow G_{x,u}$. Then our given mapping can be realized by a system of the same type which has property (C) and is observable. Moreover, any two such realizations are isomorphic.</u>

The following trivial example shows that Theorem 5 is no longer true in the non-analytic case. Let M_1 be two-dimensional Euclidean space, and let the output space N be M_1 itself, and the output map be the identity. The control system on M_1 is taken to be

$$\dot{x}_1 = u^2 \quad , \quad \dot{x}_2 = v^2 \qquad (u , v \quad \text{reals}) .$$

The initial state is taken to be the origin. It is clear that this system is observable, and that property (C) holds. We now define a new system which is not isomorphic to the given one, but such that there is a state which is indistinguishable from the origin considered as a state in the first system. Let $h(x_1,x_2)$ be an infinitely differentiable function which is identically equal to 1 on the nonnegative quadrant, but such that $h(-1,-1) = 0$. Let M_2 be M_1 with the point $(-1,-1)$ removed, and consider, on M_2, the system

$$\dot{x}_1 = h(x_1,x_2)u^2 \quad , \quad \dot{x}_2 = h(x_1,x_2)v^2 \quad .$$

The initial state is again the origin and the output map is simply the inclusion of M_2 in N. It is clear that the new system is complete, satisfies (C) and is observable, and that, for any control, the outputs obtained from the given initial states are the same for both systems Yet, they are not isomorphic.

We now give a brief discussion of equivalence of realizations of systems without a fixed initial state. Given such a system

$$\dot{x} = f(x,u) \quad , \quad x \in M \quad , \quad y = g(x) \quad , \quad y \in N \quad ,$$

we can define an input-output correspondence $u \longrightarrow G_u$ by assigning to each input $u(t)$ the set of all outputs that correspond to this input and all possible initial states. We call two such systems <u>equivalent</u> if the associated input-output correspondences are the same. This says that for every state of one of the systems, and every input, there is a state of the other system which, for the given input, gives the same output as the first state. Notice that, in principle, this does not mean that we can associate, with a given state of the first system, a state of the second that gives the same outputs for all inputs. In other words, equivalence does not seem to imply strong equivalence. However, we have been able to show that this implication holds if some additional conditions are satisfied. The detailed proof is given in Sussmann 5. The assumptions required by this proof are: (i) that the set of values of the control variable u is a connected subset U of some Euclidean space, and that U has a nonempty interior; (ii) that, for both systems, the functions $f(x,u)$ are real analytic as functions of both variables x and u, and (iii) that the class of admissible controls consist of <u>all</u> bounded, measurable, U-valued functions defined on intervals $[0,T]$. However, we are convinced that the result is valid under fairly more general conditions (and we do not know of any counterexample).

Let us consider systems for which (i),(ii) and (iii) hold. It is then possible to develop a reasonable theory, provided some rather harmless precautions are taken. First, we must enlarge slightly our class of systems, by allowing as state spaces manifolds whose connected components have different dimensions. Second, we must be satisfied with weak isomorphism, rather than isomorphism, as our formalization of "being the same system". Given an arbitrary system with a state space M, we can first modify M by changing its topology and differentiable structure. The new M is taken to be the manifold whose connected components are the orbits of the associated vector field system. It is then possible to take a quotient and get an observable system. Finally, Theorem 3 implies that this observable system is unique up to weak isomorphism. It is clear that this is all that can be hoped for, because it is impossible to tell, by looking at inputs and outputs, how the orbits are "glued together".

In all the preceding considerations the role of analyticity and completeness is essential. As shown by the example following Theorem 5, the reason why analyticity is important is that it implies "rigidity", i.e. that the behavior of the system "in the large" is completely determined by its behavior in an arbitrarily small open set. Analyticity and completeness are conditions that will be satisfied for any class of systems defined by sets of algebraic equations, and having a reasonable amount of homogeneity. Examples are: linear systems, systems on Lie groups and coset spaces, bilinear systems. Minimal realization theory for linear systems is, of course, classical. Results on minimal realizations of bilinear systems have been obtained recently by Brockett. Our results here can be viewed as "metatheorems" that cover all the classes of systems enumerated above, and any other class of systems that anybody might possibly think of, as long as they are finite dimensional and "algebraic". A detailed description of minimal-realization theory for subcategories of the category of all complete analytic systems is given in Sussmann 5.

Acknowledgement

This work owes its existence to Roger Brockett, both because he initiated this line of research, and because of his helpful comments and suggestions.

References

Brockett, R.W., System Theory on Group Manifolds and Coset Spaces, SIAM J. Control 10 (1972), 265-284.

Elliott, D.L., A Consequence of Controllability, J. Diff. Equations 10, 1971,364-370.

Haynes, G.W. and Hermes, H., Nonlinear Controllability via Lie Theory, SIAM J. Control 8,(1970), 450-460.

Hermann, R., On the Accessibility Problem in Control Theory, in "International Symposium on Nonlinear Differential Equations and Nonlinear Mechanics, 325-332, Academic Press, New York, 1963.

Jurdjevic, V. and Sussmann, H.J., Control Systems on Lie Groups, J. Diff. Equations 12 (1972), 313-329.

Krener, A., A Generalization of Chow's Theorem and the Bang-bang Theorem to Nonlinear Control Problems, SIAM J. Control, to appear.

Lobry, C., Contrôlabilité des Systèmes Non-linéaires, SIAM J. Control 8(1970), 573-605.

Serre,J.P., Lie Groups and Lie Algebras, Benjamin Press, New York, 1965.

Sussmann, H.J. and Jurdjevic, V., Controllability of Nonlinear Systems, J. Diff. Equations 12 (1972),95-116.

Sussmann, H.J.,
1.-Orbits of Families of Vector Fields and Integrability of Systems with Singularities,Bull.Amer.Math.Soc.79 (1973), 197-199.
2.-Orbits of Families of Vector Fields and Integrability of Distributions, Trans.Amer.Math.Soc., to appear.
3.-On Quotients of Manifolds: a Generalization of the Closed Subgroup Theorem, submitted.
4.-Observable Realizations of Nonlinear Systems,submitted.
5.-Equivalence of Realizations of Nonlinear Systems, submitted.

CAUSAL DYNAMICAL SYSTEMS: IRREDUCIBLE REALIZATIONS

Velimir Jurdjevic

University of Toronto

Toronto, Canada.

I. INTRODUCTION

Recently, much attention has been devoted to control systems described by the following data:

$$(1) \qquad \frac{dx}{dt} = F(x,u)$$

$$(2) \qquad y = h(x)$$

where the state of the system x belongs to a differentiable manifold X, u belongs to a class of admissible inputs I, and the output map h takes values in a manifold Y ([2], [5], [9]). In this context it is generally assumed that for a variety of physical considerations, the state of the system x is not known directly but that it is only observed through the output $h(x)$. From this point of view, it appears natural to regard the system as an input - output relation S where (1) and (2) along with the set X where (1) is defined represent a state description for S. More precisely, if we denote the solution of (1) which corresponds to $x \in X$, and $u \in I$ by $\pi(x,u,t)$ {i.e., $\pi(x,u,t) = x$, and $\frac{\partial \pi}{\partial t} = F(\pi,u)$}, then $(u,y) \in S$ if and only if $y(t) = h\pi(x,u,t)$ for some $x \in X$, and all $t \geq 0$. Thus, a state representation defines a system S, but corresponding to an S

there may be many state representations. In this framework, the following questions have been of considerable interest ([1], [7]):

(a) Given a state description for S does the knowledge of the input-output data uniquely determine the state of the system, or in the control theory terminology, is S observable in this representation?

(b) Does every system S have observable representations, and if so what is the relationship among various such representations?

In regard to (a), the intuitive idea of observability leads to several plausible definitions. For instance, we may define S to be observable if for *a given input* u there is a 1-1 correspondence between the elements of X and the outputs y such that $(u,y) \in S$ ([5], [8]); or, we may say that S is observable if there exists a function from S onto X such that $(u,y) \to x$ if and only if $y(t) = h\pi(x,u,t)$ for all $t \leq 0$ ([2], [6]).

In this paper we adopt the latter definition, and we address question (b). We will proceed axiomatically, partly for the sake of generality, but mainly we feel that such an approach illuminates the conceptual aspects of the problem. We start with two classes of functions I and O called inputs and outputs respectively. $S \subset I \times O$ is causal if it admits a state representation (X, π, h) where π is an abstract generalization of solutions of (1). On that level of generality we show that every causal system has irreducible (observable) state representations which are isomorphic in the sense of finite automata.

We specialize this result to systems which admit a "linear" state representation: we show that any system which a linear state representation in a Banach space, has an irreducible state representation which is also linear; furthermore, we show that any two reduced state spaces are both algebraically and topologically equivalent. Incidentally, an attempt to generalization of the "observability rank condition" ([8]) to infinite dimensional spaces leads to difficulty, for it touches upon a long outstanding problem of analysis; namely, whether each linear transformation has a nontrivial closed invariant subspace. With the exception of the last remark, the observability theory of linear systems in R^n extends without much additional difficulty to infinite dimensional spaces. In view of [4], and [6], this suggests that the duality between observability and controllability is restricted to R .

In the case of nonlinear systems the situation is in general different: even simple differential systems give rise to reduced state spaces which fail to be differentiable manifolds.

Finally, a few remarks concerning the choice of terminology.

The notion of causality as used in this context was originated in [10]. Otherwise, the terminology is a mixture of terms from the control theory and the theory of finite machines. While this approach may have not resulted in the best possible choice, it is hoped that it will at least suggest the need for the long overdue terminology of general dynamical systems.

II. DEFINITIONS AND BASIC CONCEPTS

Throughout the sequel we will assume the existence of two classes of functions I and O which will be called the class of inputs and the class of outputs respectively. We will make the following overall assumptions about I and O :

(i) I is a non-empty class of functions defined on the set of non-negative reals R^+ and having values in an abstract set Σ . We further assume that if $f \in I$, and $s \in R^+$, then f_s defined as $f_s(t) = f(t+s)$ for all $t \in R^+$ is in I . Also, if f and g are in I , and $s \geq 0$ then the map h defined by

$$h(t) = \begin{cases} f(t) & 0 \leq t \leq s \\ g(t-s) & t > s \end{cases}$$

is an element of I . We shall denote such a concatenation of maps by $h = f(*t)g$.

(ii) O is a non-empty set of functions defined on R^+ and having values in an abstract set Γ which will be called the output space.

If X is a non-empty set, then a map

$\pi : X \times I \times R^+ \to X$ is a *dynamical system in* X if the following properties hold:

(1°) $\pi(x,f,0) = x$ for all $x \in X$, $f \in I$.

(2°) $\pi(x,f,t+s) = \pi(\pi(x,f,s),f_s,t)$ for all $x \in X$, $f \in I$, and all s,t in R^+.

(3°) If for some x and y in X, f and g in I, and $s > 0$, $\pi(x,f,s) = \pi(y,g,s)$ and if $f(t) = g(t)$ for $t \in (s,v)$, then $\pi(x,f,v) = \pi(y,g,v)$.

Axioms (1°), (2°) and (3°) are called the identity, the homomorphism and the nonanticipation axiom respectively. We will say that a system $S \subset I \times 0$ is a *causal system* if there exists a triple (X,π,h) where π is a dynamical system in X, $h : X \to \Gamma$, with the property that $(f,g) \in S$ if and only if $g(t) = h\pi(x,f,t)$ for some $x \in X$, and all $t \geq 0$. (X,π,h) will be referred to as a *state representation* for S, and h will be called the *output map*: Evidently, a causal system may have more than one state representation. If (X,π,h) and (X,π,g) are two state representations for S, then we will say that (X,ψ,g) is *related to* (X,ψ,h) if there exists a map $F : X \to Y$ such that $F\pi(x,f,t) = \psi(F(x),f,t)$ for all $x \in X$, $t \in R$. If, furthermore, F is one-one and onto, then we will say that (X,π,h) and (Y,ψ,g) are *isomorphic*. It follows that the notion of isomorphism partitions the set of all state representations of S.

If (X,π,h) is a state representation for S, then x and y in X are *indistinguishable* in this representation whenever $h\pi(x,f,t) = h\pi(y,f,t)$ for all $f \in I$, $t \in R$. (X,π,h) is an *irreducible state representation* if no distinct states are indistinguishable in it. Clearly, the notion of indistinguishability is an equivalence relation over X.

III. IRREDUCIBLE STATE REPRESENTATIONS

Let (X,π,h) be a state representation for S.

Define the following equivalence relation $\sim$ in X : $x \sim y$ if and only if x and y are indistinguishable in (X,π,h) . Let $\tilde{X}$ be the quotient space of X under $\sim$. If $x \in X$, we denote by $\tilde{x}$ the equivalence class to which it belongs. Let $x \sim y$, $f \in I$, and $t \in R^+$. For any $u \in I$, define a map v by $v = f(*t)u$. It follows that for any $s > 0$, $h\pi(\pi(x,f,t),u,s) = h\pi(x,v,s+t) = h\pi(y,v,s+t) = = h\pi(\pi(y,f,t),u,s)$. Hence, $\pi(x,f,t) \sim \pi(y,f,t)$. Thus, if we define $\tilde{\pi} : \tilde{X} \times I \times R^+ \to \tilde{X}$ by $\tilde{\pi}(\tilde{x},f,t) = = \tilde{\pi}(x,f,t)$, the above argument shows that such a map is well defined. It is now straightforward to verify that $\tilde{\pi}$ is a dynamical system in $\tilde{X}$. If $x \sim y$, then $h\pi(x,f,t) = h\pi(y,f,t)$ for all $f \in I$, $t \geq 0$. In particular, when $t = 0$, $h(x) = h\pi(x,f,0) = = h\pi(y,f,0) = h(y)$. Thus, h is constant over each equivalence class $\tilde{x}$. It follows that $\tilde{h} : \tilde{X} \to \Gamma$ defined by $\tilde{h}(\tilde{x}) = \tilde{h}(x)$ is well defined. Now it is clear that $(\tilde{X},\tilde{\pi},\tilde{h})$ is an irreducible state representation for S . Let $P : X \to \tilde{X}$ be the natural projection map, i.e., $P(x) = \tilde{x}$. Then $P\pi(x,f,t) = \tilde{\pi}(x,f,t) = \tilde{\pi}(x,f,t) = \tilde{\pi}(P(x),f,t)$.

Thus we have proved the following

Proposition 3.1.

If (X,π,h) is a state representation for S , then there exists a state representation $(\tilde{X},\tilde{\pi},h)$ which is irreducible and which is related to (X,π,h) .

Let (X,π,h) and (Y,ψ,g) be irreducible state representations for S . Define a map $F : X \to Y$ by $x \to F(x)$ if and only if $h\pi(x,f,t) = g\psi(F(x),f,t)$ for all $f \in I$, and all $t \geq 0$. Since both state representations are irreducible, F is a well defined, one-one and onto map.

Let $x \in X$, $f \in I$, and $t \in R^+$. If $u \in I$, then let $v = f(*t)u$. We have that $g\psi(\psi(F(x),f,t),u,s) = = g\psi(F(x),v,s+t) = h\pi(x,v,s+t) = h\pi(\pi(x,f,t),u,s)$. Since u and s are arbitrary, $F(\pi(x,f,t)) = \psi(F(x),f,t)$. Thus we have proved

Proposition 3.2.

Any two irreducible state representations of a

given system S are isomorphic. Furthermore, an irreducible state representation is related to any state representation of S.

If (X,π,h) is a state representation for S, then X is π-*connected* if for every x and y in X, there exists $f \in I$ and $t \in R^+$ with $y = \pi(x,f,t)$. The notion of π-connectedness is in the context of control theory equivalent to complete controllability. An obvious corollary of the above is the following

Proposition 3.3.

Let (X,π,h) be a π-connected state representation for S. If (Y,ψ,g) is an irreducible state representation of S, then Y is ψ-connected.

IV. LINEAR SYSTEMS

This section is primarily concerned with the application of the preceding results to linear systems in infinite dimensional spaces. The main motivation for such a development comes from systems described by partial differential equations (for instance, [1], [4] and [6]).

Throughout the following we will assume that I and Γ are linear spaces over the same field of scalars $\mathcal{R}$ which in our case will be either real or complex. We will say that (X,π,h) is a *linear state representation* if X is a linear space over $\mathcal{R}$, $\pi(\alpha x + \beta y, \alpha f + \beta g, t) = \alpha\pi(x,f,t) + \beta\pi(y,g,t)$ for all α, β in $\mathcal{R}$, f, g in I, and all $t \in R^+$, and h is a linear map.

We will say that S is *linear* if it has a linear state representation. If S is linear, then the range of S $R(S)$ can be regarded as a linear space over $\mathcal{R}$. Then it follows that $(f,g) \in S$ and $(u,v) \in S$ imply that $(\alpha f + \beta u, \alpha g + \beta v) \in S$ for all α, β in $\mathcal{R}$.

If (X,π,h) is a state representation for S, then for each constant input f, the map $\pi_f : X \times R^+ \to X$ satisfies $\pi_f(x,0) = x$, and $\pi_f(x,s+t) = \pi_f(\pi_f(x,s),t)$. If (X,π,h) is linear, then for each $t \in R^+$ the map $x \to \pi_f(x,t)$ is an affine map, which is also linear just in case $f = 0$.

In such a case, let $U = \{x : h\pi_0(x,t) = 0$ for all $t \in R^+\}$. U is a linear subspace of X, and $x \underset{\sim}{\sim} y$ if and only if $x - y \in U$. If $\tilde{X} = X/\sim$, then $\tilde{X}$ is a factor space modulo U. Thus $\tilde{X}$ is a linear space. Since the projection map from X onto $\tilde{X}$ is linear, it follows that $(\tilde{X},\tilde{\pi},h)$ is a linear state representation for S.

The above, along with the fact that irreducible state representations are isomorphic proves

Proposition 4.1.

If S is a linear causal system, then every irreducible state representation is linear.

We will now assume that the output space Γ is a topological linear space. A linear state representation (X,π,h) is said to be *continuous* if X is a topological linear space, h is continuous, and for each $f \in I$, the map $(x,t) \to \pi(x,f,t)$ is continuous in the product topology of $X \times R^+$. If (X,π,h) is a continuous linear state representation, then for each $f \in I$, π_f is a continuous map. In particular, π_0 is continuous. Hence, $U = \{x : h\pi_0(x,t) = 0$, for all $t \in R^+\}$ is a closed subspace of X. If X is a Banach space then $\tilde{X} = X/U$ becomes a Banach space with $\|\tilde{x}\| = \inf\{\|x+y\| : y \in U\}$. Furthermore, the projection map P from X onto $\tilde{X}$ is continuous, and hence, open. Therefore, it easily follows that $(\tilde{X},\tilde{\pi},\tilde{h})$ is a continuous linear state representation of S. Now assume that (Y,ψ,g) is any other continuous linear state representation of S with Y being a Banach space. We know that $(\tilde{X},\tilde{\pi},\tilde{h})$ and (Y,ψ,g) are isomorphic. Let $F : \tilde{X} \to Y$ be such that $F(\tilde{\pi}(x,f,t)) = \psi(F(\tilde{x}),f,t)$. F is one-one, onto, and it follows by direct verification that F is linear. Let $\{x_n\}$ be a convergent sequence in $\tilde{X}$ with $\lim x_n = x$. Assume that $\{F(x_n)\}$ converges. Let $y = \lim F(x_n)$. We have that $\tilde{h}\tilde{\pi}(x,f,t) = \lim \tilde{h}\tilde{\pi}(x_n,f,t) = \lim g\psi(F(x_n),f,t) = g\psi(y,f,t)$. Since f and t are arbitrary, it follows that $y = F(x)$. Hence, F is a closed map, and therefore continuous.

Analogously, F^{-1} is continuous. Thus we have proved the following

Proposition 4.2.

Let (X,π,h) be a linear continuous state representation for S with X a Banach space. Then, $\tilde{X}$ can be normed so that it becomes a Banach space, and in this topology $(\tilde{X},\tilde{\pi},\tilde{h})$ is continuous. Furthermore, if (Y,ψ,g) is any other irreducible linear continuous representation for S with Y a Banach space, then $\tilde{X}$ and Y are both algebraically and topologically equivalent.

Let (X,π,h) be a continuous linear state representation with X a Banach space. Then $\pi(x,f,t) = \pi(x,0,t) + \pi(0,f,t) = \pi_0(x,t) + \pi(0,f,t)$. If we regard π_0 as a family of operators $\{T(t)\}$ on X defined by $T(t)x = \pi_0(x,t)$, then it follows that $\{T(t)\}$ is a strongly continuous semi-group of operators in X [3]. For $h > 0$, let A_h be the linear operator defined by the formula $A_h = h^{-1}(T(h) - I)$. Let D be the set of all $x \in X$ for which $\lim A_h(x)$ exists as $h \to 0$. Define an operator A on D by $Ax = \lim A_h(x)$. It is well known that D is a dense linear subspace of X , and that A is a closed linear operator on D [3]. A is bounded if and only if $\lim T(t) = I$ as $t \to 0$ in the operator topology. A is called the infinitesimal generator of T . Now let (X,π,h) and (Y,ψ,g) be any two linear continuous irreducible state representations for S . Let $P : X \to Y$ be such that $P\pi(x,f,t) = \psi(P(x),f,t)$. In particular, $P\pi_0(x,t) = \psi_0(P(x),t)$. If $x \in D(A)$ then for any $h > 0$ $h^{-1}[P\pi_0(x,h) - P(x)] = h^{-1}[\psi_0(P(x),h) - P(x)]$. Since P is continuous, the left hand side of the above equality has a limit as $h \to 0$, and this limit is $PA(x)$.

Let B be the infinitesimal generator of ψ_0 . It follows that $B(Px) = P(Ax)$. Analogous argument applied to P^{-1} shows that for any $y \in D(B)$, $P^{-1}(By) = AP^{-1}(y)$. Hence, $P(D(A)) = D(B)$, and $PA = BP$. Since $h\pi_0(x,t) = g\psi_0(P(x),t)$ for all $(x,t) \in X \times R^+$, we get that in particular $h = g \circ P$. Thus we have shown the following

Proposition 4.3.

Let (X,π,h) and (Y,ψ,g) be linear continuous irreducible state representations for S with X and Y Banach spaces. Then there exists an invertible bounded transformations $P : X \to Y$ such that:

(i) If A is the infinitesimal generator of π_0, and B is the infinitesimal generator of ψ_0, then $P(D(A)) = D(B)$, and $PA = BP$.

(ii) $h = g \circ P$, and $P\pi(0,f,t) = \psi(0,f,t)$ for all $f \in I$, $t \in R^+$.

Actually, (i) and (ii) in the above proposition are also sufficient conditions for (Y,ψ,g) to be an irreducible representation for S ([7]).

V. REMARKS

In view of the preceding results the following question seems very natural: If S is such that it has a state representation (X,π,h) with X a differentiable manifold and π and h sufficiently smooth in the manifold structure of X, then find conditions under which the reduced state space $\tilde{X}$ is a smooth manifold and $\tilde{\pi}$ and $\tilde{h}$ are differentiable in $\tilde{X}$.

In order to suggest the complexities associated with this problem consider the following simple example: let $X = R^2$, $\pi(x,u,t) = (x_1, x_2 + t)$, and $h(x_1, x_2) = x_1(x_2^2 + 1)$. Since h maps vertical lines of X into non-degenerate parabolas with the exception of the x_2-axis, which is mapped onto the origin, it follows that $(0,x) \sim (0,y)$ for any x, y in R, but otherwise, no distinct points are equivalent to each other. If we assign to $\tilde{X}$ the quotient topology (for such topology is the weakest topology under which the projection map is continuous), then $\tilde{0}$ has no neighborhood homeomorphic to an open subset of R^n. Hence, $\tilde{X}$ is not a manifold.

In regard to linear systems we showed that (X,π,h) is irreducible if and only if $U = \{x : h\pi_0(x,t) = 0 \text{ for all } t \geq 0\}$. If X is of dimension n, then, as it is well known, $U = 0$ if and only if $\operatorname{rank}[h\ Ah \ldots A^{n-1}h] = n$.

An attempt to generalizations of the above result to infinite dimensional spaces leads to difficulty; for in general it is not even known if every operator has a nontrivial invariant closed subspace.

REFERENCES

[1] A.V. Balakrishnan, On the controllability of nonlinear system, Proc. N.A.S., Vol. 55 (1966), 465-468.

[2] R.W. Brockett, On the algebraic structure of bilinear systems, submitted for publication.

[3] N. Dunford and L. Schwartz, *Linear Operators:* Part I, Interscience Publishers Inc., New York, 1958.

[4] H.O. Fattorini, On complete controllability of linear systems, J. Diff. Equations 3 (1967), 391-402.

[5] E.W. Griffith and K.S. Kumar, On the observability of nonlinear systems: I, J. Math. Anal. and Appl. 35 (1971), 135-147.

[6] V. Jurdjevic, Abstract control systems; controllability and observability, SIAM J. on Control, 3 (1970), 424-439.

[7] V. Jurdjevic, On the structure of irreducible state representations of a causal system, to appear in the Math. Systems Theory.

[8] L. Markus and B. Lee, *Foundations of Optimal Control Theory*, John Wiley, New York, 1967.

[9] T.G. Windeknecht, Mathematical systems theory; causality, Math. Systems Theory, 1 (1967), 279-289.

[10] R. Triggiani, Controllability, observability and stabilizability of systems in Banach spaces, Ph.D. thesis, 1972, University of Minnesota.

ON THE INTERNAL STRUCTURE OF BILINEAR INPUT-OUTPUT MAPS

E. Fornasini - G. Marchesini

Dept. of Electrical Eng., Univ. of Padua

INTRODUCTION

This paper deals with some observations arising in the realization of bilinear input-output maps. It concerns essentially with some internal structural properties which are consequences of the definition of the state by the most natural way i.e. Nerode equivalence classes.

Most questions are yet to be studied in depth but the results are sufficient to give a picture of the problems.

Let K be a field and $\mathbb{Z}$ the ring of integers. A bilinear zero-state discrete-time input-output map is a map $f : U \times U \longrightarrow Y$ defined as follows :

i) $U = \{u : u \in K^{\mathbb{Z}}, \text{ card (supp } u) \leqq \aleph_0\}$; is naturally endowed with the structure of K-module.

ii) $Y = \{y : y \in K^{\mathbb{Z}}\}$

iii) $f : U \times U \rightarrow Y$ (input-output map) has the following properties :

i) $\min \text{supp } f(u_1,u_2) > \max \text{supp}(u_1,u_2) \triangleq \text{supp } u_1 \cup \text{supp } u_2$

ii) f is bilinear :

$$f(ku_1,u_2) = kf(u_1,u_2) \text{ , } k \in K \text{ , } u_1,u_2 \in U$$

$$f(u_1,hu_2) = hf(u_1,u_2) \text{ , } h \in K \text{ , } u_1,u_2 \in U$$

$$f(u_1+v_1,u_2) = \left[f(u_1,u_2) + f(v_1,u_2)\right]_{\text{supp } f(u_1+v_1,u_2)}$$

This work has been supported by CNR-GNAS

$$f(u_1,u_2+v_2) = \left[f(u_1,u_2) + f(u_1,v_2)\right]_{\text{supp } f(u_1,u_2+v_2)}$$

$u_1,u_2,v_1,v_2 \in U$

The zero-state response of a bilinear map to an arbitrary pair of input sequences (u_1,u_2) with left compact support, is given by:

(1) $$y(r) = f(T_r u_1,\ T_r u_2)(r) \quad , \quad r \in \mathbb{Z}$$

where:

(2) $$T_r u_i(t) = \begin{cases} u_i(t) & , \quad t < r \\ 0 & , \quad t \geqslant r \end{cases} \quad , \quad i = 1,2$$

Denoting by σ the shift operator on $K^{\mathbb{Z}}$, the bilinear input-output map is stationary if it satisfies the condition :

(3) $$\sigma f(u_1,u_2) = f(\sigma u_1,\ \sigma u_2) \quad , \quad u_1,u_2 \in U$$

In this case $t = 0$ can be assumed as the max supp (u_1,u_2).

CHARACTERIZATION OF THE BILINEAR MAP

The input-output map defined in the previous section can be represented via a sequence of infinite matrices.

In particular the representation of the K-bilinear stationary input-output maps onto the infinite K-valued matrices is then biunique.

Actually it is immediate to observe that

$$f(u_1,u_2)(r) \quad , \quad r > \text{max supp } (u_1,u_2)$$

can be considered as a bilinear functional

$$f(u_1,u_2)(r) \ : \ T_r[U] \times T_r[U] \longrightarrow K \ .$$

<u>Proposition 1</u> : Let $\{\beta_i\}$ $i = r-1, r-2, \ldots$ the usual basis in $T_r[U]$, where β_i is the sequence:

(4) $$\beta_i = (\delta_{ik})_{k = r-1, r-2, \ldots}$$

and

(5) $$f(\beta_i,\ \beta_j)(r) \triangleq w_{ij}(r) \quad ,$$

then

(6) $f(u_1,u_2)(r)=\sum_{ij} w_{ij}(r)\, u_1(i)\, u_2(j) \quad , \quad u_1,u_2 \in T_r[U]$.

The sequence w_{ij} comes out naturally depending on r since stationarity has not been assumed.

Obviously if we assume the bilinear i-o map to be stationary the w_{ij} dependence on r fails and the following Proposition holds.

Proposition 2 Let f be stationary. Then

$$(7) \qquad w_{i,j}(r) = w_{i-r,j-r}(0) \quad , \quad \forall\, i,j,r \in \mathbb{Z}$$

Proof. Direct application of invariancy definition to (6).

The stationarity assumption allows us to consider input sequences (u_1,u_2) having $t = 0$ as max $\operatorname{supp}(u_1,u_2)$. This has as consequence the possibility of evaluating the output of the bilinear map from the infinite matrix $W = (w_{ij})_{i,j<0}$. In this way the K-bilinear stationary i-o maps can be biuniquely represented onto the infinite K-valued matrices.

The input space being constituted by pairs of sequences with compact support in $\mathbb{Z}_-$, we shall adopt the usual polynomial representation :

$$(8) \quad u_1=(a^1_{-p}, a^1_{-p+1},\dots, a_0) \longmapsto p_1(z_1^{-1})=\sum_0^p{}_i\, a_i z_1^{-i}, \quad a_h= a^1_{-h}$$

$$(9) \quad u_2=(b^1_{-q}, b^1_{-q+1},\dots,b_0) \longmapsto p_2(z_2^{-1})=\sum_0^q{}_j\, b_j z_2^{-j} \quad , \quad b_k= b^1_{-k}$$

Analogously the output space, constituted by sequences with support $\mathbb{Z}_{++}$ can be represented in $K[[z]]$.

It is now possible to give a global representation for the input-output map in terms of series and polynomials.

Proposition 3 Let (8), (9) and

$$(10) \qquad S(z_1, z_2) = \sum_1^\infty{}_{h,k}\, w_{-h,-k}(0)\, z_1^h\, z_2^k \quad =$$

$$= \sum_{1}^{\infty}{}_{h,k}\, s_{h,k}\, z_1^h z_2^k \quad , \quad w_{-h,-k}(0) \overset{\Delta}{=} s_{h,k}$$

Then :

$$(11) \quad f(p_1(z_1^{-1}),\ p_2(z_2^{-1})) = \underset{>0}{\text{diag}}\ S(z_1,z_2)p_1(z_1^{-1})\ p_2(z_2^{-1})$$

<u>Proof.</u>

$$(12) \qquad y(m) = \sum_{0}^{-\infty}{}_{i,j}\, w_{ij}(m)\, a_i^1\, b_j^1 =$$

$$= \sum_{0}^{-\infty}{}_{i,j}\, w_{i-m,\ j-m}(0)\ a_i^1\ b_j^1 = \sum_{0}^{+\infty}{}_{h,k}\, s_{h+m,k+m}\, a_h b_k$$

But

$$(13) \quad \underset{>0}{\text{diag}}\ S(z_1,z_2)\ p_1(z_1^{-1})\ p_2(z_2^{-1}) =$$

$$= \sum_{1}^{\infty}{}_{m}\Big(\sum_{0}^{\infty}{}_{ij}\, s_{i+m,\,j+m}\, a_i\, b_j\Big)(z_1\, z_2)^m$$

<u>INTERNAL PROPERTIES</u>

The most natural way to attack the realization problem is by introducing the Nerode equivalence [2,3]:

$$(14) \qquad (p_1,p_2) \underset{N}{\sim} (\hat{p}_1,\hat{p}_2) \iff f(p_1 \circ \sigma_1, p_2 \circ \sigma_2) =$$

$$= f(\hat{p}_1 \circ \sigma_1, \hat{p}_2 \circ \sigma_2)\ , \quad \forall \sigma_1 \in K[z_1^{-1}],\ \forall \sigma_2 \in K[z_2^{-1}]$$

where

$$(15) \qquad p_1(z_1^{-1}) \circ \sigma_1(z_1^{-1}) = p_1(z_1^{-1})\ z_1^{-(c+1)} + \sigma_1(z_1^{-1})$$

$$p_2(z_2^{-1}) \circ \sigma_2(z_2^{-1}) = p_2(z_2^{-1})\ z_2^{-(c+1)} + \sigma_2(z_2^{-1})$$

$$c = \max\ (\ \deg \sigma_1,\ \sigma_2\ )$$

and similar expressions for $\hat{p}_1 \circ \sigma_1$ and $\hat{p}_2 \circ \sigma_2$

It is known that [4] :

(16) $$(p_1,p_2) \underset{N}{\sim} (\hat{p}_1,\hat{p}_2) \Longleftrightarrow (1),2),3))$$

1) $f(p_1 z_1^{-c}, p_2 z_2^{-c}) = f(\hat{p}_1 z_1^{-c}, \hat{p}_2 z_2^{-c}) \qquad \forall c \geqslant 0$

2) $f(p_1 z_1^{-c-1}, \sigma_2) = f(\hat{p}_1 z_1^{-c-1}, \sigma_2), \forall \sigma_2 \in K[z_2^{-1}]\ c=\deg \sigma_2$

3) $f(\sigma_1, p_2 z_2^{-c-1}) = f(\sigma_1, \hat{p}_2 z_2^{-c-1}), \forall \sigma_1 \in K[z_1^{-1}] \quad c=\deg \sigma_1$

The N-state space (Nerode-state space) is the set of Nerode equivalence classes :

(17) $$X = (U \times U)/\underset{N}{\sim} = \left\{ [u_1,u_2] : (u_1,u_2) \in U \times U \right\}$$

In the linear case the N-state space X can be endowed with the linear vector space structure. If we represent the i-o linear map by a formal power series $\sum_{1}^{} {}_i\, a_i z^i$ in $K[z]$ the following facts are equivalent [5,6] :

i) $\dim X < \infty$

ii) card ([0]) > 1

iii) $\sum_{1} {}_i\, a_i z^i = \dfrac{P(z^{-1})}{Q(z^{-1})}$

iv) $(a_i)_{i>0}$ is a H-sequence (°)

v) X is a $K[z^{-1}]$ torsion module

In the bilinear case the N-state space X cannot be endowed with a K-module structure and consequently there is no way to extend some of the previous statements to the bilinear case.

Nevertheless it is still interesting to investigate how the structure of the bilinear i-o map and the structure of the N-state space are correlated.

(°) by "H-sequence" we intend a sequence whose Hankel matrix has finite rank [7] .

We recall now some properties of H-sequences and introduce some notations to simplify the formalism.

1 - (s_i) is a H-sequence iff there exist $(r+1)$ numbers $a_0, \dots, a_r$ not all zero such that :

$$\sum_{0}^{r}{}_i \; a_i \, s_{q-i} = 0 \;, \qquad q = r+1, \dots\dots \tag{18}$$

we call the ordered set $(a_0, \dots a_r)$ an annihilating set of the H-sequence.

2 - The annihilating sets constitute a principal ideal $((a_0, \dots, a_r))$ in the ring of finite sequences with Cauchy product. This ideal is generated by a "minimal annihilating set" (i.e. of minimal length).

3 - Consider a j-indexed family of H-sequences (s_{ij}), $j \in \mathbf{N}$ with the corresponding annihilating ideals I_j; the principal ideal $\bigcap_j I_j$ is the "annihilating ideal of the family".

4 - Given a double sequence (s_{ij}), a finite matrix $A = (a_{rs})$ is a (p,q)-annihilating matrix for (s_{ij}) if

$$\sum_{0}^{n,m}{}_{rs} \; a_{rs} \; s_{p+r,q+s} = 0$$

<u>Lemma 1.</u> Let S as in (10). The annihilating ideal of the j-indexed family (s_{ij}), $j \geqslant 1$, $i > j$ is different from (0) iff there exists $p_1(z_1^{-1}) \in K[z_1^{-1}]$, $p_1 \neq 0$, such that $(p_1(z_1^{-1}), 0) \in [0,0]$.

<u>Proof.</u> Let $p_1(z_1^{-1}) = \sum_{0}^{n}{}_r \; a_r \, z_1^{-r}$ and $(p_1, 0) \in [0,0]$. For every $\sigma_2(z_2^{-1})$, (16) and (11) give

$$\underset{>0}{\text{diag}} \sum_{1}^{\infty}{}_{ij} \; s_{ij} z_1^i \, z_2^j \; \sum_{0}^{n}{}_r \; a_r \, z_1^{-r} z_1^{-c-1} \sum_{0}^{c}{}_s b_s z_2^{-s} = 0, \forall b_s, \forall c \geqslant 0 \tag{19}$$

$$\sum_{0}^{c}{}_s\, b_s \operatorname*{diag}_{>0} \sum_{ijr} s_{ij}\, a_r\, z_1^{i-r-c-1}\, z_2^{j-s} = 0$$

$$\sum_{0}^{n}{}_r\, s_{t+r+c+1,t+s}\, a_r = 0 \quad , \ \forall t>0, \ \forall c\geqslant 0, \forall s, \ 0\leqslant s\leqslant c$$

It is then immediate that

$$(20) \quad \sum_{0}^{n}{}_r\, s_{t+1+c+r,t}\, a_r = 0 \ , \quad \forall c\geqslant 0 \ , \quad \forall t > 0$$

Conversely, let (20) hold and assume $p_1(z_1^{-1}) = \sum_{0}^{n}{}_r a_r\, z_1^{-r}$.

Then recalling (16) and straightforward checking that

$$(21)\ f(p_1(z_1^{-1})\ z_1^{-c-1} \ , \ \sigma_2(z_2^{-1})) = 0, \forall \sigma_2 \in K[z_2^{-1}], \quad c = \deg \sigma_2$$

it follows that the input pair $(p_1(z_1^{-1}),0) \in [0,0]$.

<u>Lemma 2.</u> Let S as in (10). The annihilating ideal of the i-indexed family (s_{ij}), $i \geqslant 1$, $j > i$ is different from (0) iff there exists $p_2(z_2^{-1}) \in K[z_2^{-1}]$, $p_2 \neq 0$, such that $(0,p_2(z_2^{-1})) \in [0,0]$.

<u>Proof.</u> As in Lemma 1.

<u>Theorem 1.</u> Let S as in (10) and $p_1(z_1^{-1}) = \sum_{0}^{n}{}_r a_r z_1^{-r}$, $p_2(z_2^{-1}) = \sum_{0}^{m}{}_s b_s z_2^{-s}$. The input pair $(p_1(z_1^{-1}),\ p_2(z_2^{-1})) \in [0,0]$ iff

i) $(a_0,\ldots,a_n)$ is an annihilating set for the j-indexed family (s_{ij}), $j \geqslant 1$, $i>j$,

ii) $(b_0,\ldots,b_m)$ is an annihilating set for the i-indexed family (s_{ij}), $i \geqslant 1$, $j>i$

iii) the matrix $A = (a_i b_j)$ is pp-annihilating for (s_{ij}), $\forall p \geqslant 1$

<u>Proof.</u> By definition $(p_1,p_2) \in [0,0]$ iff

$$(22) \quad f(p_1 \circ \sigma_1, p_2 \circ \sigma_2) = f(\sigma_1, \sigma_2), \forall (\sigma_1, \sigma_2) \in K[z_1^{-1}] \times K[z_2^{-1}]$$

Recalling (16), (22) implies :

$$(23) \qquad f(p_1(z_1^{-1})\, z_1^{-c-1}, \sigma_2(z_2^{-1})) = 0 , \quad c = \deg \sigma_2$$

$$(24) \qquad f(\sigma_1(z_1^{-1}), p_2(z_2^{-1}) z_2^{-c-1}) = 0 \quad , \quad c = \deg \sigma_1$$

$$(25) \qquad f(p_1(z_1^{-1})\, z_1^{-c} , p_2(z_2^{-1})\, z_2^{-c}) = 0 , \quad \forall c \geq 0 .$$

(23) and (24) imply i) and ii) by Lemma 1 and Lemma 2.

(25) implies

$$(26) \ \operatorname{diag}_{>0} \sum_{1}^{\infty}{}_{ij} s_{ij} z_1^i z_2^j \sum_{0}^{n}{}_{r} a_r z_1^{-r} \sum_{0}^{m}{}_{s} b_s z_2^{-s} z_1^{-c} z_2^{-c} = 0 , \ \forall c \geq 0$$

$$\sum_{1}^{\infty}{}_{t} \Big(\sum_{0}^{n}{}_{r} \sum_{0}^{m}{}_{s} a_r b_s s_{t+c+r, t+c+s} \Big) (z_1 z_2)^t = 0 \quad \forall c \geq 0$$

and

$$(27) \ \sum_{0}^{n}{}_{r} \sum_{0}^{m}{}_{s} a_r b_s s_{t+r, t+s} = 0 \qquad \forall t > 0$$

The converse is proved assuming $p_1(z_1^{-1})$ and $p_2(z_2^{-1})$ having as coefficient sets the annihilating sets.

<u>Remark.</u> The assumption that the zero state $[0,0]$ contains at least one element of the form (p_1, p_2) with $p_1, p_2 \neq 0$ is a necessary condition for controllability to zero state. This condition has a direct implication on the operator S as proved in Theorem 1.

<u>Lemma 3.</u> Let S satisfy i) and ii) in Theorem 1.
Then

$$(28) \ S(z_1, z_2) = \frac{N(z_1^{-1}, z_2^{-1})}{p_1(z_1^{-1}) p_2(z_2^{-1})} + \frac{(z_1 z_2)^{-1}}{p_1(z_1^{-1}) p_2(z_2^{-1})} \sum_{\substack{1 \\ -m \leq h-k \leq n}}^{\infty}{}_{h,k} s^1_{h,k} z_1^h z_2^k$$

Conversely if S has the structure (28) then i) and ii) hold.

Proof. $S(z_1,z_2)p_1(z_1^{-1})p_2(z_2^{-1}) = N(z_1^{-1},z_2^{-1}) + S^1(z_1,z_2)$.

It is immediate to check that $S^1(z_1,z_2)$ is a double series with $s^1_{h,k}=0$, $-m > h-k > n$ and that $N(z_1^{-1}, z_2^{-1}) \in K[z_1^{-1},z_2^{-1}]$, with $\deg N(z_2^{-1})(z_1^{-1}) < \deg p_1$, $\deg N(z_1^{-1})(z_2^{-1}) < \deg p_2$

The converse is a consequence of the fact that for every polynomial $p(z^{-1})$, the sequence corresponding to the series $1/p(z^{-1}) = \sum_1^{i} s_i z^i$ has the coefficients of $p(z^{-1})$ as minimal annihilating set.

Theorem 2. $(p_1,p_2) \in [0,0]$, $p_1,p_2 \neq 0$ iff S can be represented as in (28) and $s^1_{h,h} = 0$, $\forall h > 1$.

Proof. S represented by (28) implies i) and ii) of Theorem 1; using (25) and (26); $s^1_{h,h} = 0$, $\forall h > 1$ gets iii).

Conversely if $(p_1,p_2) \in [0,0]$, $p_1,p_2 \neq 0$ by Theorem 1, i) and ii) are satisfied and by Lemma 3 the structure (28) holds. $s_{h,h} = 0$ $\forall h > 1$ follows from $\operatorname{diag}_{>0} S\, p_1 p_2 = 0$.

Corollary. Consider now the two minimal annihilating sets (possibly zero)

(29) $(a^1_0,\dots,a^1_N)$, for (s_{ij}), $j \geq 1$, $i > j$

(30) $(b^1_0,\dots,b^1_M)$, for (s_{ij}), $i \geq 1$, $j > i$

with the associated polynomial $p_{1m}(z_1^{-1})$ and $p_{2m}(z_2^{-1})$. The state $[0,0]$ belongs to the ideal $(p_{1m}) \oplus (p_{2m})$ and contains the ideals $(p_{1m}) \oplus (0)$ and $(0) \oplus (p_{2m})$. If $p_{1m}, p_{2m} \neq 0$, S can be represented as in (28), with -of course- $p_1 = p_{1m}$ and $p_2 = p_{2m}$.

As recalled before, in the linear case the rational structure of the i-o map is equivalent to other fundamental properties of the system that can be evidentiated from the i-o map. In the bilinear case the rationality condition does not come out so directly from i-o characteristics. It is still possible to give an interpretation in this direction referring to a finite space repetition of the system [1] .

Theorem 3. Let S as in (28) and $p_1 = p_{1m}$, $p_2 = p_{2m}$. Assume S be the series expansion of a rational function $L(z_1^{-1},z_2^{-1})/D(z_1^{-1},z_2^{-1})$, $\deg L(z_1^{-1})(z_2^{-1}) < \deg D(z_1^{-1})(z_2^{-1})$, $\deg L(z_2^{-1})(z_1^{-1}) < \deg D(z_2^{-1})(z_1^{-1})$. Then

$$(31)\quad S(z_1,z_2) = \frac{N(z_1^{-1},z_2^{-1})}{p_{1m}(z_1^{-1})p_{2m}(z_2^{-1})} + \frac{(z_1z_2)^{-1}}{p_{1m}(z_1^{-1})p_{2m}(z_2^{-1})}\left(\frac{P_0((z_1z_2)^{-1})}{Q_0((z_1z_2)^{-1})} + \frac{1}{z_1^{-1}}\frac{P_1((z_1z_2)^{-1})}{Q_1((z_1z_2)^{-1})} + \dots + \frac{1}{z_1^{-N}}\frac{P_N((z_1z_2)^{-1})}{Q_N((z_1z_2)^{-1})} + \frac{1}{z_2^{-1}}\frac{\hat{P}_1((z_1z_2)^{-1})}{\hat{Q}_1((z_1z_2)^{-1})} + \dots + \frac{1}{z_2^{-N}}\frac{\hat{P}_N((z_1z_2)^{-1})}{\hat{Q}_N((z_1z_2)^{-1})}\right),$$

$\deg P_i < \deg Q_i$, $\deg \hat{P}_i < \deg \hat{Q}_i$. (by semplicity of notations M=N)

Proof. (Hint) An infinite matrix (s_{ij}) is associated to a series expansion of a proper rational function

$$\frac{R(z_1^{-1},z_2^{-1})}{Q(z_1^{-1},z_2^{-1})} = \sum_{1}^{}{}_{ij}\, s_{ij}\, z_1^i\, z_2^j$$

iff there exists a pq-annihilating matrix for any (p,q), p or q $\geqslant 1$.

Consider the series $\sum_{\substack{1 \\ |h-k|\leqslant n}}^{\infty}{}_{h,k}\, s^1_{hk}\, z_1^h\, z_2^k$ which is now a rational function.

By assigning appropriate values to p and q one proves that the side located diagonals in the matrix $(s^1_{h,k})$ are H-sequences and then can be represented by rational functions in $(z_1z_2)^{-1}$.

The procedure is repeated for the infinite matrix obtained deleting the above mentioned diagonals.

An i–o map having the structure (31) can be of course realized by connecting a finite number of linear maps and multipliers.

<u>Theorem 4.</u> Let S as in (31). Then there exist $(p_1,p_2) \in [0,0]$; $p_1,p_2 \neq 0$.

<u>Proof.</u> See [1] .

Theorem 4 shows that the rationality condition with i) and ii) of Theorem 1 are sufficient to guarantee that there exist $p_1,p_2 \neq 0$ such that $(p_1,p_2) \in [0, 0]$.

Before concluding we give a rough sketch of some consequences implied by the introduction of a state basing on the structure (31).

Assume S as in (31) and introduce the following equivalence relation on $K[z_1^{-1}] x\, K[z_2^{-1}]$: two input polynomial pairs are L–equivalent, $(p_1,p_2) \underset{L}{\sim} (\hat{p}_1,\hat{p}_2)$, iff every linear subsystem, appearing in the structure (31), reaches in t = 0 the same state, starting from zero state.

<u>Theorem 5.</u> $(p_1,p_2) \underset{L}{\sim} (\hat{p}_1,\hat{p}_2) \Longrightarrow (p_1,p_2) \underset{N}{\sim} (\hat{p}_1,\hat{p}_2)$

<u>Proof.</u> See [1] .

The set of L–classes can be obviously embedded in the direct sum of the state spaces of the linear subsystems. The L–classes constitute a finer partition of the input space than Nerode equivalence classes.

REFERENCES

1. E. Fornasini - G. Marchesini : "Abstract realization theory of multilinear input-output maps." UPee, 1973.

2. J.C. Willems : "Minimal Realization in State Space Form from input-output Data". Int.Rept.Dept.of Math.,Univ.of Groningen,1973.

3. G. Marchesini - G. Picci : "Some Results of the Abstract Realization Theory of Multilinear Systems. Theory and Applications of Variable Structure Systems". Academic Press 1972.

4. M.A. Arbib : "Decomposition Theory for Automata and Biological Systems", in System Structure (Ed. by A.S. Morse), N.Y., IEEE Control Systems Society, pp. 1-56, 1971.

5. R.E. Kalman : "Lectures on Controllability and Observability", CIME, Bologna Italy 1968.

6. R.E. Kalman - P.L. Falb - M.A. Arbib : "Topics in Mathematical System Theory", McGraw-Hill, 1969.

7. F.R. Gantmacher : "The Theory of Matrices", Chelsea P.C., N.Y. 1964.

OPTIMAL CONTROL OF DISCRETE BILINEAR SYSTEMS

K.N. Swamy and T.J. Tarn

Washington University

St. Louis, Missouri, U.S.A.

I. INTRODUCTION

The problem of optimal control of single-input discrete bilinear systems with unbounded control and the cost criterion quadratic in state is investigated in this paper.

Closed-form solutions have been obtained for the deterministic and stochastic optimization problems, via the method of dynamic programming. The work reported here represents the first successful attempt at obtaining an explicit solution to the optimization problem for a system which is not linear.

The deterministic problem has been studied in Section 2. Several features of bilinear systems are discovered which are curiously different from the corresponding features of linear systems. The optimal control is seen to be a nonlinear function of the state, being a ratio of quadratic forms in the state. The optimal cost is a quadratic form in the state as in linear systems. Because of the dependence of the existence of certain inverses on the state, regular and singular paths satisfying the dynamic programming functional equation are seen to exist simultaneously, unlike the situation in linear systems. It is shown, then, that if the regular path exists, it is optimal.

The stochastic control problem with complete state information, where the state equation is corrupted by additive and multiplicative noise, has been examined in Section 3. The expression for optimal control is seen to be similar to that in the deterministic case, but with certain points of difference.

Firstly, the presence of noise is seen to make the calculation of optimal control much easier now (than in the deterministic case) by virtue of integration over certain sets of measure zero, which annihilate the effects of singularities! Secondly, the presence of noise is seen to reduce the several possible recursive equations in the deterministic case to a single recursive equation. The multiplicative noise influences the recursive equation directly by contributing an additive term -- unlike the additive noise.

II. DETERMINISTIC OPTIMAL CONTROL

Free-end-point problem is considered in this section for the deterministic system. A closed-form solution has been obtained for the optimal control problem by the method of dynamic programming. The optimal control is seen to be a ratio of quadratic forms in state. It is shown that the dynamic programming functional equation may have nonunique solution, with a multiplicity of paths satisfying the functional equation. The order of the system n is assumed to be ≥ 2, since the solution for the case n = 1 is trivial.

Before introducing the main result, we state the following definitions.

Definition 2.1: Path of a dynamical system generated by applying at each time the uniquely determined optimal control is called the regular path of the system.

Definition 2.2: A singularity is said to exist at k if the optimal control at k is not unique.

Definition 2.3: Optimal path of a dynamical system which has at least one singularity is called a singular path of the system.

It is shown that for the bilinear systems considered, a unique regular path, and singular paths may exist simultaneously. The main result of this section is that if the regular path and singular paths exist simultaneously, it is the regular path that is optimal.

2.1 Problem Statement

Consider the bilinear discrete system described by

$$\underline{x}_{k+1} = (\Phi + u_k B)\underline{x}_k \ , \ \underline{x}_k = \underline{x}(t_k) \ , \ 0\leq k\leq N-1 \tag{1}$$

where $\underline{x}_k = (x_k^1, x_k^2, \ldots , x_k^n)^T \in R^n$ is the state vector, $u_k \in R^1$ is the scalar control. The $(n \times n)$ constant matrices Φ and B have the form

$$\Phi = \begin{bmatrix} 0 & 1 & 0 & 0 & \dots & 0 & 0 \\ 0 & 0 & 1 & 0 & \dots & 0 & 0 \\ \cdot & \cdot & \cdot & \cdot & & \cdot & \cdot \\ \cdot & \cdot & \cdot & \cdot & & \cdot & \cdot \\ \cdot & \cdot & \cdot & \cdot & & \cdot & \cdot \\ 0 & 0 & 0 & 0 & \dots & 0 & 1 \\ a_1 & a_2 & \dots & & & & a_n \end{bmatrix}, \quad B = \begin{bmatrix} 0 & 0 & \dots & 0 \\ 0 & 0 & \dots & 0 \\ \cdot & \cdot & & \cdot \\ \cdot & \cdot & & \cdot \\ \cdot & \cdot & & \cdot \\ 0 & 0 & \dots & 0 \\ b_1 & b_2 & \dots & b_n \end{bmatrix} \quad (2)$$

The problem is to choose the sequence of controls $u_0, u_1, \dots, u_{N-1}$ so that the cost functional

$$J = \sum_{k=1}^{N-1} \underline{x}_k^T Q \underline{x}_k + \underline{x}_N^T Q' \underline{x}_N \quad (3)$$

is a minimum, subject to (1) and the condition $\underline{x}_0 = \underline{x}(0)$. Furthermore, $\underline{x}_N$ may be specified (fixed-end-point problem), or may not be specified (free-end-point problem). Q and Q', the (n×n) constant cost matrices, are symmetric, and non-negative. The superscript T denotes transpose. N is a fixed, finite positive integer.

2.2 Optimal Control

Theorem 1: The solution to the problem stated in Section 2.1 with $\underline{x}_N$ free is given by

$$u_k = \begin{cases} -\dfrac{\underline{x}_k^T \Phi^T Q_k B \underline{x}_k}{\underline{x}_k^T B^T Q_k B \underline{x}_k}, & \text{if } Q_k B \underline{x}_k \neq \underline{0} \\ \text{nonunique}, & \text{if } Q_k B \underline{x}_k = \underline{0} \end{cases} \quad 0 \le k \le N-1 \quad (4)$$

where Q_k is given recursively by

$$Q_k = \begin{cases} Q + F(Q_{k+1}), & \text{if } Q_{k+1} B \underline{x}_{k+1} \neq \underline{0} & (5a) \\ Q + \Phi^T Q_{k+1} \Phi, & \text{if } Q_{k+1} B \underline{x}_{k+1} = \underline{0} & (5b) \end{cases}$$

with the initial condition $Q_{N-1} = Q'$. Q_k is nonnegative, symmetric for all k, $0 \le k \le N-1$. The optimal cost for the last (N-k) stages is given by

$$\underset{0 \le k \le N-1}{V_{N-k}} = \begin{cases} \underline{x}_k^T \{F(Q_k)\} \underline{x}_k, & \text{if } Q_k B \underline{x}_k \neq \underline{0} \\ \underline{x}_k^T (\Phi^T Q_k \Phi) \underline{x}_k, & \text{if } Q_k B \underline{x}_k = \underline{0}. \end{cases} \quad (6)$$

$F(S)$ is defined for any symmetric $S = (s_{ij})$ as

$$F(S) \triangleq \begin{bmatrix} 0 & \underline{0}^T \\ \underline{0} & \tilde{F}(S) \end{bmatrix} \tag{7}$$

where $\tilde{F}(S) = (s'_{ij})$

$$\underset{1 \le i,j \le n-1}{s'_{ij}} = \begin{cases} s_{ij} - \dfrac{s_{in}s_{nj}}{s_{nn}} , & s_{nn} \neq 0 \\ s_{ij} , & s_{nn} = 0 . \end{cases}$$

Proof: The proof proceeds inductively, using the functional equations of Dynamic Programming [1,3].

Remark (i) From (4), the optimal control u_k is uniquely determined at each k, if $Q_k B \underline{x}_k \neq \underline{0}$, $0 \le k \le N-1$. In view of Definition 2.1, it follows that the regular path, if it exists, is uniquely determined.

(ii) From (4) u_k is nonunique when $Q_k B \underline{x}_k = \underline{0}$. In view of Definitions 2.2 and 2.3, a singularity is said to exist at k if $Q_k B \underline{x}_k = \underline{0}$ (i.e., $q_{nn}^{(k)} (\underline{b}^T \underline{x}_k) = 0$ where $Q_k = (q_{ij}^{(k)})$, $1 \le i, j \le n$ and $\underline{b}^T = (b_1, \cdots , b_n)$) and the path is singular if $Q_k B \underline{x}_k = \underline{0}$ at least for one value of k, $0 \le k \le N-1$.

(iii) From (5) and (6) it is clear that regular and singular paths have different sequences of V_k's associated with them.

(iv) Since the existence of the inverse in the calculation of control (i.e., $Q_k B \underline{x}_k \neq \underline{0}$ or $= \underline{0}$) depends on $\underline{x}_k$, a situation in which the inverse both exists and does not exist at a given k can occur in bilinear systems depending on $\underline{x}_k$. This implies the simultaneous existence of regular and singular paths. This feature is nonexistent in linear systems since the inverse required in the calculation of control, being independent of $\underline{x}_k$, either exists or does not at a given k.

2.3 Main Result

The central result on optimality is contained in the following theorem.

Theorem 2 (Optimality Theorem): For the free-end-point problem stated in Section 2.1, if regular path and singular paths exist simultaneously, the regular path is optimal.

Proof: The idea of the proof is as follows. For any singular path S, there corresponds a set of Q_k's from which new sets of Q_k's (denoted $S_1, S_2, \ldots$) can be constructed. Comparison of the sets of Q_k's for S, $S_1, \ldots$ and R permits a comparison of V_N's for R and S. It can be shown that $V_N^R \leq V_N^S$, which establishes the optimality of the regular path [1,3].

Remark: Equation (1) can be expressed as

$$\underline{x}_{k+1} = \Phi \, \underline{x}_k + \underline{c} \, v_k \tag{1a}$$

$$v_k = u_k \, (\underline{b}^T \, \underline{x}_k) \tag{1b}$$

where, $\underline{c}^T = (0,0, \cdots ,0,1)$. It can be shown that (5a) is the simplified version of the Riccati equation for the optimal control problem for the linear system (1a) with cost functional (3). When $q_{nn}^{(k+1)} = 0$ for some k, the inverse does not exist in the Riccati equation. The corresponding v_{k+1} is arbitrary and the recursive equation takes the form of (5b) for the k in question.

When $q_{nn}^{(k)} \neq 0$ and $\underline{b}^T \, \underline{x}_k \neq 0$, $0 \leq k \leq N-1$, the optimal control u_k (1b) obtained by solving the linear optimization problem with (1a) and (3) corresponds to the regular case of this paper. However, the analysis of the linear optimization problem does not give any information about the simultaneous existence of the regular path and singular paths. Hence, optimality of the regular path is not clear from the analysis of the linear optimization problem.

When $q_{nn}^{(k)} = 0$ for some k, the regular path does not exist. There may be several singular paths satisfying the functional equation. Though the linear optimization problem (1a) and (3), and (1b) may yield $\{u_k\}$, its optimality has to be investigated. No attempt is made here to investigate the optimality when the regular path does not exist. When $q_{nn}^{(k)} \neq 0$ but $\underline{b}^T \, \underline{x}_k = 0$ for some k, the analysis of linear optimization problem provides no solution.

III. STOCHASTIC OPTIMAL CONTROL

In this section, a combination of additive and multiplicative noise is considered in the state equation; but complete information about the state is assumed available. A closed-form solution has been obtained for the optimal control problem. The expression for optimal control is seen to be similar in form to that in the deterministic case. Recursive equation is obtained, of structure similar to that in the deterministic case, but now affected by the noise statistics. The principal contribution of this section consists in showing that the presence of noise simplifies the analysis compared to that in the deterministic case! This novel phenomenon, which is discovered here for the first time, is absent in linear systems.

It is hoped that this feature of "noise smoothing-out bad points" will be of some fundamental value in future studies of stochastic control of nonlinear systems and related topics.

3.1 Problem Statement

Consider the stochastic discrete bilinear system

$$\underline{x}_{k+1} = (\Phi + u_k B)\, \underline{x}_k + H\, \underline{\xi}_k + G(\underline{x}_k)\underline{\alpha}_k \quad 0 \leq k \leq N-1, \tag{8}$$

where $\underline{x}_k \in R^n$, $u_k \in R^1$, Φ and B are as in (2), H is an $(n \times d_1)$ matrix, and $G(\underline{x}_k) = \sum_{i=1}^{n} x_k^i\, G_i$, where G_i, $1 \leq i \leq n$, are $(n \times d_2)$ matrices. The initial state $\underline{x}_0$ is Gaussian with $E(\underline{x}_0) = \underline{m}_0$, cov $(\underline{x}_0) = \Sigma_0$. The noise sequence $\{\underline{\xi}_k : k = 0, 1, \ldots, N-1\}$, $\{\underline{\alpha}_k : k = 0, 1, \ldots, N-1\}$ are each Gaussian-white, and independent of each other and of $\underline{x}_0$, with mean and covariance

$$E(\underline{\xi}_k) = \underline{0}\ , \ E(\underline{\xi}_k\, \underline{\xi}_j^T) = \Xi_k\, \delta_{kj}\ ;$$

$$E(\underline{\alpha}_k) = \underline{0}\ , \ E(\underline{\alpha}_k\, \underline{\alpha}_j^T) = \Lambda_k\, \delta_{kj}$$

respectively, where δ_{kj} is the Kronecker delta function. Exact measurement of the state is assumed available at each $k, 0 \leq k \leq N-1$. The problem is to choose the control policy $\{u_i(\underline{x}_i),\ 0 \leq i \leq N-1\}$ so that the (unconditional) cost functional

$$J = E\ \{\sum_{k=1}^{N-1} \underline{x}_k^T\, Q\, \underline{x}_k + \underline{x}_N^T\, Q'\, \underline{x}_N\}$$

is a minimum, where Q and Q' are symmetric and nonnegative.

3.2 Optimal Control Policy

Theorem 3: The solution to the problem stated in Section 3.1 is given by:

$$u_k \underset{0\le k\le N-1}{} = \begin{cases} \dfrac{\underline{x}_k^T \Phi^T \tilde{Q}_k B \underline{x}_k}{\underline{x}_k^T B^T \tilde{Q}_k B \underline{x}_k}, & \text{if } \tilde{Q}_k B \underline{x}_k \ne \underline{0} \\ \text{arbitrary}, & \text{if } \tilde{Q}_k B \underline{x}_k = 0 \end{cases} \qquad (9)$$

where $\tilde{Q}_k$ is given recursively by

$$\tilde{Q}_k = Q + F(\tilde{Q}_{k+1}) + \tilde{G}_{k+1} \qquad 0 \le k \le N-2 \qquad (10)$$

with initial condition $\tilde{Q}_{N-1} = Q'$, where $\tilde{G}_{k+1}$ is given by

$$(\tilde{G}_{k+1})_{ij} = \text{tr}\,(\Lambda_{k+1} G_i^T \tilde{Q}_{k+1} G_j), \quad -1 \le k \le N-2$$

and $F(\cdot)$ is as in (7). $\tilde{G}_k$ and $\tilde{Q}_k$ are symmetric and nonnegative, $0\le k\le N-1$. The (conditional) optimal cost for the last $(N-k)$ stages given $\underline{x}_k$ is given by

$$V_{N-k} \underset{0\le k\le N-1}{} = \begin{cases} \underline{x}_k^T\{F(\tilde{Q}_k)+\tilde{G}_k\}\underline{x}_k + \text{tr} \sum_{i=k}^{N-1} (H^T\tilde{Q}_i H)\Xi_i, & \text{if } \tilde{Q}_k B\underline{x}_k \ne \underline{0} \\ \underline{x}_k^T\{\Phi^T\tilde{Q}_k\Phi)+\tilde{G}_k\}\underline{x}_k + \text{tr} \sum_{i=k}^{N-1} (H^T\tilde{Q}_i H)\Xi_i, & \text{if } \tilde{Q}_k B\underline{x}_k = \underline{0}. \end{cases}$$

Minimum value of J is given by

$$\underline{m}_0^T S_0 \underline{m}_0 + \text{tr}\,(S_0 \Sigma_0) + \text{tr} \sum_{i=0}^{N-1} (H^T \tilde{Q}_i H)\, \Xi_i,$$

where

$$S_0 = \begin{cases} F(\tilde{Q}_0) + \tilde{G}_0 & \text{if } \tilde{Q}_0 B \underline{x}_0 \ne \underline{0} \\ \Phi^T \tilde{Q}_0 \Phi + \tilde{G}_0 & \text{if } \tilde{Q}_0 B \underline{x}_0 = \underline{0}\,. \end{cases}$$

Remark: It is assumed that each component of $\underline{x}_k$, $1 \le k \le N-1$, is random.

Proof: Proof of the theorem is inductive [2,3].

3.3 Discussion

The significance of the results in the above theorem will now be discussed.

As we have seen in Section 2.2, in deterministic bilinear systems Q_k, and hence u_k, is affected by the singularities in the future. This raises the question of simultaneous existence of paths and the consequent question of optimality.

In the stochastic case, however, the singularities determined by $\tilde{Q}_k B \underline{x}_k = \underline{0}$ are located on a set of measure zero at each k. By virtue of integration over this set the $\tilde{Q}_k$'s are uniquely determined by (10) irrespective of the singularities in the future. Hence, the presence of noise simplifies the analysis (compared to the deterministic case) by smoothing out the singularities. This is a feature of bilinear systems different from that of linear systems.

Consider now the role of the additive noise only. In the deterministic problem, there are several recursive equations to consider depending on the distribution of singularities. However, addition of noise results in a single recursive equation (10) with $G_k = 0$, $0 \leq k \leq N-1$. Though the expressions for control in the deterministic and stochastic cases are similar (compare (4) and (9), Q_k and $\tilde{Q}_k$ are described, in general, by different recursive equations. This is in contrast to the behavior in linear systems where the additive noise does not affect the recursive equation or the feedback parameters in the expression for control. As in linear systems the noise results in added cost.

The multiplicative noise also results in a single recursive equation (10). It affects the recursive equation directly by contributing an additive term, unlike the additive noise, and results in added cost like the additive noise.

IV. EXAMPLES

Example 1: Described in this example is the method of examining if a path with a certain distribution of singularities exists, and finding the path if it exists.

$$\text{Let } \Phi = \begin{bmatrix} 0 & 1 & 0 \\ 0 & 0 & 1 \\ 1 & 0 & 1 \end{bmatrix}, \quad B = \begin{bmatrix} 0 & 0 & 0 \\ 0 & 0 & 0 \\ 1 & 0 & 1 \end{bmatrix}, \quad Q = Q' = \begin{bmatrix} 1 & -1 & 0 \\ -1 & 1 & 0 \\ 0 & 0 & 1 \end{bmatrix}$$

$\underline{x}_0^T = (1, 1, 1)$, $N = 4$, and $\underline{x}_4$ be free. Q_k, $0 \leq k \leq 3$ are obtained from (5) for the assumed distribution of singularities. u_k, $0 \leq k \leq 3$, $\underline{x}_k$, $1 \leq k \leq 4$, and V_4 are obtained from (4), (1) and (6) respectively.

(i) Assume that there are no singularities at k, $0 \leq k \leq 3$. Then, we can compute Q_3, Q_2, Q_1 and Q_0. It is seen that $Q_0 B \underline{x}_0 \neq 0$ (hence, no singularity at $k = 0$). We get $u_0 = -0.81$, $\underline{x}_1^T = (1,1,0.38)$. Since $Q_1 B \underline{x}_1 \neq \underline{0}$, $u_1 = -0.89$, $\underline{x}_2^T = (1,0.38,0.152)$. $Q_2 B \underline{x}_2 \neq \underline{0}$; $u_2 = -0.94$,

$\underline{x}_3^T$ = (0.38, 0.152, 0.069). Q_3 B $\underline{x}_3 \neq \underline{0}$ so that u_3 = -1 and $\underline{x}_4^T$ = (0.152.0.069,0). Since Q_k B $\underline{x}_k \neq 0$, $0 \leq k \leq 3$, is indeed satisfied, regular path exists. V_4 = 0.64.

(ii) Assume that a singular path exists with a singularity at k = 1. Then similarly as in (i), Q_3,Q_2,Q_1 and Q_0 can be computed. It is seen that Q_0 B $\underline{x}_0 \neq \underline{0}$; u_0 = -15/14, $x_1^T = (1,1,-\frac{1}{7})$. However, Q_1 B $\underline{x}_1 \neq 0$; hence there is no singularity at k = 1. The proposed singular path does not exist.

Example 2: Simultaneous existence of regular and singular paths is illustrated in this example.

$$\text{Let } \Phi = \begin{bmatrix} 0 & 1 \\ 1 & 1 \end{bmatrix}, \quad B = \begin{bmatrix} 0 & 0 \\ 3/5 & 1 \end{bmatrix}, \quad Q = Q' = \begin{bmatrix} 1 & 1 \\ 1 & 1 \end{bmatrix},$$

$\underline{x}_0^T = (0, \eta_0)$, $\eta_0 \neq 0$, and N = 2.

(i) Regular path (R) exists with $u_0 = -2$, $u_1 = -\frac{5}{2}$; $\underline{x}_1^T = (\eta_0, -\eta_0)$, $\underline{x}_2^T = (-\eta_0, \eta_0)$; $V_2 = 0$ (because $\underline{x}_1$ and $\underline{x}_2$ are in the null space of Q).

(ii) Singular path (S) exists with a singularity at k = 1, for which $u_0 = -\frac{8}{5}$, u_1 = arbitrary (hence can be taken = 0); $\underline{x}_1^T = (\eta_0, -\frac{3}{5}\eta_0)$, $\underline{x}_2^T = (-\frac{3}{5}\eta_0, \frac{2}{5}\eta_0)$; $V_2 = (\frac{1}{5})\eta_0^2$.

Remark: $V_2^R < V_2^S$

V. ACKNOWLEDGEMENT

This research was supported in part by National Science Foundation Grants GK - 36531, GK - 22905A#2.

REFERENCES

[1] Swamy, K.N., Tarn, T.J., and Zaborszky, J., "Deterministic Optimal Control of Single-Input Discrete Bilinear Systems", submitted for publication.

[2] Swamy, K.N., and Tarn, T.J., "Stochastic Optimal Control of Single-Input Discrete Bilinear Systems", submitted for publication.

[3] Swamy, K.N., "Optimal Control of Single-Input Discrete Bilinear Systems", DSc. Dissertation, December 1973, Washington University, St. Louis, Missouri.

DIFFUSIONS ON MANIFOLDS
ARISING FROM CONTROLLABLE SYSTEMS

David L. Elliott

Washington University

St. Louis, Missouri

1. INTRODUCTION

We shall consider the relationships between a control system, on a connected real-analytic manifold M,

$$\dot{x} = X_0 x + u_1(t)\, X_1 x + \cdots + u_r(t) X_r x \tag{1}$$

and a corresponding stochastic differential equation

$$dx = Lx dt + \Sigma_1^r \; X_i x \, dw_i(t) \;, \tag{2}$$

where

$$L = X_0 + (1/2)\, \Sigma_1^r \, X_i^2 \;;$$

the X_i are real-analytic vector fields (differential operators), the u_i are piecewise real-analytic controls, and $w_1, \cdots, w_r$ are independent Wiener processes.

Since the completion of a first attempt at this [1], others have obtained much more information about these two systems, and most of it has been presented here in previous lectures. However, the main thread of [1], the relation between controllability of (1) and the existence of transition densities for (2), remains to be discussed.

We will first give an easy way to study Eq. (2) on M; return to Lie-algebraic facts about Eq. (1), and Hörmander's Theorem for L; then derive the existence of transition densities; and finally explore some consequences for stochastic optimal control and other problems.

2. HOW TO LIVE ON A MANIFOLD

At first we will suppose Eq. (1) is given on euclidean space R^n. We will assume that if X is in the real Lie algebra generated by the vector fields $X_0, \ldots, X_r$, then X is complete (solutions of $\dot{x} = Xx$ do not escape from R^n in finite time). Then given a control vector $\underline{u} = (u_1, \ldots, u_r)$ and an initial point y, Eq. (1) has a unique solution $x(t) = \sigma(y, t; \underline{u})$, $t \geq 0$. The class U^r of admissible controls is the class of piecewise real-analytic functions $\underline{u}: R^1 \to R^r$.

Let M be an m-dimensional real-analytic closed submanifold of R^n. It is known [15] that this is not a restriction on the abstract manifold M if $n \geq 2m + 1$. Then system (1) lives on M (has M for an invariant manifold) if every real-analytic function f defined on a neighborhood $N_f \subset R^n$ and constant on $N_f \cap M$ satisfies $X_i f = 0$ on N_f, $i = 0, \ldots, r$. For example, M may be given as a differentiable variety:

$$\{x \in R^n: f_k(x)=0,\ k=1, \ldots ,n-m;\ \text{rank}\ (\partial f_k(x)/\partial x_j)=n-m\},$$

with $X_i f_k = 0$, $i = 0, \ldots, r$. Such an invariant M is then an acceptable state space for (1).

<u>Example 1</u>: the pendulum

$$\ddot{\theta} + a\dot{\theta} + b\sin\theta = cu(t),\quad (\theta,\dot{\theta}) \in S^1 \times R^1 ;$$

in R^3, we have $M = \{x: x_1^2 + x_2^2 - 1 = 0\}$,

$$X_0 = (-x_2 x_3)\partial/\partial x_1 + (x_1 x_3)\partial/\partial x_2 + (-bx_2 - ax_3)\partial/\partial x_3 ,$$

$$X_1 = (c)\partial/\partial x_3 .$$

Let $\underline{w} = (w_1, \ldots, w_r)$ be an r-dimensional Brownian motion with associated σ-algebra F and probability $\underline{P}$. Using the method of Secs. 3.2-3.3 of McKean [2] we can construct, <u>globally</u> on R^n , a degenerate diffusion $\mathcal{X}$ from Eq. (2); as we shall point out below, the solutions of (2) will not explode. By Ito's Lemma ([2]) a function f constant on M will have stochastic differential

$$df = Lf\ dt + \Sigma_1^r X_i f\ dw_i(t) = 0;$$

if $x(0) \varepsilon M$ with probability one, so we can with the condition $x(0) \varepsilon M$ define a Markov process $\mathcal{D} = (x(\cdot),F,\underline{P})$ on the state space M. This works when L is not elliptic, unlike Ch. 4 of McKean. See [3] for details of the case $M = S^n$.

If γ is a trajectory through $x(0)$ of a vector field X, a similar argument shows that the one-dimensional diffusion

$$dx = (1/2)\; X^2 x dt + Xx\; dw(t),\; x(0) \;\varepsilon\; \gamma$$

moves on the curve γ w.p.1. This equation is independent of coordinates, and furnishes a strong argument that (2) is a good way to describe what we intuitively mean by saying "replace the controls u_i by independent white noises." Wong and Zakai have given other arguments, based on differentiable approximations of $w(\cdot)$, for associating (2) with (1). The vector version of their result is discussed by Stroock and Varadhan [4] while showing (loosely speaking) that in the space of continuous curves $[0,\infty) \to R^n$, starting at x_0, the probability associated with (2) is concentrated on the closure of the set of curves obtained from (1) with piecewise constant controls. Since (1) doesn't have exploding solutions, neither does (2).

We call (1) <u>controllable</u> on M if for every pair of points y, y^1 of M there exists $t > 0$ and $\underline{u} \;\varepsilon\; U^r$ for which $\sigma(y,t;\underline{u}) = y^1$. In that case the results of Stroock and Varadhan imply that if G is an open neighborhood of y^1 in M, then for $\mathcal{D}$ we have $\underline{P}\{x(t) \;\varepsilon\; G \mid x(0) = y\} > 0$.

> Example 2: again $M = S^1 \times R^1 \quad R^3$,
> $X_0 = (-x_2)\partial/\partial x_1 + (x_1)\partial/\partial x_2$, $X_1 = \partial/\partial x_3$;
> this is controllable, but $\sigma(y,t;U^1)$ for fixed (y,t), is only one-dimensional, i.e. a line on the cylinder.

3. CONTROLLABILITY AND HYPOELLIPTICITY

System (1) has the <u>strong accessibility</u> property from $y \;\varepsilon\; M$ if for all $t > 0$ the set $\sigma(y,t;U^r)$ has non-empty interior in M. If this holds for all y, we say the property holds on M. Example 1 has the strong accessibility property on $S^1 \times R^1$; example 2 does not.

Nagano's [5] Theorem 1 shows that M can be partitioned into maximal immersed submanifolds $\{M_\alpha\}$ invariant under (1); these submanifolds constitute a "foliation with singularities" induced by the Lie algebra A^o generated by $X_0, \ldots, X_r$. If $y \in M_\alpha$, $\dim(M_\alpha) = \dim(A_y)$. If system (1) is controllable on M, then M is the only invariant set and $\dim(A_x) = m$ everywhere on M; we say A^o has rank M. Looking the same way at the (x,t) space $M \times R$, we see that the strong accessibility property implies rank $(A^+) = m + 1$, where A^+ is the Lie algebra generated by

$$\partial/\partial t + X_0,\ X_1, \ldots, X_r.$$

It was shown in [6] that if the fundamental group (first homotopy group) $\pi_1(M)$ has no element of infinite order, then controllability implies rank $(A^+) = m + 1$. My second argument (in the Remark), attempting to avoid Haefliger's Lemma, was incorrect; it assumed that the leaves of the foliation induced by $B = A^o \cap A^+$ are closed, and it takes said Lemma to prove that. Stefan and Taylor [7] have given a simple proof, which is briefly paraphrased here (with thanks):

If rank (B) = m, we are through. Otherwise rank (B) = m-1 and B induces a foliation F on M of codimension one. From fundamental properties of F there exists a differential 1-form ω such that $\omega(B) = 0$, $\omega(X_0) = 1$ as a normalization, and $d\omega = 0$. Then if $x(t) = \sigma(y,t;\underline{u})$, $\omega_{x(t)} \cdot \dot{x}(t) = 1$. By controllability we can find $T > 0$ and $\underline{u}$ such that $\sigma(y,T;\underline{u}) = y$; i.e., we have a closed loop trajectory. The line integral of ω around this loop is T, but since ω is a closed form the integral depends only on the homotopy class of the loop through a group homomorphism $\pi_1(M) \to R$. Since $T \neq 0$ the loop represents an element of $\pi_1(M)$ of infinite order.

In many important problems (systems on cylinders, tori, etc.) the condition on $\pi_1(M)$ is not satisfied. Then strong accessibility may not follow from controllability (Ex. 2 above). On the other hand:

> Proposition 1. If system (1) is controllable on M and $\sigma(y,t;U^r)$ has non-empty interior in M for <u>one</u> $y \in M$ and some $t > 0$, then (1) has the strong accessibility property on M.

The proof is just to point out that if $\dim(A^+_{y,0}) = m + 1$, then by controllability and analyticity this is true everywhere on M, since (1) is time-invariant. Sussmann and Jurdjevic [8] show that this rank condition is equivalent to strong accessibility.

The nice thing about this simple result is that at one point y(say y = (1,0,0) in Example 1) we can check the strong accessibility property by looking at the linearization of (1) along the trajectory of X_0 through y. It is sufficient (but not necessary) that the linearized, time-varying, system satisfy Kalman's complete controllability condition, in order to guarantee strong accessibility at y.

A linear differential operator K is called hypoelliptic if given Schwartz distributions ϕ, ψ such that $K\phi = \psi$ then ϕ is C^∞ (essentially) off the support of ψ. Thus, if ψ is the Dirac $\delta(x,t)$ then ϕ is C^∞ except at (0,0). For a readable account of Hörmander's Theorem on hypoelliptic second-order operators, and related work, see Oleinik and Radkevich [9]. In our context the theorem and its converse (due to M. Derridj) give us the following facts.

(A) Provided $X_0, \ldots, X_r$ do not simultaneously vanish anywhere on M, L is hypoelliptic on M if and only if rank $(A^o) = m$;

(B) $-\partial/\partial t + L$ is hypoelliptic on M if and only if rank $(A^o) = m + 1$.

That is to say,

(C) The Kolmogorov backward operator $-\partial/\partial t + L$ of the diffusion $\mathcal{D}$ induced by (2) is hypoelliptic if and only if (1) has the strong accessibility property.

4. TRANSITION FUNCTIONS AND DENSITIES ON M

<u>Theorem</u>: If the real-analytic system (1) has the strong accessibility property on M, then the (<u>formally</u>) degenerate diffusion $\mathcal{D}$ defined by (2) on M has smooth transition densities on M, and has the strong Feller property.

The proof is sketched below; the R^n part appears in [1]. First we will deal with the degenerate diffusion $\mathcal{X}$ induced on R^n by (2). The usual technical assumption of a global Lipschitz condition on the coefficients can be replaced by our assumption that the vector fields in A^o are complete on R^n, thus (from [4]) the escape (explosion) time for the trajectories of (1) is ∞ with probability one. This will also obviate Skorokhod's [10, Ch. 3] assumption that the first and second derivatives of the coefficients are bounded; the argument follows.

We define a sequence of Markov processes $\{X_k\}_1^\infty$ on R^n, by smoothly modifying the coefficients of (2) outside $V_k = \{x : \|x\| < k\}$ to satisfy Skorokhod's assumptions. The part of X obtained by cutting off trajectories at their first exit time from V_k agrees with the corresponding part of X_k. By non-explosiveness, $P(y,t,R^n - V_k) \to 0$ as $k \to \infty$ and $P^k(y,t,B) \to P(y,t,B)$ where $P^k(y,t,B) = \underline{P}\{x(t) \in B \mid x(0) = y\}$ defines the transition function X_k ; $P(\cdot,\cdot,\cdot)$ is the transition function for X, B is any Borel set in R^n. This kind of convergence suffices to extend the properties we need from the well-behaved X_k to X itself.

The notion of weak convergence we require ([10], Ch. 5) is $f_k \overset{w}{\to} f$ if the $f_k: R^n \times R^1 \to R^1$ (or $M \times R^1 \to R^1$) are uniformly bounded and $f_k(x,t) \to f(x,t)$ for all x,t as $k \to \infty$.

L is the differential generator of X; its domain includes $C_0^\infty(R^n)$, the smooth functions with compact support, by Ch. 11 of Dynkin [11]. If $g \in C_0^\infty(R^n)$ and $t \geq 0$ then the conditional expectation $f(y,t) = E[g(x(t)) \mid x(0) = y]$ satisfies $f(y,0) = g(y)$ and, by Ch. (3) of [10], the forward equation

$$\partial f/\partial t = Lf, \quad t \geq 0; \tag{3}$$

then f is smooth by (C) in Sec. 3 above. For fixed y,t the functional $T_{t,y} : g \to f(y,t) = E_y\, g(x(t))$ is linear and bounded (norm 1), so it has a (Schwartz) distribution kernel $\hat{p}(y,t,\cdot)$, which is a function of its first two arguments, a generalized function of the third. Of course

$$f(y,t) = \int_{R^n} g(x)\ \hat{p}(y,t,x)\,dx = \int_{R^n} g(x)\ P(y,t,dx).$$

Now let us restrict y to M, obtaining the Markov process $\mathcal{D}$ on M as before; $P(y,t,M) = 1$. From (C) and our hypothesis, $\partial/\partial t - L$ is hypoelliptic on M. The Lebesgue measure on R^n induces a natural m-dimensional measure μ (arc-length, area, ...) on M; with $y \in M$, the support of $P(y,t,\cdot)$ and $\hat{p}(y,t,\cdot)$ is on M, so we can define a distribution $p(y,t,\cdot)$ on M by

$$\int_M g(x)\ p(y,t,x)\,d\mu\ (x) = \int_{R^n} g(x)\ \hat{p}(y,t,x)\,dx \tag{4}$$

for all $g \in C_0^\infty(R^n)$, $y \in M$, $t \geq 0$.

Let $\mathcal{H}$ be the class of functions $g: M \to R^1$ such that $f(y,t) - T_{t,y}g$ satisfies Eq. (3). $\mathcal{H}$ is a linear space, includes $C_o^\infty(M)$, and is closed under weak limits: if $g_i \overset{W}{\to} g$, by dominated convergence $T_{t,y}g_i \to T_{t,y}g = f(t,y)$, hence $T_. g_i \overset{W}{\to} f(\cdot)$; since $f_i = T_. g_i$ is a weak (distribution) solution of (3), $i = 1,2, \ldots$, then so is f (see next paragraph); by hypoellipticity f is a C^∞ solution of (3), so $g \in \mathcal{H}$. The C^∞ version of Lemma 5.12 of [11] shows $\mathcal{H} \supset \mathcal{B}(M)$, the class of bounded Borel functions. We conclude that if $g \in \mathcal{B}(M)$, for $t > 0$ $T_t g \in C^\infty(M)$. A fortiori, $T_t\mathcal{B}(M) \supset C_b(M)$, the class of bounded continuous functions; that is the strong Feller Property [11]. If g is the indicator of a Borel set B, we see that $P(\cdot,\cdot,B)$ is smooth.

L has a formal adjoint L^* defined by

$$\int_{R^n} vLu\, dx = \int_{R^n} uL^*v\, dx, \quad u,v \in C_o^\infty(R^n),$$

and the Schwartz distribution version of (3) is:

$$\int_0^\infty \int_{R^n} f(y,t) \quad (\partial v/\partial t + L^*v)dxdt = 0,$$

for all $v \in C_o^\infty(R^n)$. Vector fields X on R^n define functions $\underline{a} : R^n \to R^n$ such that $X = \underline{a} \cdot \nabla$. Then if $u, v \in C_o^\infty(R^n)$, from Green's formula

$$\int_{R^n} [v\underline{a} \cdot \nabla u + u\nabla \cdot (\underline{a}v)]dx = \int_{R^n} \nabla \cdot (u\underline{a}v)dx = 0;$$

that is an integration-by-parts formula by which we calculate

$$L^* = -X_0 + Y + (1/2)\, \Sigma_1^r X_i^2 + q_0, \text{ where } Y = \Sigma_1^r q_i X_i$$

and $q_0, \ldots, q_r$ are in $C^\omega(R^n)$; $q_i|_M \in C^\omega(M)$. It follows that $L^*f = 0$ whenever f is constant on M; and $\partial/\partial t - L^*$ is hypoelliptic (q_0 doesn't affect that) on $M \times R^1$.

Let $v \in C_o^\infty(M \times R^1)$; v can be extended to $C_o^\infty(R^n \times R^1)$. For fixed $s > 0$, $v(\cdot,s)$ is in the domain of L. Using the Markov property, for $y \in M$ and all test functions v we have

$$\int_M v(x,s)\ p(y,t + s,x)d\mu(x) = \int_{R^n} [T_{t,x}v(\cdot,s)]\ \hat{p}(y,s,x)dx.$$

Differentiating, at $t = 0$,

$$\int_M v(x,s) \frac{\partial}{\partial s}\ p(y,s,x)d\mu(x) = \int_{R^n} [Lv(s,x)]\ \hat{p}(y,s,x,)dx;$$

$$\int_0^\infty \int_M v(x,s)\frac{\partial}{\partial s}\ p(y,s,x)d\mu(x)ds =$$

$$= \int_0^\infty \int_{R^n} v(x,s)\ L^*\ \hat{p}(y,s,x)dxds$$

$$= \int_0^\infty \int_M v(x,s)\ L^*\ p(y,s,x)d\mu(x)ds$$

where L^* acts on the x variables. That is, $p(y,\cdot,\cdot)$ is a Schwartz distribution solution of the forward (Fokker-Planck) equation smooth for $t > 0$ by the hypoellipticity of $\partial/\partial t - L^*$. (Compare [2], Sec. 3.5) Therefore, $p(\cdot,\cdot,\cdot)$ is a transition density, smooth on $M \times R^+$.

5. CONCLUSION

First of all, why are smooth densities nice to know about? For an answer see Kushner [12] and Rishel [13], who have shown that if smooth transition densities exist for (2) then one can solve by transformation of measures a similar stochastic differential equation with an added discontinuous control term. This solution is used to establish the existence of optimal feedback controls for certain stochastic systems; so far this has been done for $M = R^n$. Where are the papers on deterministic optimal control on manifolds? (some answers are available for Ex. 1).

If rank $(A^+) < m + 1$, the situation is like Ex. 2. The transition function $P(y,t,B)$ is singular and the diffusion is really degenerate.

There are additional results in [1] obtained by usual probabilistic methods and extendable to $M \neq R^n$:

(i) if $g \varepsilon C_b$ (bounded continuous), $\lim_{t \downarrow 0} f(y,t) \to g(y)$ uniformly on compacta;

(ii) if $g \varepsilon C_b$ then f is the unique bounded solution of $\partial f/\partial t = Lf$ satisfying $\lim_{t \downarrow 0} f(y,t) \to g(y)$.

(iii) if ρ is an invariant measure for $\mathcal{D}$ and L is hypo-elliptic, then ρ has a density p such that $L^* p = 0$ (steady-state Fokker-Planck equation).

For an introduction to Dirichlet problems, see [4] and [14] and their references, especially the work of Bony.

6. ACKNOWLEDGEMENTS

Finally, my thanks to the many people who have discussed these problems with me, especially A.V. Balakrishnan who supervised my dissertation.

The research reported here was supported in part by NSF Grants GK-27873 and GK-386924.

REFERENCES

1. D.L. Elliott, "Controllable Nonlinear Systems Driven by White Noise," Dissertation, U.C.L.A., September 1969.

2. H.P. McKean, Jr., Stochastic Integrals, Academic Press, 1969.

3. R.W. Brockett, "Lie Theory and Control Systems Defined on Spheres," preprint, Aiken Laboratory, Harvard University (1973).

4. D.W. Stroock and S.R.S. Varadhan, "On the Support of Diffusion Processes, with Applications to the Strong Maximum Principle," in 6th Berkeley Symposium on Mathematical Statistics and Probability, Vol. III, U. of California, 1973.

5. T. Nagano, "Linear Differential Systems with Singularities and an Application to Transitive Lie Algebras," J. Math. Soc. Japan 18, 398-404 (1966).

6. D.L. Elliott," A Consequence of Controllability," J. Diff. Eqs. 10, 364-370 (1971).

7. P. Stefan and J.B. Taylor, "A Remark on a Paper of D.L. Elliott," preprint, School of Mathematics and Computer Science, University College of North Wales, Bangor (1973).

8. H.J. Sussmann and V. Jurdjevic, "Controllability of Nonlinear Systems," J. Diff. Eqs. 12, 95-116 (1972).

9. O.A. Oleinik and E.V. Radkevich, "On Local Smoothness of Generalized Solutions and Hypoellipticity of Second Order Differential Equations," Russ. Math Surveys 26, 139-156 (1971).

10. A.V. Skorokhod, Studies in the Theory of Random Processes, Addison-Wesley, Reading, Mass., 1965 (Ch. 3).

11. E.B. Dynkin, Markov Processes, Springer-Verlag, Berlin, 1965.

12. H.J. Kushner, "Stability and Existence of Diffusions with Discontinuous or Rapidly Growing Drift Terms," J. Diff. Eqs. 11, 156-168 (1972).

13. R.W. Rishel, "Weak Solutions of a Partial Differential Equation of Dynamic Programming," SIAM J. Control, 9, 519-528 (1971).

14. D. Stroock and S.R.S. Varadhan, "On Degenerate Elliptic-Parabolic Operators of Second Order and Their Associated Diffusions," Comm. Pure and Appl. Math 25, 651-713 (1972).

15. C.B. Morrey," The Analytic Embeddings of Abstract Real-Analytic Manifolds," Ann. of Math. (2), 68, 159-201 (1958).

SIGNAL DETECTION ON LIE GROUPS

James Ting-Ho Lo

Division of Mathematics and Physics, University of Maryland Baltimore County, Baltimore, Maryland 21228

1. Introduction.

In a series of recent papers (e.g., [1] through [6]), control, estimation, and detection problems in which the natural state space is a Lie group have been studied. Among practical problems of this kind are frequency modulation and demodulation, frequency stability, gyroscopic analysis, satellite attitude control, power conversion, etc. It is the purpose of this paper to formulate and study a class of detection problems on an arbitrary matrix Lie group.

The trick used in this paper to formulate an observation process is borrowed from McKean [7], where he constructed the Brownian sample paths on a Lie group. Thus we will inject the differentials of a skew observation process on the Lie algebra into the Lie group via the exponential map and then piece them together as a so-called product integral. This product integral represents our observation process on the Lie group, and satisfies a bilinear matrix stochastic differential equation, when the skew observation process is linear. We will then show that given an arbitrary bilinear matrix observation process, the corresponding skew observation process can be obtained by "reversing" the above injecting procedure. Furthermore, these two procedures will be seen to induce two "almost sure" bijective mappings between a vector-valued and a matrix-valued function spaces, one being the inverse of the other. It is well known that the study of a Lie group may be greatly simplified by considering the tangent space (the Lie algebra) of the Lie group at its identity. In fact, the local study of a Lie group is entirely equivalent to the study of the finite dimensional linear algebraic structures of the Lie algebra. In this paper, the above bijective mappings facilitate similar simplification. It enables us to evaluate the likelihood ratio in a finite dimensional linear space--the Lie algebra!

In view of the above construction, the null and the alternative hypotheses that the signal is respectively absent and present in the observation on a Lie group can be written as a pair of bilinear matrix stochastic differential equations. The observational noise can be viewed as entering multiplicatively. Using the bijective mappings, we may transform these hypotheses on a Lie group into those on the corresponding Lie algebra. There the likelihood ratio can be expressed by the well-known formula due to Duncan [8] and Kailath [9]. Thus the evaluation of the likelihood ratio on a Lie group also hinges on the least-squares estimation.

When the signal is a linear diffusion process, the idea of using the bijective mappings to work in the Lie algebra also leads to a finite dimensional filtering equation for evaluating the least-squares estimate. This equation is indeed an immediate extension of the Kalman-Bucy filter to the case with observation on Lie groups.

When the signal is on a Lie group and only its skew form enters the observation directly (such as in an integrating gyroscope), the above techniques can be applied immediately to the detection problem. However, the estimation and the controls problems, in which the cost criteria are expressed in connection with the Riemannian metric, are more involved. The special case where the state space of the signal is an abelian Lie group will be treated elsewhere.

The reader is referred to [1] for a brief introduction to Lie algebra and Lie groups. The differential geometric methods used here are elementary.

2. Almost Sure Representation of Continuous Curves on Lie Groups.

Let $R^{n\times n}$ denote the set of real $n \times n$ matrices, and $\{R_1,\ldots,R_m\}$ a basis of a Lie algebra L in $R^{n\times n}$. Then the set

$$G = \{M : M = \exp(A_1)\exp(A_2)\ldots\exp(A_k);\ A_i \in L\ ,\quad i = 1,\ldots,k;\ k = 0,1,\ldots\}$$

is a m-dimensional Lie group related to L by a one-to-one map ϕ_M from a neighborhood of $0 \in L$ onto a neighborhood of $M \in G$, which is defined by $\phi_M(A) = \exp(A)M$, $A \in L$.

A continuous curve in G is usually represented by a $n\times n$-matrix-valued continuous function on a closed interval $T = [o,s]$ of the real line. In this section we will see that, under certain assumptions, a continuous curve in G starting from the identity element $I \in G$ can also be represented by a m-vector-valued function on T in a certain "almost sure" sense.

We will need the following notations:

C_ℓ = the family of the continuous m-vector-valued functions, a, on T with initial value $a(o) = 0$,
C_g = the family of the continuous $m \times m$-matrix-valued functions, A, on T with initial value $A(o) = I$ such that A represents a continuous curve on G, the matrix Lie group associated with L,
B_ℓ = the Borel σ-field of C_ℓ in the uniform topology,
B_g = the Borel σ-field of C_g in the uniform topology,
w = a standard m-vector Brownian motion on $(\Omega, ,P)$,

$$(\sum_j,\sum_i) = (\sum_{j=1}^m, \sum_{i=1}^m).$$

Let y be an m-vector stochastic process on T satisfying the following Ito differential equation:

$$dy(t) = f(t)dt + Q^{\frac{1}{2}}(t)dw(t) \tag{1}$$

where f is an m-vector stochastic process on T and $Q^{\frac{1}{2}} : T \to R^{m\times n}$ is Borel-measurable and bounded, i.e.

$$||Q^{\frac{1}{2}}(t)||^2 \triangleq \operatorname{tr} Q^{\frac{1}{2}}(t)(Q^{\frac{1}{2}}(t))' \leqslant C_1^2\ , \quad \text{for } t \in T\ . \tag{2}$$

Let $H_n : C_\ell \to C_g$ be defined by

$$\begin{aligned}(H_n(a))(t) &= I & (t = 0)\\ &= \exp[\sum_j(a_j(t) - a_j(\ell 2^{-n}))R_j](H_n(a))(\ell 2^{-n}) \\ & & (t \geqslant 0,\ \ell = [2^n t])\end{aligned}$$

for $a = [a_1,a_2,\ldots,a_m]' \in C_\ell$. Let $K(\Delta) \triangleq \sum_j y_j(\Delta)R_j \triangleq \sum_j(y_j(t) - y_j(\ell 2^{-n}))R_j$ and $Y_n(t) = (H_n(y))(t)$. Then

$$Y_n(t) - Y_n(\ell 2^{-n}) = (\exp(K(\Delta)) - I)Y_n(\ell 2^{-n})$$

Recalling the oscillation property ([7], p. 57), it is clear that up to terms involving $K^3(\Delta)$ (of magnitude $< 2^{-4n/3}$),

$$Y_n(t) - Y_n(\ell 2^{-n}) = (K(\Delta) + \frac{1}{2} K^2(\Delta))Y_n(\ell 2^{-n})\ .$$

By simple calculations,

$$K^2(\Delta) \cong \sum_i \sum_j Q_{ij}(t)R_iR_j\Delta t\ .$$

Thus the definition of the Ito integral leads at once to the conjecture that $Y \triangleq \lim_{n\to\infty} Y_n$, a.s. is the solution of

$$dY(t) = [\sum_j R_j dy_j(t) + M(t)dt]Y(t) \tag{3}$$

$$M(t) = \frac{1}{2} \sum_i \sum_j Q_{ij}(t)R_iR_j \tag{4}$$

$$Y(0) = I\ .$$

This conjecture can be proven by following closely the six steps taken in Section 4.8 of [7] and keeping in mind the assumption (2) and the oscillation property.

The operator $H = \lim_{n\to\infty} H_n$ is the so-called product integral operator, which is usually used to solve matrix differential equations.

In the following, we will construct the inverse operator, J, of H by defining appropriate "inverse" operator J_n of H_n.

Let C_m denote the family of $n \times n$-matrix-valued continuous functions which are representations of continuous curves on L.

Let $A \in C_g$ and n_1 be the smallest integer such that for all $n \geq n_1$ and $0 \leq i \leq [s2^n]$,

$$||A(i+1)2^{-n})A^{-1}(i2^{-n}) - I|| \leq 1$$

Define $K_n : C_g \to C_m$ by

$$(K_n(A))(t) = 0 , \qquad (t \in T) ,$$

for $n < n_1$, and

$$\begin{aligned}(K_n(A))(t) &= 0 , & (t = 0) ,\\ &= (K_n(A))(\ell 2^{-n}) + \lg(A(t)A^{-1}(\ell 2^{-n})) & (t \geq 0, \ell = [t2^n])\end{aligned}$$

for $n \geq n_1$.

Letting $(K_n(A))(t) = \sum_j R_j[(K_n(A))(t)]_j$, we define $J_n : C_g \to C_\ell$ by

$$(J_n(A))(t) = [[(K_n(A))(t)]_1, \ldots, [(K_n(A))(t)]_m] ,$$

for $A \in C_g$. It can be shown that a subsequence of $\{J_n(A)\}$ converges uniformly on T to a continuous function $J(A) \in C_\ell$, for each element A of a B_g-measurable set $B_2 \subset C_g$ such that $\nu_y(B_2) = 1$. Furthermore, for each $a \in B_1$, $H(a)$ is an element of $B_2 (B_1 \subseteq J(B_2))$ and $J(H(a)) = a (J = J^{-1})$. We now come to the conclusion summarized as the following theorem.

<u>Theorem 1</u>. A continuous curve on a matrix Lie group G, which is represented by an element A of C_g, can almost surely be represented by an element a of C_ℓ with respect to the measure ν_y induced by Y. These two representations are related by the almost sure bijective mapping $J = H^{-1} : B_2 \to B_1 = J(B_2)$ with $\nu_y(B_2) = \mu_y(B_1) = 1$ or equivalently the Ito stochastic integral equation

$$\sum_i R_i a_i(t) = -\frac{1}{2}\sum_i\sum_j R_iR_j \int_o^t Q_{ij}(s)ds + \int_o^t (dA(s))A^{-1}(s)$$

where the second term on the right side denotes the value the Ito integral $\int_0^t (dY(s))Y^{-1}(s)$ takes on at $Y = A$.

3. Hypotheses on Lie Groups and Evaluation of Likelihood Ratios.

In view of the results of the previous section, it is easy to see that, given a matrix Lie group G , we may formulate a signal detection problem on G by injecting a skew observation process y, which satisfies $dy = mdt + Q^{\frac{1}{2}}dw$ under the hypothesis H_1 and $dy = Q^{\frac{1}{2}}dw$ under the hypothesis H_o , into G via the bijection H. The differential equation satisfied by $H = H(y)$ under either H_1 , or H_o, is bilinear in form. Thus a natural question arises as to the detection for arbitrary bilinear $g \times g$-matrix Ito differential equations as the following:

$$H_1 : dY = [\sum_{i=1}^{\alpha} A_i m_i dt + \sum_{i=1}^{\beta} B_i dw_i + \sum_{i=1}^{\gamma} C_i q_i dt]Y \tag{5}$$

$$H_o : dY = [\sum_{i=1}^{\beta} B_i dw_i + \sum_{i=1}^{\gamma} C_i q_i dt]Y \tag{6}$$

$$Y(o) = I$$

where we assume:

I. simply for simplicity in illustration of our approach, A_i, B_i, and C_i are constant matrices,

II. $\{A_i, \ i = 1,\ldots,\alpha\}$ are linearly independent,

III. $w = [w_1,\ldots,w_\beta]'$ is a standard β-vector Wiener process,

IV. $m = [m_1,\ldots,m_\alpha]'$ is a measurable α-vector stochastic process $\sum_{i=1}^{\alpha} \int_o^s m_i^2(t)dt < \infty$, a.s.

V. $q = [q_1,\ldots,q_\gamma]'$ is a measurable γ-vector stochastic process such that $\sum_{i=1}^{\gamma} \int_o^s q_i^2 (t)dt < \infty$, a.s., and $q(t)$ is Y^t-measurable,

VI. w is independent of m .

Let L_n denote the Lie algebra generated by $\{A_1,\ldots,A_\alpha,B_1,\ldots,B_\beta\}$. Assume that L_n is an n-dimensional linear space of which $\{R_1,\ldots,R_n\}$ is a basis such that $R_i = A_i$, for $i = 1,\ldots,\alpha$. Since $B_i \in L_n$ we may write

$$B_i = \sum_{j=1}^{n} R_j Q_{ji}^{\frac{1}{2}}, \quad i = 1,\ldots,\beta , \tag{7}$$

where $Q_{ji}^{\frac{1}{2}}$, $j = 1,\ldots,n$ are the coordinates of B_i in the basis $\{R_1,\ldots,R_n\}$. Let $Q^{\frac{1}{2}}$ denote the $n \times \beta$ matrix, $[Q_{ij}^{\frac{1}{2}}]$.

Assume that the Lie algebra L generated by $\{A_1,\ldots,A_\alpha,B_1,\ldots,B_\beta,C_1,\ldots,C_\gamma, R_iR_j , \ i = 1,\ldots,n, \ j = 1,\ldots,\beta\}$ is an m-dimensional. In view of the results in Section 2, we may presume that a skew observation $\underline{y} = J(Y)$ in the following form exists

$$d\underline{y} = \underline{f}\, dt + \underline{Q}^{\frac{1}{2}} dw \ , \tag{8}$$

where an n-vector stochastic process $\underline{f}$ and an $n \times \beta$-matrix $\underline{Q}^{\frac{1}{2}}$ are to be determined.

Substituting (8) into (3), we obtain

$$\underline{f} = \underline{m} + \underline{r} \tag{9}$$

$$\underline{m} = [m_1, \ldots, m_\alpha, 0, \ldots, 0]' \tag{10}$$

$$\underline{r} = [r_1, \ldots, r_m]' \tag{11}$$

$$\sum_{i=1}^{\gamma} C_i q_i - \frac{1}{2} \sum_{i=1}^{n} \sum_{j=1}^{\beta} Q_{ij} R_i R_j = \sum_{i=1}^{m} R_i r_i \tag{12}$$

(Q_{ij} is the ixj^{th} component of the $n \times n$-matrix $Q = Q^{\frac{1}{2}}(Q^{\frac{1}{2}})'$ and hence $Q_{ij} \neq (Q^{\frac{1}{2}}_{ij})^2$)

$$\underline{Q}^{\frac{1}{2}} = \begin{bmatrix} Q^{\frac{1}{2}} \\ \hline 0 \end{bmatrix} \begin{matrix} \updownarrow n \\ \updownarrow n-m \end{matrix} \quad (\leftrightarrow \beta) \tag{13}$$

Hence $\underline{y} = J(Y)$ under H_1 satisfies

$$H_{11} : d\underline{y} = \underline{m}dt + \underline{Q}^{\frac{1}{2}} dw + \underline{r}\, dt \tag{14}$$

and similarly $\underline{y} = J(Y)$ under H_o satisfies

$$H_{10} : d\underline{y} = \underline{Q}^{\frac{1}{2}} dw + \underline{r}\, dt. \tag{15}$$

This completes showing that the hypotheses, (5) and (6), are on the Lie group G of L . Therefore, Y(t) is invertible and, by Theorem 1, $\underline{y}$ and Y are causally equivalent.

Let C_g and B_g be defined as in the previous sections, and let the measures on (C_g, B_g) induced by Y under H_1 and H_o be denoted by μ_1 and μ_o respectively. In the following, we will evaluate the Radon-Nikodym derivative $d\mu_1/d\mu_o$, if it exists.

Let C^k ($k = \alpha$ or $m - \alpha$) denote the family of the continuous $k \times k$-matrix-valued functions, A , on T with initial value $A(o) = I$, and let B^k denote the Borel σ-field of C^k .

Let $y = [\underline{y}_1, \ldots, \underline{y}_\alpha]'$ and ν_i^1 be the measure on (C^α, B^α) induced by y under H_i . Let $z = [\underline{y}_{\alpha+1}, \ldots, \underline{y}_n]'$ and ν_i^2 be the measure on $(C^{m-\alpha}, B^{m-\alpha})$ induced by z under H_i . Then the measure on (C^m, B^m) induced by $\underline{y}$ under H_i is equal to product measure

$\nu_i^1 \times \nu_i^2$. It is easy to see that $\nu_1^2 = \nu_o^2$ and thus $\frac{d\nu_1^2}{d\nu_o^2} = 1$. Using a well-known lemma (Lemma 2, p. 99 in [10]), we have, for $t\varepsilon T$,

$$\frac{d(\nu_1^1 \times \nu_1^2)}{d(\nu_o^1 \times \nu_o^2)}(\underline{y}^t) = \frac{d\nu_1^1}{d\nu_o^1}(y^t)\frac{d\nu_1^1}{d\nu_o^2}(z^t) = \frac{d\nu_1^1}{d\nu_o^1}(y^t) , \tag{16}$$

provided $\frac{d\nu_1^1}{d\nu_o^1}$ exists.

We need the following special notations:

$$I_j^i = \begin{cases} \begin{bmatrix} I_i^i , & 0_{i-j}^i \end{bmatrix} & \text{if } i < j \\ \begin{bmatrix} I_j^j \\ \\ 0_j^{i-j} \end{bmatrix} & \text{if } i > j \end{cases} \tag{17}$$

where I_i^i is the $i \times i$ identity matrix and 0_j^i is the $i \times j$ zero matrix.

We note that y under H_1 satisfies $H_{21} : dy = mdt + I_n^\alpha Q^{\frac{1}{2}}dw + rdt$ and under H_o it satisfies $H_{20} : dy = I_n^\alpha Q^{\frac{1}{2}}dw + rdt$, where $r = I_m^\alpha \underline{r}$. It is well known ([8] and [9]) that if $\det(I_n^\alpha Q I_\alpha^n) \neq 0$, the likelihood ratio of H_{21} to H_{20} can be written as:

$$\frac{d\nu_1^1}{d\nu_o^1}(y) = \exp[-\frac{1}{2}\int_o^s \hat{m}'(t)(I_m^\alpha Q I_\alpha^m)^{-1}\hat{m}(t)dt$$
$$- \int_o^s \hat{m}'(t)(I_m^\alpha Q I_\alpha^m)^{-1}r(t)dt + \int_o^t \hat{m}'(t)(I_m^\alpha Q I_\alpha^m)^{-1}dy(t)] \tag{18}$$

where $\hat{m}(t) = E(m(t)|y^t , H_{21})$.

Since J is bijective, it can be shown that

$$\frac{d\mu_1}{d\mu_o}(Y) = \frac{d(\nu_1^1 \times \nu_1^2)}{d(\nu_o^1 \times \nu_o^2)}(J(Y)) = \frac{d\nu_1^1}{d\nu_o^1}(J(Y)) . \tag{19}$$

Let e_{ij} be the $m \times m$-matrix of which the ixj^{th} component is one and the other components are zero and let $\{R_{m+j}, j = 1,\ldots,g^2 - m\}$ be $g \times g$-matrices such that $\{R_j, j = 1,\ldots,g^2\}$ form a basis of $R^{g\times g}$. Now we may write $e_{ij} = \sum_{k=1}^{g^2} R_k e_{ij}^k$, for some constants $\{e_{ij}^k\}$

Let

$$e_k = [e^k_{ij}] = \text{the matrix of which the } ixj^{th} \text{ component is } e^k_{ij} \quad (20)$$

Since $\int_o^t (dY(\tau))Y^{-1}(\tau) - \int_o^t M(\tau)d\tau$ belongs to L spanned by $\{R_1,\ldots,R_m\}$, we have

$$tr\left\{e_j'\left[\int_o^t (dY(\tau))Y^{-1}(\tau) - \int_o^t M(\tau)d\tau\right]\right\} = 0 \, , \quad \text{for } j > m \, .$$

From (5), it can be shown by simple calculation that

$$y(t) = (J(Y))(t) = [tr\{e_1'[\int_o^t (dY(\tau))Y^{-1}(\tau) - \int_o^t M(\tau)d\tau]\} \, ,$$
$$\ldots, tr\{e_m'[\int_o^t (dY(\tau))Y^{-1}(\tau) - \int_o^t M(\tau)d\tau]\}]' \quad (21)$$

where M is defined by (4).

Substituting (18) into (19), in view of (21), leads to the theorem:

<u>Theorem 2.</u> Consider the hypotheses H_1 and H_o described by (5) and (6) Let the constant matrix Q be defined (7) and assume that $\det(I^\alpha_m Q I^m_\alpha) \neq 0$. Then the likelihood ratio for H_1 against H_o given a realization of Y can be expressed as follows:

$$\frac{d\nu_1}{d\nu_o}(Y^t) = \exp[-\frac{1}{2}\int_o^t \hat{m}_s'(I^\alpha_m Q I^m_\alpha)^{-1}\hat{m}_s ds$$
$$- \int_o^t \hat{m}_s'(I^\alpha_m Q I^m_\alpha)^{-1} r(s)ds + \int_o^t \hat{m}_s'(I^\alpha_m Q I^m_\alpha)^{-1} dy(s)$$

where $\hat{m}_s = E(m(s)|Y^s,H_1)$

$$dy(s) = [tr\{e_1'[(dY(s))Y^{-1}(s) - Mds]\} \, , \ldots,$$
$$tr\{e_\alpha'[(dY(s))Y^{-1}(s) - Mds]\}]'$$

$$M = \frac{1}{2}\sum_{i=1}^{n}\sum_{j=1}^{\beta} Q_{ij}R_iR_j$$

and I^α_m , e_k , and $r = I^\alpha_m \underline{r}$ are defined by (17), (20), and (11) respectively.

Thus the evaluation of the likelihood ratio hinges on the evaluation of the conditional expectation $\hat{m}_s$. We recall that under H_1 , Y satisfies (5) . So we have a nonlinear filtering problem. However, we note that $E(m(t)|Y^t,H_1) = E(m(t)|\underline{y}^t,H_{11}) = E(m(t)|y^t,H_{21})$, a.s., and that if, for $t \in T$,

$$m(t) = H(t)X(t) \tag{23}$$

$$dX(t) = F(t)X(t)dt + G(t)dv(t) \ , \quad X(o) = X_o \tag{24}$$

where v is a standard vector Wiener process, X_o is a normal vector, and w, v_t, X_o are statistically independent, then it is well known $E(m(t)|y^t, H_{21})$ satisfies the Kalman-Bucy filtering equations. Hence, under the assumptions (23) and (24), some reflection shows that

$$\hat{m}_t = H(t)\hat{X}_t$$

$$d\hat{X}_t = F\hat{X}_t dt + rdt + PH'(I_m^\alpha Q I_\alpha^m)^{-1}(dy - H\hat{X}_t dt)$$

$$dy(t) = [tr\{e_1'[(dY(t))Y^{-1}(t) - Mdt]\} \ , \ \ldots,$$
$$tr\{e_\alpha'[(dY(t))Y^{-1}(t) - Mdt]\}]' \ ,$$

$$\dot{P} = FP + PF' - PH'(I_m^\alpha Q I_\alpha^m)^{-1}HP + GG'$$

$$P(o) = E(X_o X_o')$$

where I_m^α, e_k, r, M, Q are determined by (17), (20), (11), (22), respectively.

ACKNOWLEDGEMENTS

This work was supported in part by a 1973 UMBC Summer Research Fellowship at the Division of Mathematics and Physics, University of Maryland at Baltimore County, Baltimore.

The author wishes to thank Professor Richard C. Roberts of UMBC for his support during the preparation of this paper.

REFERENCES

[1] R. W. Brockett, "System Theory on Group Manifolds and Coset Spaces", SIAM J. on Control, May, 1972.

[2] H. J. Sussmann, "The Bang-Bang Problem for Linear Control Systems in Lie Groups", SIAM J. on Control,

[3] V. Jurdjevic and H. J. Sussmann, "Control Systems in Lie Groups", J. Differential Equations, 12, No. 2., 1972.

[4] J. T. Lo and A. S. Willsky, "Estimation for Rotational Processes with One Degree of Freedom", Harvard University Technical Report No. 635, July 1927.

[5] J. T. Lo and A. S. Willsky, "Stochastic Control of Rotational Processes with One Degree of Freedom", submitted for publication.

[6] J. T. Lo, "Signal Detection of Rotational Processes and Frequency Demodulation", presented at the 7th Annual Princeton Conference on Information Sciences and Systems, March, 1973.

[7] H. P. McKean, Jr., Stochastic Integrals, Academic Press, New York, 1969.

[8] T. E. Duncan, "Evaluation of Likelihood Functions", Information and Control, Vol. 13, 1968.

[9] T. Kailath, "A General Likelihood-Ratio Formula for Random Signals in Gaussian Noise", IEEE Trans. on Information Theory, Vol. IT-15, No. 3, May 1969.

[10] A. V. Skorokhod, Studies in Theory of Random Processes, Addison-Wesley, Reading, Massachusetts, 1965.

SOME ESTIMATION PROBLEMS ON LIE GROUPS

Alan S. Willsky

Dept. of Electrical Engineering,

M.I.T., Cambridge, Massachusetts

1. INTRODUCTION

Recently a number of results [1]-[8] related to stochastic bilinear systems have been obtained. In particular, the work of Willsky and Lo [1]-[6] on optimal estimation on certain Lie groups leads to concrete and simple solutions to several bilinear estimation and stochastic control problems. The major limitation of Willsky and Lo's work is that it deals only with abelian Lie groups. Although there are a number of interesting problems in that setting [4], this limitation prevents their techniques from being directly applicable to a number of problems.

In this paper we consider several problems on nonabelian Lie groups. We define a class of bilinear stochastic systems with the aid of an extension of an injection procedure of McKean [9]. Several associated estimation and analysis problems are formulated, and solutions are obtained for a number of special cases, including the case in which the bilinear system matrices obey certain Lie bracket relations, and also several other cases involving optimal angular velocity estimation for rigid bodies.

2. SIGNALS, OBSERVATIONS, AND ESTIMATION ON MATRIX LIE GROUPS

In this section we construct signal and observation processes on matrix Lie groups. We then briefly discuss several general estimation problems and the difficulties encountered.

Let $x(t)$ be an n-dimensional process satisfying

$$dx(t) = F(t)x(t)dt + B(t)dw(t) \tag{1}$$

where w is an m-dimensional Brownian motion with strength $Q(t)$

(i.e. $E(dw(t)dw(t)') = Q(t)dt$). Let $A_0, A_1, \ldots, A_n$ be linearly independent k x k matrices, and define $L = \{A_0, \ldots, A_n\}_A$ to be the Lie algebra generated by the A_i. Let $G = \{\exp L\}_G$ be the associated Lie group. We then inject x into G [w.p.1] in one of two ways

$$X_1(t) = \prod_{s \leq t} \exp[A_0 ds + \sum_{i=1}^{n} A_i dx_i(s)] \tag{2}$$

$$X_2(t) = \prod_{s \leq t} \exp[A_0 ds + \sum_{i=1}^{n} A_i x_i(s) ds] \tag{3}$$

where "$\prod$" denotes the product integral [1],[9]. Note that unless $F(t) \equiv 0$, X_1 is <u>not</u> a Markov process, and even in that case X_2 is not Markov. Later in this paper we present examples that allow us to interpret these processes physically. One can show [1] that X_1 and X_2 satisfy

$$dX_1(t) = [A_0 dt + \sum_{i=1}^{n} A_i dx_i(t) + \frac{1}{2} \sum_{i,j=1}^{n} S_{ij}(t) A_i A_j dt] X_1(t) \tag{4}$$

$$dX_2(t) = [A_0 dt + \sum_{i=1}^{n} A_i x_i(t) dt] X_2(t) \tag{5}$$

where $S = BQB'$. Depending upon the problem, we regard one of the processes x, X_1, or X_2 as the signal process to be tracked.

We now define an observation process (see Section 4 for several others). Let H be another Lie group with Lie algebra M generated by the linearly independent set $B_0, \ldots, B_p$. Let z be the p-dimensional process satisfying

$$dz(t) = C(t)x(t)dt + dv(t) \tag{6}$$

where v is a Brownian motion, independent of x, with strength $R(t) > 0 \ \forall t$. Define the H-valued function

$$Z(t) = \prod_{s \leq t} \exp(B_0 ds + \sum_{i=1}^{p} B_i dz_i(s)) \tag{7}$$

which satisfies

$$dZ(t) = [B_0 dt + \sum_{i=1}^{p} B_i dz_i(t) + \frac{1}{2} \sum_{i,j=1}^{p} R_{ij}(t) B_i B_j dt] Z(t) \tag{8}$$

Depending upon the problem, we regard either z or Z as the observation.

As discussed in [1]-[6] and in [11], the processes x, X_1, and X_2 are equivalent if we assume $x(0) = 0$ - - i.e. knowledge of $x^t \triangleq \{x(s) | 0 \leq s \leq t\}$ is equivalent to knowledge of either $X_1{}^t$ or $X_2{}^t$. We can see this from (4) and (5), the invertibility of X_1 and X_2, and the linear independence of the A_i. We can make the same statement about z and Z. Thus if we are given the bilinear observation process Z, we can nonlinearly transform it to the causally equivalent linear process z (see [1]-[6]). Therefore, we assume we have z.

In this case the optimal estimate of x(t) given z^t is produced by a Kalman-Bucy filter. As shown in [1]-[6], one can write $X_1(t)$ and $X_2(t)$ as explicit functions of x(t) and $y(t) = \int_0^t x(s)ds$ only, if G is abelian. Thus the optimal conditional estimates of $X_1(t)$ and $X_2(t)$ can be computed once we compute the conditional densities for x(t) and y(t) given z^t, which, by linearity, are normal. Thus we need only compute means and covariances (as the Kalman-Bucy filter does) to determine $\hat{X}_1$ and $\hat{X}_2$. However, as discussed in [1], the nonabelian problem is much more difficult, since X_1 and X_2 are more complicated functions of x, and, in general, we need more information about z^t than $p(x(t)|z^t)$ or $p(y(t)|z^t)$. As the next section indicates, there are special cases other than the abelian one for which we can obtain finite-dimensional solutions.

Before looking at special cases, we comment on the problem of estimation on homogeneous spaces [10]. As discussed in [1] and [10], one can study processes on various homogeneous spaces by examining equations such as

$$dy(t) = \{A_0 dt + \sum_{i=1}^{n} A_i dx_i(t) + \frac{1}{2} \sum_{i,j=1}^{n} s_{ij}(t) A_i A_j dt\} y(t) \qquad (9)$$

where y is a k-vector. However, we note that

$$y(t) = X_1(t) y(0) \qquad (10)$$

where X_1 is given in (2), and thus the problem of estimating y(t) can be solved if we can estimate $X_1(t)$ (see Example 2).

3. OPTIMAL ESTIMATION FOR A CLASS OF BILINEAR SYSTEMS

In this section we prove that the optimal estimator for a certain class of bilinear systems can be realized by a finite dimensional system. To do this we define the operator ad_A on the space $M(k,R)$ of k x k real matrices, where A is also k x k .

$$\mathrm{ad}_A(B) \stackrel{\Delta}{=} AB - BA \stackrel{\Delta}{=} [A,B] \tag{11}$$

We use the notation ad_A^i to denote the ith power of ad_A. In addition, we assume we are given a cost function ϕ: $G \times G \rightarrow [0,\infty)$.

Theorem 1: Consider the linear equations (1) and (6) with $x(0)=0$, and the bilinear equations (4) and (5). Suppose

$$[\mathrm{ad}_{A_0}^i A_j, A_\ell] = 0 \qquad \text{for } j,\ell = 1,\ldots,n \quad i = 0,\ldots,k^2 - 1 \tag{12}$$

Then the z^t-measurable estimates $\hat{X}_1(t|t)$ and $\hat{X}_2(t|t)$ that minimize

$$J_i(M) = E[\phi(X_i(t),M)|z^t] \qquad i = 1,2 \tag{13}$$

over the class of all z^t-measurable, G-valued functions M, can be computed by finite-dimensional linear filters followed by nonlinear postprocessors.

Proof: We prove the result for X_1. The proof for X_2 is quite similar. The key to the proof is a linear transformation used by Brockett [10] and Krener [12]. Let

$$Y(t) = e^{-A_0 t} X_1(t) \tag{14}$$

As in [3], if we let $H_1,\ldots,H_r$ be a basis for the abelian Lie algebra generated by $\mathrm{ad}_{A_0}^i (A_j)$ $j = 1,\ldots,n$; $i = 0,\ldots, k^2 - 1$, we have

$$e^{-A_0 t} A_i e^{A_0 t} = \sum_{j=1}^{r} e_{ij}(t) H_j \qquad i = 1,\ldots,n \tag{15}$$

for some functions e_{ij}. If we let $E(t) = [e_{ij}(t)]$ and define

$$dy(t) = E'(t)dx(t) \qquad y(0) = 0 \tag{16}$$

we can compute

$$dY(t) = [\sum_{i=1}^{r} H_i dy_i(t) + \frac{1}{2}\sum_{i,j=1}^{r} T_{ij}(t)H_iH_j dt]Y(t) \tag{17}$$

where $T = E'SE$.

Since the H_i commute

$$Y(t) = \bigcap_{s\leq t} \exp(\sum_{i=1}^{r} H_i dy_i(s)) = \exp(\sum_{i=1}^{r} H_i y_i(t)) \tag{18}$$

Also, because y is a linear function of x, the conditional density for y(t) given z^t is normal with mean $\hat{y}(t|t)$ and covariance $P_y(t)$ determined by solving the linear filtering problem associated with

$$\begin{bmatrix} dx(t) \\ dy(t) \end{bmatrix} = \begin{bmatrix} F(t) & 0 \\ E'(t)F(t) & 0 \end{bmatrix} \begin{bmatrix} x(t) \\ y(t) \end{bmatrix} dt + \begin{bmatrix} B(t) \\ E'(t)B(t) \end{bmatrix} dw(t) \tag{19}$$

$$dz(t) = \begin{bmatrix} C(t) & 0 \end{bmatrix} \begin{bmatrix} x(t) \\ y(t) \end{bmatrix} + dv(t) \tag{20}$$

We also have

$$J_1(M) = E[\phi(e^{A_0 t} \exp \sum_{i=1}^{r} H_i y_i(t), M)|z^t] \tag{21}$$

which is solely a function of $\hat{y}(t|t)$, $P_y(t)$, and t. Thus we have the desired structure of the optimal filter: a linear filter (perhaps preceded by a nonlinear preprocessor to generate z from Z) followed by a nonlinear postprocessor. ■

Note that if $A_0 = 0$, we obtain the abelian group results of Lo and Willsky. Also, the linear system (19),(20) is not observable; however, since y is an explicit function of x, we expect no problems in the behavior of the linear filter -- i.e. if the original x-z problem leads to a well-behaved filter, so will the augmented problem (see Ex. 1). Finally, note that even if the original system is time-invariant, the optimal filter is inherently time varying; however, such a complication seems well worth it when we can produce a finite dimensional optimal filter with only linear dynamics.

Example 1: We present a simple 2 x 2 affine group example that can also be solved using linear theory, but which illustrates the important aspects of our approach. Let x and z be scalar processes satisfying

$$dx(t) = Fx(t)dt + Q^{1/2}\, dw(t), \quad x(0) = 0 \tag{22}$$

$$dz(t) = Cx(t)dt + R^{1/2}\, dv(t) \tag{23}$$

where v and w are independent standard Brownian motion processes. Let X be the 2 x 2 process defined by

$$dX(t) = [A_0 dt + A_1 dx(t)]X(t), \quad X(0) = I \tag{24}$$

$$A_0 = \begin{bmatrix} A & 0 \\ 0 & 0 \end{bmatrix} \qquad A_1 = \begin{bmatrix} 0 & B \\ 0 & 0 \end{bmatrix} \tag{25}$$

Defining Y as in (14) and letting $y(t) = B\int_0^t e^{-As} dx(s)$, we have

$$Y(t) = \begin{bmatrix} 1 & y(t) \\ 0 & 1 \end{bmatrix} \tag{26}$$

and the equations generating the conditional means are

$$\begin{bmatrix} d\hat{x}(t|t) \\ d\hat{y}(t|t) \end{bmatrix} = \begin{bmatrix} F & 0 \\ e^{-At}BF & 0 \end{bmatrix} \begin{bmatrix} \hat{x}(t|t) \\ \hat{y}(t|t) \end{bmatrix} dt + \begin{bmatrix} P_x(t)C/R \\ P_{xy}(t)C/R \end{bmatrix} (dz(t) - C\hat{x}(t|t)dt) \tag{27}$$

where P_x, P_{xy}, and P_y are the appropriate elements of the error covariance satisfying the Riccati equation:

$$\dot{P}_x(t) = 2FP_x(t) - C^2P_x^2(t) + Q, \qquad P_x(0) = 0 \tag{28}$$

$$\dot{P}_{xy}(t) = [F - C^2P_x(t)]P_{xy}(t) + e^{-At}BFP_x(t) + e^{-At}BQ, \; P_{xy}(0)=0 \tag{29}$$

$$\dot{P}_y(t) = [2e^{-At}BF - C^2P_{xy}(t)]P_{xy}(t) + e^{-2At}B^2Q, \; P_y(0) = 0 \tag{30}$$

Note that $\hat{y}$ and P_y can be computed as quadratures involving $\hat{x}$, P_x and z only, and the well-posedness of (28) guarantees the same for (29), (30).

If we choose the cost function

$$\phi(M,N) = \sum_{i,j=1}^{2} (M_{ij} - N_{ij})^2 \tag{31}$$

we can show that the optimal estimate is

$$\hat{X}(t|t) = \begin{bmatrix} e^{At} & e^{At}\hat{y}(t|t) \\ 0 & 1 \end{bmatrix} \tag{32}$$

<u>Example 2</u>: Consider the dynamical equations

$$d\xi_1(t) = [\xi_1(t) - \xi_2(t)][dt + dx(t)], \quad \xi_1(0) = 1 \tag{33}$$

$$d\xi_2(t) = [\xi_1(t) - \xi_2(t)]dx(t), \qquad \xi_2(0) = 0 \tag{34}$$

where x and the observation process z satisfy (22) and (23). If we define X by

$$dX(t) = [A_0dt + A_1dx(t)]X(t), \quad X(0) = I \tag{35}$$

$$A_0 = \begin{bmatrix} 1 & -1 \\ 0 & 0 \end{bmatrix} \qquad A_1 = \begin{bmatrix} 1 & -1 \\ 1 & -1 \end{bmatrix}. \tag{36}$$

we have $[\mathrm{ad}_{A_0}^{i}(A_1), A_1] = 0$ and $\xi(t) = X(t)\xi(0)$. Thus the problem of estimating $\xi(t)$ given z^t is solved by estimating $X(t)$, and this problem falls into the framework of Theorem 1.

4. OTHER MEASUREMENT PROCESSES AND THE SPECIAL CASE OF RIGID BODY ROTATION

In this section we consider the Lie group SO(3), consisting of all 3 x 3 orthogonal matrices with positive determinant. Such a matrix can be thought of as determining the orientation of a rigid body in R^3. In this setting, we will interpret the various processes described in Section 2 and will introduce two other measurement processes that have special physical motivation for SO(3), although they can be formulated for the general matrix Lie group problem as well.

The Lie algebra so(3) associated with SO(3) has the basis

$$A_1 = \begin{bmatrix} 0 & 0 & 0 \\ 0 & 0 & -1 \\ 0 & 1 & 0 \end{bmatrix} \quad A_2 = \begin{bmatrix} 0 & 0 & 1 \\ 0 & 0 & 0 \\ -1 & 0 & 0 \end{bmatrix} \quad A_3 = \begin{bmatrix} 0 & -1 & 0 \\ 1 & 0 & 0 \\ 0 & 0 & 0 \end{bmatrix} \tag{37}$$

Note that so(3) (and thus SO(3)) is not abelian. Consider the differential equation on SO(3)

$$dX(t) = \left(\sum_{i=1}^{3} du_i(t) A_i \right) X(t) \tag{38}$$

If we interpret X(t) as the orientation of a rigid body with respect to some inertial frame, then the $du_i(t)$ are infinitessimal rotations about orthogonal axes -- either body or inertially fixed, depending upon our coordinatization. Referring to the definitions of X_1 and X_2 in (4) and (5), we see that if we let $du_i(t) = dx_i(t)$, x(t) has the interpretation as the total angle swept (up to time t) about 3 orthogonal axes. On the other hand, if we let $du_i(t) = x_i(t)dt$, x(t) is an angular velocity vector. It is easy to imagine cases in which the angular velocity history (or the history of total angle swept) should be modeled as a correlated process. In such a case, as mentioned in Section 2, X is <u>not</u> a Markov process, but it is the physical motivation provided by problems such as rigid body rotation that leads us to study such processes.

In the SO(3) setting we can also find a physical interpretation for the observation process given by (6) and (8). In strapdown navigation systems, one receives information about either angular velocity or incremental angle changes -- i.e. either $\dot{z}(t)$ or dz(t), where in this case x must be thought of as an angular velocity. This information is then processed by a "direction cosine computer," which produces the quantity Z(t), defined in (7) and (8). If we wish to interpret x, given by (1), as total angle swept, the strapdown observation process should be

$$\begin{aligned} dz(t) &= C(t)dx(t)dt + dv(t) \\ &= C(t)F(t)x(t)dt + C(t)B(t)dw(t) + dv(t) \end{aligned} \quad (39)$$

The problem of estimating x(t) given z(t) specified in (39), is a linear estimation problem with correlated observation and process noise, and it can be handled with little added complexity compared to the standard problem.

Finally, we consider an observation process related to inertial navigation systems. Suppose X, representing the relative orientation of a rigid body with respect to inertial space, satisfies (38) with $du_i = x_i dt$. In addition, suppose the rigid body is equipped with an inertial platform that is to be kept fixed with respect to inertial space. However, because of drifts in the gyros, the platform drifts relative to inertial space. A reasonable assumption is to model the relative orientation V(t) of inertial space with respect to the platform as an SO(3) Brownian motion process.

$$dV(t) = V(t)\left\{\sum_{i=1}^{3} A_i dv_i(t) + \frac{1}{2}\sum_{i,j=1}^{3} R_{ij}(t)A_iA_j dt\right\} \quad (40)$$

where v is a 3-dimensional Brownian motion, independent of X, with strength R(t).

We take as our observation the relative orientation M(t) of the rigid body and the inertial platform, which can be determined by reading off gimbal angles, and which is given by M(t) = X(t)V(t) Note that for abelian Lie groups the multiplicative form (45) and the bilinear form (8) are interchangeable, [1]-[6].

We now compute

$$dM(t) = \left\{\sum_{i=1}^{3} A_i x_i(t)\right\} M(t)dt + M(t)\left\{\sum_{i=1}^{3} A_i dv_i(t) + \frac{1}{2}\sum_{i,j=1}^{3} R_{ij}(t)A_iA_j dt\right\} \quad (41)$$

Since M(t) is conditionally known, (41) is a linear measurement of x(t), where M serves as a gain matrix. Thus the error covariance cannot be precomputed, but one can show [1] that the minimum variance angular velocity estimate is produced by a Kalman-Bucy filter where the filter gain is computed by solving the Riccati equation on-line using the incoming values of M.

5. CONCLUSIONS

In this paper we have examined several estimation problems on Lie groups and have obtained results that extend those developed by Willsky and Lo and also that indicate that the structure of bilinear systems can be exploited in order to derive easily implemented estimation equations. Of course the general Lie group problem has not been solved, but these results provide hope that the general case may in some sense yield under further analysis.

Acknowledgements

The author gratefully acknowledges Professor Roger W. Brockett of Harvard University for providing the original motivation for this work. This research was conducted in part while the author was a Fannie and John Hertz Foundation Fellow in the Dept. of Aeronautics and Astronautics at M.I.T. and in part at the Decision and Control Sciences Group of the M.I.T. Electronic Systems Laboratory with partial support provided by AFOSR under grant 72-2273.

References

1. A.S. Willsky, Dynamical Systems Defined on Groups: Structural Properties and Estimation, Ph.D. Thesis, Dept. of Aeronautics and Astronautics, M.I.T., June 1973.

2. J.T. Lo and A.S. Willsky, "Estimation for Rotational Processes with One Degree of Freedom I:Introduction and Continuous Time Processes," submitted for publication.

3. A.S. Willsky and J.T. Lo, "Estimation for Rotational Processes with One Degree of Freedom II:Discrete Time Processes," submitted for publication.

4. A.S. Willsky and J.T. Lo, "Estimation for Rotational Processes with One Degree of Freedom III:Applications and Implementation," submitted for publication.

5. J.T. Lo and A.S. Willsky, "Stochastic Control of Rotational Processes with One Degree of Freedom," submitted for publication.

6. J.T. Lo and A.S. Willsky, Estimation for Rotational Processes with One Degree of Freedom, Harvard Tech. Dept. No. 635, Div. of Eng. and Appl. Phys. , Harvard, July 1972.

7. C. Bruni, G. DiPillo, and G. Koch, Bilinear Systems:An Appealing Class of "Nearly Linear" Systems in Theory and Applications, Universita' di Roma, Istituto di Automatica, R.2-29, December 1972.

8. R.W. Brockett, "Lie Theory and Control Systems on Spheres," SIAM J. Appl. Math., to appear.

9. H.P. McKean Jr., Stochastic Integrals, Academic Press, N.Y., 1969.

10. R.W. Brockett, "System Theory on Group Manifolds and Coset Spaces," SIAM J. on Control, Vol. 10, No. 2, May 1972, pp. 265-284.

11. J.T. Lo, "Signal Detection on Lie Groups," The Proceedings of this Conference.

12. A.J. Krener, "On the Equivalence of Control Systems and the Linearization of Nonlinear Systems," Preprints of the 14th JACC, Columbus, Ohio, June 1973, pp. 759-764.